Annual Reviews of
Computational Physics VIII

ANNUAL REVIEWS OF COMPUTATIONAL PHYSICS

Series Editor: Dietrich Stauffer (*Cologne University*)

Published:

Vol. I: ISBN 981-02-1881-8

Vol. II: ISBN 981-02-2176-2

Vol. III: ISBN 981-02-2427-3
            ISBN 981-02-2506-7 (pbk)

Vol. IV: ISBN 981-02-2728-0
            ISBN 981-02-2753-1 (pbk)

Vol. V: ISBN 981-02-3181-4
            ISBN 981-02-3182-2 (pbk)

Vol. VI: ISBN 981-02-3563-1

Vol. VII: ISBN 981-02-4080-5

Vol. VIII: ISBN 981-02-4524-6

Annual Reviews of Computational Physics VIII

Theme Issue: Scaling and Disordered Systems

Guest editors: M.R.H. Khajehpour, M.R. Kolahchi and M. Sahimi

edited by

Dietrich Stauffer

Cologne University

World Scientific

Singapore • New Jersey • London • Hong Kong

Published by

World Scientific Publishing Co. Pte. Ltd.

P O Box 128, Farrer Road, Singapore 912805

USA office: Suite 1B, 1060 Main Street, River Edge, NJ 07661

UK office: 57 Shelton Street, Covent Garden, London WC2H 9HE

British Library Cataloguing-in-Publication Data
A catalogue record for this book is available from the British Library.

ANNUAL REVIEWS OF COMPUTATIONAL PHYSICS VIII

ISBN 981-02-4524-6

Printed in Singapore by FuIsland Offset Printing

PREFACE

For the first time in this series of reviews, a volume is dedicated to one theme: Scaling and Disordered Systems. This is possible because it presents the proceedings of the Regional Summer School on this subject, held in July 1999 in Zanjan, Iran, at the Institute for Advanced Studies in Basic Sciences. We are all grateful to the organizers of this international meeting, M. R. H. Khajehpour, M. R. Kolahchi and M. Sahimi, for making it possible. In addition, I thank M. Hashemi and M. Sahimi for guiding me through Tehran, and N. Jan and NSERC of Canada for hospitality and supporting me during August and September 1999 when the articles were collected.

As usual in this series, all texts were submitted electronically and were with the publisher the day after the October 2, 1999, deadline. No author missed it. The cover picture shows an aperiodic tiling of the Friday mosque in Isfahan, Iran, as discussed in the quasicrystalline article of Rivier, Fig. 1.

St. F. X. University
Antigonish, Nova Scotia, Canada *Dietrich Stauffer*

CONTENTS

Shorter Talks

INTRODUCTION

The Regional Summer School on Scaling and Disordered Systems was held at the Institute for Advanced Studies in Basic Sciences (IASBS) in Zanjan, Iran, during July 3–14, 1999. The motivation for organizing this international school was twofold: (1) To bring to Iran some of the leading researchers in the field of scaling and disordered media in order to provide a forum in which a wide range of problems in this active research area can be discussed, and (2) to introduce to the physics community the recent progress that has been made in physics education and research in Iran, and in particular at IASBS. Our motivation is perhaps best described by the Inauguration Address of the School, delivered by Professor Y. Sobouti, the Director of IASBS, from which we quote:

> Mohammad ibn Musa al-Khwarazmi, mathematician (d. 846), Mohammad ibn Zakariyya al-Razi, physician and chemist (d. 925), al-Farabi, philosopher and musicologist (d. 950), Avicenna, philosopher and man of medicine (d. 1037), al-Biruni, astronomer and chronologist (d. 1046), Omar Khayyam, mathematician and poet (d. 1131), and Khaajeh Nasiruddin Tusi, astronomer, mathematician, and statesman (d. 1274) are but a few among the many top-ranking thinkers that the people and culture of Iran have offered the world in the course of their long history. The same potential still exists in this land. The key to rekindle this long forgotten tradition is to provide the youth of Iran with the proper environment and means that trigger their ingenuity and imagination.

The papers in this volume represent the lectures that were delivered at the School. Computer simulations and algorithms played an important role in most of the lectures and discussions that took place in the School, which is why these papers are published in the Annual Reviews of Computational Physics VIII. (The word "algorithm" has its root in this name al-Khwarazmi.) The School was highly successful in introducing to many talented students, who had been selected from amongst universities of the region, the enormous success that the ideas of scaling and universality have enjoyed in describing disordered media, while at the same time making it clear that many of the theoretical approaches and models for such media have found practical applications in the industry.

The School was sponsored by IASBS, the Abdus Salam International Centre for Theoretical Physics (ICTP) in Trieste, Italy, and the Physical Society of Iran. We are also grateful to the Office of External Activities at ICTP and Professor Y. Sobouti for their support. We would like to thank all the lecturers for their efforts to make their presentations of the highest quality, and Dietrich Stauffer who made publication of these Proceedings possible, and who pushed everybody to deliver his/her paper on time.

The organizers and Guest Editors,
M. R. H. Khajehpour
M. R. Kolahchi
M. Sahimi

Annual Reviews of Computational Physics VIII (pp. 1–47)
Edited by Dietrich Stauffer

STOCHASTIC DYNAMICS OF GROWING FILMS

MEHRAN KARDAR

*Department of Physics, Massachusetts Institute of Technology,
Cambridge, Massachusetts 02139, USA*

These notes describe a set of pedagogical lectures delivered in July 1999 at the *Regional Summer School on Scaling and Disordered Systems* in Zanjan, Iran. **Lecture 1** is an introduction to the Langevin method for stochastic dynamics. The general ideas are first introduced in the context of the Brownian motion of a single particle, and then elaborated for the case of a fluctuating surface. **Lecture 2** provides a brief review of *dynamic scaling* phenomena for growing surfaces. A number of simple discrete growth models are described, and compared with continuum Langevin equations that capture their underlying symmetries and conservation laws. Applications of the method to other *nonequilibrium* situations such as moving flux-lines or drifting polymers are explained in **lecture 3**. These examples involve the coupling of fields describing fluctuations parallel and perpendicular to the direction of motion. The coupling of different fields is also encountered in **lecture 4** which is a brief account of our ongoing research on the growth of "ordered" films. The coupling of surface roughness to the order parameter leads to interesting new phenomena and textures.

1. Introduction

These lectures provide both an introduction to stochastic nonequilibrium phenomena in growing surfaces, and some of the current problems of research. These notes are intended mainly as a *supplement* to the lectures, and will probably not serve the reader well as a review article. They are likely to include many errors and omissions, in particular when it comes to references. I shall attempt to partially alleviate this deficiency by listing some general sources for further study. The research described here has involved many collaborators, principally Y.-C. Zhang, G. Parisi, T. Hwa, E. Medina, D. Ertas, and B. Drossel. Partial support was provided by the NSF through grant number DMR-98-05833. Many thanks are due to Professors M. R. H. Khajehpour, M. Kolahchi, and M. Sahimi for their efforts in organizing a most pleasant and productive summer school at the Institute for Advanced Studies in Basic Sciences (IASBS) in Zanjan.

In order to set the stage, the introductory chapter describe the standard approach to dynamics that starts from an energy functional. The simplest example is the Brownian motion of a single particle which is reviewed in the next section. Many of the classic articles on stochastic processes can be found in Ref. 1, and are described in standard texts.[2] I then go on to generalize to many degrees of freedom by examining the dynamics of a field. The standard reference for dynamic critical phenomena is the review article by Hohenberg and Halperin,[3] and more details can be found in Refs. 4 and 5. The next chapter deals with the roughness of growing films, employing a general approach to open systems which is described in more detail in Refs. 6 and 7. It serves as a rather incomplete review of roughness of growing surfaces; more detailed reviews can be found in Refs. 8–10.

The third chapter is more mathematical in nature. We first review some special properties of the KPZ equation which facilitate the analysis of its scaling properties. We then go on to introduce generalizations of this equation which occur in the contexts of moving flux lines and drifting polymers. The final chapter is a brief account of our ongoing research on the coupling of the roughness of the growing surface with an ordering field.

1.1. *Brownian motion of a particle*

Observations under a microscope indicate that a dust particle in a liquid drop undergoes a random jittery motion. This is because of the random impacts of the much smaller fluid particles. The theory of such (*Brownian*) motion was developed by Einstein in 1905 and starts with the equation of motion for the particle. The displacement $\vec{x}(t)$, of a particle of mass m is governed by,

$$m\ddot{\vec{x}} = -\frac{\dot{\vec{x}}}{\mu} - \frac{\partial \mathcal{V}}{\partial \vec{x}} + \vec{f}_{\text{random}}(t)\,. \tag{1}$$

The three forces acting on the particle are

(i) A friction force due to the viscosity of the fluid. For a spherical particle of radius R, the mobility in the low Reynolds number limit is given by $\mu = (6\pi\bar{\eta}R)^{-1}$, where $\bar{\eta}$ is the specific viscosity.

(ii) The force due to the external potential $\mathcal{V}(\vec{x})$, e.g. gravity.

(iii) A random force of zero mean due to the impacts of fluid particles.

The viscous term usually dominates the inertial one (i.e. the motion is over-damped), and we shall henceforth ignore the acceleration term. Equation (1) now reduces to a *Langevin equation*,

$$\dot{\vec{x}} = \vec{v}(\vec{x}) + \vec{\eta}(t) \,, \tag{2}$$

where $\vec{v}(\vec{x}) = -\mu \partial \mathcal{V}/\partial \vec{x}$ is the *deterministic* velocity. The *stochastic* velocity, $\vec{\eta}(t) = \mu \vec{f}_{\text{random}}(t)$, has zero mean,

$$\langle \vec{\eta}(t) \rangle = 0 \,. \tag{3}$$

It is usually assumed that the probability distribution for the noise in velocity is Gaussian, i.e.

$$\mathcal{P}[\vec{\eta}(t)] \propto \exp\left[-\int d\tau \frac{\eta(\tau)^2}{4D} \right]. \tag{4}$$

Note that different components of the noise, and at different times, are independent, and the covariance is

$$\langle \eta_\alpha(t)\eta_\beta(t') \rangle = 2D\delta_{\alpha,\beta}\delta(t - t') \,. \tag{5}$$

The parameter D is related to *diffusion* of particles in the fluid. In the absence of any potential, $\mathcal{V}(\vec{x}) = 0$, the position of a particle at time t is given by

$$\vec{x}(t) = \vec{x}(0) + \int_0^t d\tau \, \vec{\eta}(\tau) \,.$$

Clearly the separation $\vec{x}(t) - \vec{x}(0)$ which is the sum of random Gaussian variables is itself Gaussian distributed with mean zero, and a variance

$$\langle (\vec{x}(t) - \vec{x}(0))^2 \rangle = \int_0^t d\tau_1 d\tau_2 \langle \vec{\eta}(\tau_1) \cdot \vec{\eta}(\tau_2) \rangle = 3 \times 2Dt \,.$$

For an ensemble of particles released at $\vec{x}(t) = 0$, that is, with $\mathcal{P}(\vec{x}, t = 0) = \delta^3(\vec{x})$, the particles at time t are distributed according to

$$\mathcal{P}(\vec{x}, t) = \left(\frac{1}{\sqrt{4\pi Dt}} \right)^{3/2} \exp\left[-\frac{x^2}{4Dt} \right] \,,$$

which is the solution to the diffusion equation

$$\frac{\partial \mathcal{P}}{\partial t} = D\nabla^2 \mathcal{P} \,.$$

A simple example is provided by a particle connected to a Hookian spring, with $\mathcal{V}(\vec{x}) = Kx^2/2$. The deterministic velocity is now $\vec{v}(\vec{x}) = -\mu K \vec{x}$, and the Langevin equation, $\dot{\vec{x}} = -\mu K \vec{x} + \vec{\eta}(t)$, can be rearranged as

$$\frac{d}{dt}[e^{\mu K t}\vec{x}(t)] = e^{\mu K t}\vec{\eta}(t)\,. \tag{6}$$

Integrating the equation from 0 to t yields

$$e^{\mu K t}\vec{x}(t) - \vec{x}(0) = \int_0^t d\tau\, e^{\mu K \tau}\vec{\eta}(\tau)\,, \tag{7}$$

and

$$\vec{x}(t) = \vec{x}(0)e^{-\mu K t} + \int_0^t d\tau\, e^{-\mu K(t-\tau)}\vec{\eta}(\tau)\,. \tag{8}$$

Averaging over the noise indicates that the mean position

$$\langle \vec{x}(t) \rangle = \vec{x}(0)e^{-\mu K t}\,, \tag{9}$$

decays with a characteristic *relaxation time* $\tau = 1/(\mu K)$. Fluctuations around the mean behave as

$$\langle (\vec{x}(t) - \langle \vec{x}(t)\rangle)^2 \rangle = \int_0^t d\tau_1 d\tau_2\, e^{-\mu K(2t-\tau_1-\tau_2)} \overbrace{\langle \vec{\eta}(\tau_1) \cdot \vec{\eta}(\tau_2)\rangle}^{2D\delta(\tau_1-\tau_2)\times 3}$$

$$= 6D \int_0^t d\tau\, e^{-2\mu K(t-\tau)}$$

$$= \frac{3D}{\mu K}[1 - e^{-2\mu K t}] \xrightarrow{t\to\infty} \frac{3D}{\mu K}\,. \tag{10}$$

However, once the dust particle reaches equilibrium with the fluid at a temperature T, its probability distribution must satisfy the normalized Boltzmann weight

$$\mathcal{P}_{\text{eq.}}(\vec{x}) = \left(\frac{K}{2\pi k_B T}\right)^{3/2} \exp\left[-\frac{Kx^2}{2k_B T}\right]\,, \tag{11}$$

yielding $\langle x^2 \rangle = 3k_B T/K$. Since the dynamics is expected to bring the particle to equilibrium with the fluid at temperature T, Eq. (10) implies the condition

$$D = k_B T \mu\,. \tag{12}$$

This is the Einstein relation connecting the *fluctuations* of noise to the *dissipation* in the medium.

Clearly the Langevin equation at long times reproduces the correct mean and variance for a particle in equilibrium at a temperature T in the potential $\mathcal{V}(\vec{x}) = Kx^2/2$, provided that Eq. (12) is satisfied. Can we show that the whole probability distribution evolves to the Boltzmann weight for any potential? Let $\mathcal{P}(\vec{x},t) \equiv \langle \vec{x}|\mathcal{P}(t)|0\rangle$ denote the probability density of finding the particle at $\vec{x}$ at time t, given that it was at 0 at $t = 0$. This probability can be constructed recursively by noting that a particle found at $\vec{x}$ at time $t + \epsilon$ must have arrived from some other point $\vec{x}'$ at t. Adding up all such probabilities yields

$$\mathcal{P}(\vec{x}, t + \epsilon) = \int d^3\vec{x}'\mathcal{P}(\vec{x}', t)\langle \vec{x}|T_\epsilon|\vec{x}'\rangle, \tag{13}$$

where $\langle \vec{x}|T_\epsilon|\vec{x}'\rangle \equiv \langle \vec{x}|\mathcal{P}(\epsilon)|\vec{x}'\rangle$ is the transition probability. For $\epsilon \ll 1$,

$$\vec{x} = \vec{x}' + \vec{v}(\vec{x}')\epsilon + \vec{\eta}_\epsilon, \tag{14}$$

where $\vec{\eta}_\epsilon = \int_t^{t+\epsilon} d\tau \vec{\eta}(\tau)$. Clearly $\langle \vec{\eta}_\epsilon \rangle = 0$, and $\langle \eta_\epsilon^2 \rangle = 2D\epsilon \times 3$, and following Eq. (4),

$$p(\vec{\eta}_\epsilon) = \left(\frac{1}{4\pi D\epsilon}\right)^{3/2} \exp\left[-\frac{\eta_\epsilon^2}{4D\epsilon}\right]. \tag{15}$$

The transition rate is simply the probability of finding a noise of the right magnitude according to Eq. (14), and

$$\langle \vec{x}|T(\epsilon)|\vec{x}'\rangle = p(\eta_\epsilon) = \left(\frac{1}{4\pi D\epsilon}\right)^{3/2} \exp\left[-\frac{(\vec{x} - \vec{x}' - \epsilon\vec{v}(\vec{x}'))^2}{4D\epsilon}\right]$$

$$= \left(\frac{1}{4\pi D\epsilon}\right)^{3/2} \exp\left[-\epsilon\frac{(\dot{\vec{x}} - \vec{v}(\vec{x}))^2}{4D}\right]. \tag{16}$$

By subdividing the time interval t, into infinitesimal segments of size ϵ, repeated application of the above evolution operator yields

$$\mathcal{P}(\vec{x}, t) = \langle \vec{x}|T(\epsilon)^{t/\epsilon}|0\rangle = \int_{(0,0)}^{(\vec{x},t)} \frac{\mathcal{D}\vec{x}(\tau)}{\mathcal{N}} \exp\left[-\int_0^t d\tau \frac{(\dot{\vec{x}} - \vec{v}(\vec{x}))^2}{4D}\right]. \tag{17}$$

The integral is over all paths connecting the initial and final points; each path's weight is related to its deviation from the classical trajectory, $\dot{\vec{x}} = \vec{v}(\vec{x})$. The recursion relation [Eq. (13)],

$$P(\vec{x}, t) = \int d^3\vec{x}' \left(\frac{1}{4\pi D\epsilon}\right)^{3/2} \exp\left[-\frac{(\vec{x} - \vec{x}' - \epsilon\vec{v}(\vec{x}'))^2}{4D\epsilon}\right] P(\vec{x}', t - \epsilon), \quad (18)$$

can be simplified by the change of variables,

$$\vec{y} = \vec{x}' + \epsilon\vec{v}(\vec{x}') - \vec{x} \implies d^3\vec{y} = d^3\vec{x}'(1 + \epsilon\nabla \cdot \vec{v}(\vec{x}'))$$

$$= d^3\vec{x}'(1 + \epsilon\nabla \cdot \vec{v}(\vec{x}) + \mathcal{O}(\epsilon^2)). \quad (19)$$

Keeping only terms at order of ϵ, we obtain

$$P(\vec{x}, t) = [1 - \epsilon\nabla \cdot \vec{v}(\vec{x})] \int d^3\vec{y} \left(\frac{1}{4\pi D\epsilon}\right)^{3/2} e^{-\frac{y^2}{4D\epsilon}} P(\vec{x} + \vec{y} - \epsilon\vec{v}(\vec{x}), t - \epsilon)$$

$$= [1 - \epsilon\nabla \cdot \vec{v}(\vec{x})] \int d^3\vec{y} \left(\frac{1}{4\pi D\epsilon}\right)^{3/2} e^{-\frac{y^2}{4D\epsilon}} \left[P(\vec{x}, t) + (\vec{y} - \epsilon\vec{v}(\vec{x})) \cdot \nabla P\right.$$

$$\left. + \frac{y_i y_j - 2\epsilon y_i v_j + \epsilon^2 v_i v_j}{2} \nabla_i\nabla_j P - \epsilon\frac{\partial P}{\partial t} + \mathcal{O}(\epsilon^2)\right]$$

$$= [1 - \epsilon\nabla \cdot \vec{v}(\vec{x})] \left[P - \epsilon\vec{v} \cdot \nabla + \epsilon D\nabla^2 P - \epsilon\frac{\partial P}{\partial t} + \mathcal{O}(\epsilon^2)\right]. \quad (20)$$

Equating terms at order of ϵ leads to the *Fokker–Planck equation*,

$$\frac{\partial P}{\partial t} + \nabla \cdot \vec{J} = 0, \quad \text{with} \quad \vec{J} = \vec{v}P - D\nabla P. \quad (21)$$

The Fokker–Planck equation is simply the statement of conservation of probability. The probability current has a deterministic component $\vec{v}P$, and a stochastic part $-D\nabla P$. A *stationary distribution*, $\partial P/\partial t = 0$, is obtained if the net current vanishes. It is now easy to check that the Boltzmann weight, $P_{\text{eq.}}(\vec{x}) \propto \exp[-V(\vec{x})/k_B T]$, with $\nabla P_{\text{eq.}} = \vec{v}P_{\text{eq.}}/(\mu k_B T)$, leads to a stationary state as long as the fluctuation–dissipation condition in Eq. (12) is satisfied.

1.2. Equilibrium dynamics of a field

The next step is to generalize the above formalism to a collection of degrees of freedom, most conveniently described by a continuous field. The procedure will be described in terms of the dynamics of a surface, although it is in fact quite general. Small fluctuations of the surface can be described by a height $h(\mathbf{x}, t)$. Specific examples are the distortions of a soap film or the fluctuations on the surface of water in a container. In both cases the minimum energy configuration is a flat surface (ignoring the small effects of gravity on the soap film). The energy cost of small fluctuations for a soap film comes from the increased area and *surface tension* σ. Expanding the area in powers of the slope results in

$$\mathcal{H}_\sigma = \sigma \int d^D\mathbf{x}[\sqrt{1 + (\nabla h)^2} - 1] \approx \frac{\sigma}{2} \int d^d\mathbf{x}(\nabla h)^2 \,. \tag{22}$$

For the surface of water there is an additional gravitational potential energy, obtained by adding the contributions from all columns of water as

$$\mathcal{H}_g = \int d^d\mathbf{x} \int_0^{h(\mathbf{x})} dh' \, \rho g h' = \frac{\rho g}{2} \int d^d\mathbf{x}\, h(\mathbf{x})^2 \,. \tag{23}$$

The total (potential) energy of small fluctuations is thus given by

$$\mathcal{H} = \int d^d\mathbf{x} \left[\frac{\sigma}{2}(\nabla h)^2 + \frac{\rho g}{2}h^2\right], \tag{24}$$

with the second term absent for the soap film.

To construct a Langevin equation governing the dynamics of height fluctuations, first calculate the *force* on each surface element from the variations of the potential energy. The *functional derivative* of Eq. (24) yields

$$F(\mathbf{x}) = -\frac{\delta \mathcal{H}[h]}{\delta h(\mathbf{x})} = -\rho g h + \sigma \nabla^2 h \,. \tag{25}$$

The straightforward analog of Eq. (2) is

$$\frac{\partial h(\mathbf{x}, t)}{\partial t} = \mu F(\mathbf{x}) + \eta(\mathbf{x}, t) \,, \tag{26}$$

with a random velocity η, such that

$$\langle \eta(\mathbf{x}, t) \rangle = 0 \quad \text{and} \quad \langle \eta(\mathbf{x}, t)\eta(\mathbf{x}', t') \rangle = 2D\delta(\mathbf{x} - \mathbf{x}')\delta(t - t') \,. \tag{27}$$

The Langevin equation,

$$\frac{\partial h(\mathbf{x}, t)}{\partial t} = -\mu\rho g h + \mu\sigma\nabla^2 h + \eta(\mathbf{x}, t) \,, \tag{28}$$

is most easily solved by examining the Fourier components,

$$h(\mathbf{q}, t) = \int d^d\mathbf{x}\, e^{i\mathbf{q}\cdot\mathbf{x}} h(\mathbf{x}, t)\,, \tag{29}$$

which evolve according to

$$\frac{\partial h(\mathbf{q}, t)}{\partial t} = -\mu(\rho g + \sigma q^2)\, h(\mathbf{q}, t) + \eta(\mathbf{q}, t)\,. \tag{30}$$

The Fourier transformed noise,

$$\eta(\mathbf{q}, t) = \int d^d\mathbf{x}\, e^{i\mathbf{q}\cdot\mathbf{x}} \eta(\mathbf{x}, t)\,, \tag{31}$$

has zero mean, $\langle \eta(\mathbf{q}, t)\rangle = 0$, and correlations

$$\langle \eta(\mathbf{q}, t)\eta(\mathbf{q}', t')\rangle = \int d^d\mathbf{x} d^d\mathbf{x}'\, e^{i\mathbf{q}\cdot\mathbf{x}+i\mathbf{q}'\cdot\mathbf{x}'} \overbrace{\langle \eta(\mathbf{x}, t)\eta(\mathbf{x}', t')\rangle}^{2D\delta^d(\mathbf{x}-\mathbf{x}')\delta(t-t')}$$

$$= 2D\delta(t - t') \int d^d\mathbf{x}\, e^{i\mathbf{x}\cdot(\mathbf{q}+\mathbf{q}')}$$

$$= 2D\delta(t - t')(2\pi)^d \delta^d(\mathbf{q} + \mathbf{q}')\,. \tag{32}$$

Each Fourier mode in Eq. (30) now behaves as an independent particle connected to a spring as in Eq. (6). Introducing a decay rate

$$\gamma(\mathbf{q}) \equiv \frac{1}{\tau(\mathbf{q})} = \mu(\rho g + \sigma q^2)\,, \tag{33}$$

the evolution of each mode is similar to Eq. (8), and follows

$$h(\mathbf{q}, t) = h(\mathbf{q}, 0)e^{-\gamma(\mathbf{q})t} + \int_0^t d\tau\, e^{-\gamma(\mathbf{q})(t-\tau)}\eta(\mathbf{q}, \tau)\,. \tag{34}$$

Fluctuations in each mode decay with a different *relaxation time* $\tau(\mathbf{q})$; $\langle h(\mathbf{q}, t)\rangle = h(\mathbf{q}, 0)\exp[-t/\tau(\mathbf{q})]$. The competition between surface tension and gravity introduces a *capillary length*, $\lambda_c \approx \sqrt{\sigma/\rho g}$. For most liquids λ_c is of the order of a few millimeters. (It is λ_c that sets the characteristic size of rain drops or ripples on the surface of a pond.) On length, for scales larger than λ_c (or $q \ll 1/\lambda_c$), the relaxation time saturates to $\tau_{\max} = 1/(\mu\rho g)$. On the other hand, for the soap film where gravity is not important, the characteristic time scale grows with wavelength as $\tau(q) \approx (\mu\sigma q^2)^{-1}$. The divergence of the time scale is usually described by a *dynamic exponent* z, as $\tau \propto \lambda^z$. The value of $z = 2$ for the soap film is characteristic of diffusion processes.

The connected height–height correlation functions are obtained from

$$\langle h(\mathbf{q},t)h(\mathbf{q}',t)\rangle_c = \int_0^t d\tau_1 d\tau_2 \, e^{-\gamma(\mathbf{q})(t-\tau_1)-\gamma(\mathbf{q}')(t-\tau_2)} \overbrace{\langle \eta(\mathbf{q},\tau_1)\eta(\mathbf{q}',\tau_2)\rangle}^{2D\delta(\tau_1-\tau_2)(2\pi)^d\delta^d(\mathbf{q}+\mathbf{q}')}$$

$$= (2\pi)^d\delta^d(\mathbf{q}+\mathbf{q}')2D\int_0^t d\tau \, e^{-2\gamma(\mathbf{q})(t-\tau)}$$

$$= (2\pi)^d\delta^d(\mathbf{q}+\mathbf{q}')\frac{D}{\gamma(\mathbf{q})}(1 - e^{-2\gamma(\mathbf{q})t})$$

$$\xrightarrow{t\to\infty} (2\pi)^d\delta^d(\mathbf{q}+\mathbf{q}')\frac{D}{\mu(\rho g + \sigma q^2)}\,. \tag{35}$$

However, direct diagonalization of the Hamiltonian in Eq. (24) gives

$$\mathcal{H} = \int \frac{d^d\mathbf{q}}{(2\pi)^d}\frac{(\rho g + \sigma q^2)}{2}|h(\mathbf{q})|^2\,, \tag{36}$$

leading to correlation functions

$$\langle h(\mathbf{q})h(\mathbf{q}')\rangle = (2\pi)^d\delta^d(\mathbf{q}+\mathbf{q}')\frac{k_BT}{\rho g + \sigma q^2}\,. \tag{37}$$

Comparing Eqs. (35) and (37) indicates that the long-time dynamics reproduce the correct equilibrium behavior if the fluctuation–dissipation condition, $D = k_BT\mu$, is satisfied. In fact it is possible to obtain the correct equilibrium weight with $\mathbf{q}$ dependent mobility and noise, as long as the generalized fluctuation–dissipation condition,

$$D(\mathbf{q}) = k_BT\mu(\mathbf{q})\,, \tag{38}$$

holds. Physically, correlations in noise at different locations are generated if the impact of particles from the surrounding fluid exerts a force over many surface elements.

Starting with a flat interface, $h(\mathbf{x}, t = 0) = h(\mathbf{q}, t = 0) = 0$, the profile at time t is

$$h(\mathbf{x},t) = \int \frac{d^d\mathbf{q}}{(2\pi)^d}e^{-i\mathbf{q}\cdot\mathbf{x}}\int_0^t d\tau \, e^{-\mu(\rho g+\sigma q^2)(t-\tau)}\eta(\mathbf{q},t)\,. \tag{39}$$

The average height of the surface, $\bar{H} = \int d^d\mathbf{x}\langle h(\mathbf{x},t)\rangle/L^d$ is zero, while its overall width is defined by

$$w^2(t,L) \equiv \frac{1}{L^d}\int d^d\mathbf{x}\langle h(\mathbf{x},t)^2\rangle = \frac{1}{L^d}\int \frac{d^d\mathbf{q}}{(2\pi)^d}|h(\mathbf{q},t)|^2\,, \tag{40}$$

where L is the linear size of the surface. Using Eq. (35), we find that the width grows as

$$w^2(t,L) = \int \frac{d^d\mathbf{q}}{(2\pi)^d}\frac{D}{\gamma(\mathbf{q})}\left(1 - e^{-2\gamma(\mathbf{q})t}\right). \tag{41}$$

There are a range of time scales in the problem, related to characteristic length scales through Eq. (33). The shortest time scale, $t_{\min} \propto a^2/(\mu\sigma)$, is set by an atomic size a. The longest time scale is set by either the capillary length (λ_c) or the system size (L). For simplicity we shall focus on the soap film where the effects of gravity are negligible and $t_{\max} \propto L^2/(\mu\sigma)$. We can now identify three different ranges of behavior in Eq. (41):

(a) For $t \ll t_{\min}$, none of the modes has relaxed since $\gamma(\mathbf{q})t \ll 1$ for all $\mathbf{q}$. Each mode grows diffusively, and

$$w^2(t,L) = \int \frac{d^d\mathbf{q}}{(2\pi)^d}\frac{D}{\gamma(\mathbf{q})}2\gamma(\mathbf{q})t = \frac{2Dt}{a^d}\,. \tag{42}$$

(b) For $t \gg t_{\max}$, all modes have relaxed to their equilibrium values since $\gamma(\mathbf{q})t \gg 1$ for all $\mathbf{q}$. The height fluctuations now saturate to a maximum value given by

$$w^2(t,L) = \int \frac{d^d\mathbf{q}}{(2\pi)^d}\frac{D}{\mu\sigma q^2}\,. \tag{43}$$

The saturated value depends on the dimensionality of the surface, and in a general dimension d behaves as

$$w^2(t,L) \propto \frac{D}{\mu\sigma}\begin{cases} a^{2-d} & \text{for } d > 2, \quad (\chi = 0)\\[4pt] \ln(L/a) & \text{for } d = 2, \quad (\chi = 0^+)\\[4pt] L^{2-d} & \text{for } d < 2, \quad \left(\chi = \dfrac{2-d}{2}\right), \end{cases} \tag{44}$$

where we have defined a *roughness exponent* χ that governs the divergence of the width with system size via $\lim_{t\to\infty} w(t,L) \propto L^\chi$. (The symbol 0^+

is used to indicate a logarithmic divergence.) The exponent of $\chi = 1/2$ in $d = 1$ indicates that the one-dimensional interface fluctuates like a random walk.

(c) For $t_{\min} \ll t \ll t_{\max}$ only a fraction of the shorter length scale modes are saturated. The integrand in Eq. (41) (for $g = 0$) is made dimensionless by setting $y = \mu\sigma q^2 t$, and

$$w^2(t, L) \propto \frac{D}{\mu\sigma} \int dq\, q^{d-3}(1 - e^{-2\mu\sigma q^2 t})$$

$$\propto \frac{D}{\mu\sigma} \left(\frac{1}{\mu\sigma t} \right)^{\frac{d-2}{2}} \int_{t/t_{\max}}^{t/t_{\min}} dy\, y^{\frac{d-4}{2}} (1 - e^{-2y}). \tag{45}$$

The final integral is convergent for $d < 2$, and dominated by its upper limit for $d \geq 2$. The initial growth of the width is usually described by an exponent β defined through $\lim_{t \to 0} w(t, L) \propto t^\beta$, and

$$w^2(t, L) \propto \begin{cases} \dfrac{D}{\mu\sigma} a^{2-d} & \text{for } d > 2, \quad (\beta = 0) \\[2ex] \dfrac{D}{\mu\sigma} \ln(t/t_{\min}) & \text{for } d = 2, \quad (\beta = 0^+) \\[2ex] \dfrac{D}{(\mu\sigma)^{d/2}} t^{(2-d)/2} & \text{for } d < 2, \quad (\beta = (2 - d)/4). \end{cases} \tag{46}$$

The dependencies on space and time in the height–height correlation function can be summarized by the *dynamic scaling* form

$$\langle [h(\mathbf{x}, t) - h(\mathbf{x}', t')]^2 \rangle = |\mathbf{x} - \mathbf{x}'|^{2\chi} g\left(\frac{|t - t'|}{|\mathbf{x} - \mathbf{x}'|^z} \right). \tag{47}$$

Since equilibrium equal time correlations only depend on $|\mathbf{x} - \mathbf{x}'|$, $\lim_{y \to 0} g(y)$ should be a constant. On the other hand correlations at the same point can only depend on time, requiring that $\lim_{y \to \infty} g(y) \propto y^{2\chi/z}$, and leading to the exponent identity $\beta = \chi/z$.

The single particle Fokker–Planck Eq. (21) can be generalized to describe the evolution of the whole probability functional, $\mathcal{P}([h(\mathbf{x})], t)$, as

$$\frac{\partial \mathcal{P}([h(\mathbf{x})], t)}{\partial t} = -\int d^d\mathbf{x} \frac{\delta}{\delta h(\mathbf{x})} \left[\frac{\partial h(\mathbf{x}, t)}{\partial t} \mathcal{P} - D \frac{\delta \mathcal{P}}{\delta h(\mathbf{x})} \right]. \tag{48}$$

For the equilibrium Boltzmann weight

$$\mathcal{P}_{\text{eq.}}[h(\mathbf{x})] \propto \exp\left[-\frac{\mathcal{H}[h(\mathbf{x})]}{k_B T}\right] \propto \exp\left[-\frac{\sigma}{2k_B T}\int d^d\mathbf{x}(\nabla h)^2\right], \qquad (49)$$

the functional derivative results in

$$\frac{\delta\mathcal{P}_{\text{eq.}}}{\delta h(\mathbf{x})} = -\nabla\cdot\frac{\delta\mathcal{P}_{\text{eq.}}}{\delta(\nabla h)} = \frac{\sigma}{k_B T}(\nabla^2 h)\mathcal{P}_{\text{eq.}}\,. \qquad (50)$$

The total probability current,

$$J[h(\mathbf{x})] = \left[\mu\sigma\nabla^2 h - \frac{D\sigma}{k_B T}\nabla^2 h\right]\mathcal{P}_{\text{eq.}}\,, \qquad (51)$$

vanishes if the fluctuation–dissipation condition, $D = \mu k_B T$, is satisfied. Once again, the Einstein equation ensures that the equilibrium weight indeed describes a steady state.

1.3. *Dynamics of a conserved height*

The prescription for dynamics that leads to the Langevin equations (25)–(27), does not conserve the net height, $\int d^d\mathbf{x}\, h(\mathbf{x}, t)$. Although this quantity is on average zero, it undergoes stochastic fluctuations in time. In dealing with a volume of liquid, if particle exchange with the surrounding gas via evaporation and condensation is negligible, the total height of the liquid must be conserved, i.e.

$$\frac{d}{dt}\int d^d\mathbf{x}\, h(\mathbf{x}, t) = \int d^d\mathbf{x}\frac{\partial h(\mathbf{x}, t)}{\partial t} = 0\,. \qquad (52)$$

How can we construct a dynamical equation that satisfies Eq. (52)? The integral clearly vanishes if the integrand is a total divergence, i.e.

$$\frac{\partial h(\mathbf{x}, t)}{\partial t} = -\nabla\cdot\vec{j} + \eta(\mathbf{x}, t)\,. \qquad (53)$$

The noise itself must be a total divergence, $\eta = -\nabla\cdot\vec{\sigma}$, and hence in Fourier space,

$$\langle\eta(\mathbf{q}, t)\rangle = 0 \quad \text{and} \quad \langle\eta(\mathbf{q}, t)\eta(\mathbf{q}', t')\rangle = 2Dq^2\delta(t - t')(2\pi)^d\delta^d(\mathbf{q} + \mathbf{q}')\,. \qquad (54)$$

We can now take advantage of the generalized Einstein relation in Eq. (38) to ensure the correct equilibrium distribution by setting

$$\vec{j} = \mu\nabla\left(-\frac{\delta\mathcal{H}}{\delta h(\mathbf{x})}\right)\,. \qquad (55)$$

A similar question arises in dealing with the dynamics of a binary mixture undergoing phase separation. The order parameter $\phi(\mathbf{x})$, the difference between the densities of the two species, may now be conserved. The standard procedure and terminology for such situations is provided by Hohenberg and Halperin.[3] Given a Hamiltonian $\mathcal{H}[\phi]$, the Langevin dynamics of the field $\phi(\mathbf{x}, t)$ is constructed from

$$\frac{\partial \phi(\mathbf{x}, t)}{\partial t} = -\hat{\mu}\left(\frac{\delta \mathcal{H}}{\delta \phi(\mathbf{x})}\right) + \eta(\mathbf{x}, t), \tag{56}$$

with

$$\langle \eta(\mathbf{x}, t)\rangle = 0 \quad \text{and} \quad \langle \eta(\mathbf{x}, t)\eta(\mathbf{x}', t')\rangle = 2\hat{D}(\delta^d(\mathbf{x} - \mathbf{x}')\delta(t - t')). \tag{57}$$

In **model A** dynamics the field ϕ is *not conserved*, and $\hat{\mu} = \mu$ and $\hat{D} = D$ are constants. In **model B** dynamics the field ϕ is *conserved*, and $\hat{\mu} = -\mu\nabla^2$ and $\hat{D} = -D\nabla^2$.

Let us now go back to the example of a conserved volume of fluid whose surface fluctuations are subject to the Hamiltonian (24). The equation of motion constructed from model B dynamics is[a]

$$\frac{\partial h(\mathbf{x}, t)}{\partial t} = \mu\rho g \nabla^2 h - \mu\sigma\nabla^4 h + \eta(\mathbf{x}, t). \tag{58}$$

The evolution of each Fourier mode is given by

$$\frac{\partial h(\mathbf{q}, t)}{\partial t} = -\mu q^2(\rho g + \sigma q^2)h(\mathbf{q}, t) + \eta(\mathbf{q}, t) \equiv -\frac{h(\mathbf{q}, t)}{\tau(\mathbf{q})} + \eta(\mathbf{q}, t). \tag{59}$$

Because of the constraints imposed by the conservation law, the relaxation of the surface is more difficult, and slower. The relaxation times diverge even in the presence of gravity, and depending on wavelength we can define dynamic exponents z, via

$$\tau(\mathbf{q}) = \frac{1}{\mu q^2(\rho g + \sigma q^2)} \approx \begin{cases} q^{-2} & \text{for } q \ll \lambda_c^{-1} & (z = 2) \\ q^{-4} & \text{for } q \gg \lambda_c^{-1} & (z = 4). \end{cases} \tag{60}$$

---

[a]While model A dynamics provides a reasonably accurate description of the relaxation of a soap film, model B dynamics is not particularly useful for describing surface waves. As conservation of momentum in the fluid is an important constraint not included here, the following results are merely intended as an illustration.

The equilibrium behavior is unchanged, and

$$\lim_{t\to\infty} \langle |h(\mathbf{q},t)|^2 \rangle = \frac{Dq^2}{\mu q^2(\rho g + \sigma q^2)} = \frac{D}{\mu(\rho g + \sigma q^2)}, \qquad (61)$$

as before. Thus the same static behavior can be achieved by different dynamics. The static exponents (e.g., χ) are determined by the equilibrium (stationary) state and are unchanged, while the dynamic exponents may be different. As a result, dynamical critical phenomena involve many more universality classes than the corresponding static ones.

2. Dynamic Scaling in Growing Films

Inhomogeneities in a deposition process may lead to formation of rough surfaces. Fluctuations in the height $h(\mathbf{x},t)$ of the surface (at location $\mathbf{x}$ and time t) can be probed directly by scanning microscopy, or indirectly by scattering. Analytical or numerical treatments of simple growth models suggest that, quite generally, the height fluctuations have a self-similar character; their average correlations exhibiting a dynamic scaling form, $\langle [h(\mathbf{x},t) - h(\mathbf{x}',t')]^2 \rangle = |\mathbf{x} - \mathbf{x}'|^{2\alpha} f(|\mathbf{x} - \mathbf{x}'|^z / |t - t'|)$. The *roughness* and *dynamic* exponents, α and z, are expected to be *universal*, depending only on the underlying mechanism that generates self-similar scaling. Despite its ubiquitous occurrence in theory and simulations, experimental confirmation of dynamic scaling has been scarce. In some cases where such scaling has been observed, the exponents are different from those expected on the basis of analysis or numerics. I shall briefly review the theoretical foundations of dynamic scaling, and suggest possible reasons for discrepancies with experimental results. Note that in Eq. (47) we used the symbol χ to indicate the roughness exponent. Following Family and Vicsek,[8] α is the symbol used in the context of growing surfaces. The symbols ζ and H are also used to describe the same exponent.

2.1. *Dynamic scaling*

The growth of films by deposition is clearly of great technological interest. It has recently been recognized that such growth processes also pose important issues in the physics of nonlinear and complex systems. Rather crudely, we can distinguish between three types of growth morphologies:

(i) *Layer by layer* growth is naturally most desired from the technological stand-point. From a theoretical perspective, this type of growth is not stable in the presence of inhomogeneities in the deposition beam. However, this is only a statement about asymptotic behavior, as it is clearly possible to grow many layers in this mode.

(ii) *Unstable* growth occurs when the selected orientation of the substrate cannot be maintained. The instability is usually manifested by the formation of mounds, or other macroscopic features on the surface. Another type of instability results from growth that is controlled by a diffusive field. This instability may lead to formation of fractal aggregates. This is probably unrelated to growth by deposition and will not be discussed here.

(iii) *Self-affine* surfaces appear in a growth mode that is intermediate between the above. The average orientation of the surface is maintained, but it becomes rough; the amount of roughness grows with time and/or scale of observation in a self-similar fashion.

The most economical way to characterize self-affine roughness is by appealing to a dynamic scaling form[11]: If the height of the surface at location $\mathbf{x}$, at time t, is described by a function $h(\mathbf{x}, t)$, its average correlations may satisfy the scaling form

$$g(\mathbf{r}, t) \equiv \langle [h(\mathbf{x}, \tau) - h(\mathbf{x} + \mathbf{r}, \tau + t)]^2 \rangle = |\mathbf{r}|^{2\alpha} f\left(\frac{t}{|\mathbf{r}|^z}\right). \tag{62}$$

Family and Vicsek[11] first proposed such scaling in the context of interface growth based on numerical results. There is now a firm theoretical basis for such behavior. The *dynamic exponent z* describes the evolution of correlated regions with time: initially different parts of the surface are independent; but regions of correlated roughness form over time, their size growing as $\xi(t) \propto t^{1/z}$. In each correlated region, the width of surface grows as the observation scale raised to the *roughness exponent* α. Thus, the overall width of the surface initially grows as t^β with $\beta = \alpha/z$, until it saturates at L^α where L is the sample size.

2.2. *Discrete models*

A large number of numerical models of growth, of varying levels of sophistication, have been developed.[12] Here, I shall briefly describe some of the simplest versions.

Random Deposition: Particles are dropped randomly over deposition sites, and stick to the top of the pre-existing column on that site.[13] The height of each column thus performs an independent random walk, such that

$$\langle h_i(t) \rangle = vt \quad \text{and} \quad \langle (h_i(t) - vt)^2 \rangle = Dt, \tag{63}$$

where v is the average deposition rate per column, and D denotes its fluctuations. Since the columns are independent, the correlation length does not grow with time, $\xi = 1$ and $1/z = 0$. This model leads to unrealistically rough surfaces whose overall width grow with the exponent $\beta = 1/2$ without saturation.

Sedimentation: After the dropped particle reaches the top of a column, it rolls down-hill to neighboring columns until it reaches a local height minimum, where it stops. The smoothening effects of the rolling particles generate correlations in height, and lead to more realistic surfaces. Since the motion of rolling particles is more or less diffusive, it is likely that the correlation length grows as $\xi(t) \propto t^{1/2}$. This value of $z = 2$ is supported by numerical simulations,[14] which find in $1 + 1$ dimensions (i.e. for a one-dimensional substrate) the exponents $\alpha = 0.48 \pm 0.02$ and $\beta = 0.24 \pm 0.01$.

Ballistic deposition: Originally introduced for particles moving in a continuous space,[15] several variants with discretized particle locations have since been developed.[11,12] The deposited particles are now attached at the *first point of contact* with the growing aggregate. Since the first contact may be to the side of a high column, this model leaves behind a finite density of voids as it grows.

Restricted solid on solid models: There are no voids, and the height is a single-valued function of position. These models impose a restriction on the maximum height difference between neighboring columns.[16] A growth attempt that violates such a restriction is rejected. (This can be regarded as a crude form of desorption.) The last two models are expected to be in the same universality class, but better numerical values for the exponents are found from the latter. The simulations in $d = 2$ give $\alpha \approx 1/2$ and $\beta \approx 1/3$, while for $2 + 1$ dimensions they result in $\alpha \approx 0.38$ and $\beta \approx 0.24$.

Realistic models: Attempt to more closely mimic the physical processes involved in the growth of surfaces.[17] These include the detailed motion of the deposited atom, its subsequent movement on the surface; incorporation on steps or islands, island nucleation, and desorption. The main difficulty with simulating all these processes is that they take place over a wide range of time scales, making the analysis of large scale roughness quite difficult.

2.3. *Continuum equations*

Rather than describing the detailed microscopic evolution of the surface, continuum equations focus on the (hopefully universal) macroscopic aspects of its roughness, e.g., the exponents α and β. The general philosophy is to examine the evolution of coarse-grained (hydrodynamic) variables; in the example at hand, the height function $h(\mathbf{x}, t)$. Whereas in near equilibrium situations the evolution equation is obtained from variations of an energy functional, such an approach is not appropriate far away from equilibrium. Here we follow an approach described in Refs. 18 and 19: the equation of motion is decomposed as

$$\frac{\partial h}{\partial t} = \eta(\mathbf{x}, t) + \Phi[h, \nabla h, \nabla^2 h, \ldots]. \tag{64}$$

Particle deposition is described by the first term. Thus $\eta(\mathbf{x}, t)$ is a random function whose mean value gives the average particle flux at $\mathbf{x}$, and its fluctuations represent the shot noise in deposition. For simplicity, we shall assume that the noise is uncorrelated at different sites and different times, that is, it is a Gaussian process with

$$\langle \eta(\mathbf{x}, t) \rangle = v \,,$$
$$\langle \delta\eta(\mathbf{x}, t) \delta\eta(\mathbf{x}', t') \rangle = D\delta(\mathbf{x} - \mathbf{x}')\delta(t - t') \,. \tag{65}$$

Surface relaxation subsequent to deposition is described by the functional Φ. As we shall see in the following examples, it can depend on various properties of the height, such as its slope ∇h, or curvature $\nabla^2 h$. We shall assume that the relaxation is *local*, that is, it can be adequately described by the first few terms of an expansion of Φ in h and its gradients. Which terms can be included in such an expansion are then determined by the underlying symmetries and conservation laws appropriate to the dynamics. The basic idea is that any term that is not excluded for fundamental reasons of symmetry or conservation, will be generically present.

We shall now describe continuum equations for the discrete models introduced in the previous section.

Random deposition with no subsequent relaxation corresponds to $\Phi = 0$. Integrating $\partial_t h = \eta(\mathbf{x}, t)$ yields

$$h(\mathbf{x}, t) = \int_0^t dt' \, \eta(\mathbf{x}, t') , \tag{66}$$

from which we can immediately obtain

$$\langle h(\mathbf{x}, t) \rangle = vt \qquad \text{and} \qquad \langle \delta h(\mathbf{x}, t) \delta h(\mathbf{x}', t) \rangle = Dt\delta(\mathbf{x} - \mathbf{x}') , \tag{67}$$

corresponding to $\beta = 1/2$, and a zero correlation length.

Sedimentation was originally analyzed by Edwards and Wilkinson (EW) in Ref. 20. They concluded that the main relaxational process is proportional to local curvature, leading to

$$\partial_t h = \eta(\mathbf{x}, t) + \nu \nabla^2 h . \tag{68}$$

This linear (diffusion) equation is readily solved in Fourier space, leading to

$$\langle |h(\mathbf{k}, \omega)|^2 \rangle = \frac{D}{\omega^2 + \nu^2 k^4} . \tag{69}$$

Recasting the above result in real space then leads to the exponents

$$z = 2, \qquad \alpha = \frac{2 - d}{2}, \qquad \beta = \frac{2 - d}{4}, \tag{70}$$

for a d-dimensional substrate in $d + 1$ dimensional space. In particular, in $d = 2$, $\alpha = 1/2$ and $\beta = 1/4$, while for $d = 3$, the mean-square width grows logarithmically in both space and time.

Ballistic deposition is generically a nonlinear process. There is no *a priori* reason why the relaxation function Φ should not depend on the slope ∇h. (A direct dependence on h itself is ruled out by the translational symmetry $h \to h + \text{constant}$ of the underlying dynamics.) By symmetry, the relaxation should be the same for slopes $\pm \nabla h$, and hence the first term in an expansion in powers of slope starts with $(\nabla h)^2$ leading to

$$\partial_t h = \eta(\mathbf{x}, t) + \nu \nabla^2 h + \frac{\lambda}{2}(\nabla h)^2 + \cdots . \tag{71}$$

Higher order terms can also be added, but are irrelevant in that they don't change the scaling properties. There are several excellent reviews of Eq. (71), known as the KPZ equation.[21] Since the origin and properties of the nonlinear term are discussed in detail in these reviews,[9,10] I shall not elaborate on them any further. In $d = 1$, the nonlinear equation can in fact be solved exactly, and leads to $\alpha = 1/2$ and $\beta = 1/3$ (as compared to $\beta = 1/4$ for the EW equation), in excellent agreement with the numerical simulations.[11,12] There are no exact solutions in $d = 2$, but based on simulations we can estimate $\alpha \approx 0.38$ and $\beta \approx 0.24$. A characteristic signature of Eq. (71) is the exponent identity

$$\alpha + \frac{\alpha}{\beta} = 2, \tag{72}$$

obeyed in almost all simulations. The KPZ equation appears to describe the asymptotic behavior of most local, random growth processes.

2.4. *Conservative MBE models*

We may ask why the KPZ nonlinearity is not present in the model of sedimentation. The reason is that it is forbidden by a conservation law. If the growth process does not allow the formation of *overhangs* or *voids*, and there is also no *desorption*, then all the incoming flux is incorporated into the growing aggregate. This means that the net mass of the aggregate, proportional to $H(t) = \int d^d x h(\mathbf{x}, t)$, can only change due to the random deposition; the relaxation processes should conserve $H(t)$. This immediately implies that the relaxation function must be related to a surface current, i.e.

$$\Phi = -\nabla \cdot \mathbf{j}[\nabla h, \nabla^2 h, \ldots]. \tag{73}$$

The KPZ nonlinearity is thus ruled out, as it cannot be written as the divergence of another function. Following some observations of Villain,[22] many studies have focused on such conservative models in the context of MBE growth. The basic idea is that aggregates formed by the MBE process are typically free from holes and defects, and that the desorption of the adsorbed particles is negligible. It is thus argued that, at least over some sizeable preasymptotic regime, the relaxation processes should be conservative. Some examples of such conservative models are discussed in the remainder of this section.

Surface diffusion currents in equilibrium can be related to variations of a chemical potential via $\mathbf{j} \propto -\nabla \mu$. Since in the equilibrium between two phases, the chemical potential is proportional to local curvature ($\mu \propto -\nabla^2 h$), this leads to an equation of motion

$$\partial_t h = \eta(\mathbf{x}, t) - \kappa \nabla^4 h. \tag{74}$$

This equation is again linear, and as for Eq. (69) can be solved by Fourier transformation to yield,

$$\langle |h(\mathbf{k},\omega)|^2 \rangle = \frac{D}{\omega^2 + \kappa^2 k^8}.$$

(75)

The corresponding exponents in real space are,

$$z = 4, \qquad \alpha = \frac{4-d}{2}, \qquad \beta = \frac{4-d}{8}.$$

(76)

This leads to $\alpha = 3/2$ and $\beta = 3/8$ in $d = 1$, and $\alpha = 1$ and $\beta = 1/4$ in $d = 2$. Note that a surface remains self–affine, maintaining a well defined orientation, only as long as $\alpha < 1$. The above large values of the exponent α indicate a break down of validity of the above equation for $d \leq 2$.

Nonlinear MBE models have been proposed partly to remedy the break down of the linear equation for $\alpha \geq 1$. One such model starts with a nonlinear chemical potential introduced by Sun, Guo, and Grant,[23] resulting in

$$\partial_t h = \eta(\mathbf{x},t) - \kappa \nabla^4 h + \frac{\lambda'}{2}\nabla^2 (\nabla h)^2.$$

(77)

Despite its nonlinear form, Eq. (77) can in fact be analyzed to yield the exact exponents[24]

$$z = \frac{8+d}{3}, \qquad \alpha = \frac{4-d}{3}, \qquad \beta = \frac{4-d}{8+d}.$$

(78)

In particular $\alpha = 2/3$ and $\beta = 1/5$ in $d = 2$. However, in nonequilibrium circumstances, there is no reason for the surface current to be derivable from a chemical potential. Removing this restriction allows the inclusion of other nonlinearities,[24,25] such as $\nabla(\nabla h)^3$, which are in fact more relevant. More importantly, nonequilibrium currents should generically include a term proportional to ∇h, which is the dominant gradient as discussed next.

Diffusion bias refers to such generic nonequilibrium currents that are proportional to the local surface slope.[26] One possible origin of such currents is in the Schwoebel barriers[27]: Atoms on a stepped surface are easily incorporated at the step to a higher ledge, but are reflected by a barrier towards jumping to a lower ledge.[28] This sets up a net up-hill current[22] $\mathbf{j} = \nu' \nabla h$, leading to an equation of motion

$$\partial_t h = \eta(\mathbf{x},t) - \nu' \nabla^2 h + \cdots.$$

(79)

This equation leads to an unstable growth of fluctuations, and therefore higher order terms are necessary to ensure stability. For example, Johnson *et al.*[29] have proposed the following nonlinear equation

$$\frac{\partial h}{\partial t} = \eta(\mathbf{x}, t) - \nabla\left(\frac{\nu'\nabla h}{1 + (\nabla h)^2}\right) - \kappa\nabla^4 h. \tag{80}$$

The instabilities in this equation develop into a complex array of mounds dubbed as SLUGs (Super Large Unstable Growths). Finally, it is also possible for the nonequilibrium currents to be oriented down-hill (as in sedimentation), in which case the behavior is the same as the EW equation (68) discussed earlier.

2.5. *Discussion*

An excellent review of the experimental research in this subject is provided by Krim and Palasantzas.[30] The observational methods include diffraction (specular or diffuse x-rays, RHEED, LEED, HR-LEED, and helium scattering), direct imaging (STM, AFM, SEM and TEM), and surface adsorption. A variety of metallic (silver, gold, copper, iron), and other (Si, InP, NbN, polymer) surfaces grown under a host of different conditions have been examined by such probes. Some of these surfaces exhibit unstable growth, while others appear to satisfy self-similar scaling. However, there is usually no clear-cut identification of the exponents with the theoretical models. Some experiments on gold and silver give roughness exponents consistent with the KPZ value, but larger values of β. Other surfaces give larger values of α, consistent with those of the nonlinear MBE equation (78). The reader is referred to this review article[30] for the details. Perhaps the following statements at the end of the review are most revealing of the experimental situation: "Over 50% of the experimental work reported on here was published in the interval from January 1993 to August 1994. The pace of experimental work is clearly accelerating, and rapid advances in the field can be expected."

Given the discrepancies between experiment and theory, we can also ask if important elements have been left out of the analysis. The formalism presented so far deals solely with a single coarse-grained variable, the height $h(\mathbf{x}, t)$. Other variables may play an important role in the evolution of h. For example, in many cases the roughness is intimately related to formation of microcrystalline grains. Variations in crystallinity have so far been left out of the theoretical picture. In principle, one could introduce an additional "order parameter" $M(\mathbf{x}, t)$

describing the local degree of crystallinity. Surface relaxation may then depend on this order parameter, leading to $\Phi(\nabla h, \nabla^2 h, M, \nabla M, \ldots)$. We should then also include an additional dynamical equation for the evolution of M. This direction will be explored further in the final section.

Equation (64) can be regarded as representing a complex filter, converting the white input noise $\eta(\mathbf{x}, t)$ to the correlated random function $h(\mathbf{x}, t)$ through the action of Φ. So far, the focus has been on the relaxation function Φ, assuming that the input noise is uncorrelated. Not surprisingly, correlations in the input noise lead to more correlated surface roughness with larger values of the exponents α and β.[31] Maybe, in view of the rather large exponents observed experimentally, this point should be further investigated. Starting with $D(\mathbf{k}, \omega) \equiv \langle |\eta(\mathbf{k}, \omega)|^2 \rangle \propto |\mathbf{k}|^{-2\rho}$, roughness exponents may be tuned continuously by changing the parameter ρ. This is not satisfactory, as it relies on a rather arbitrarily chosen exponent for noise correlations. Here I shall propose another possibility which is less arbitrary, and may be relevant to some experiments. My choice of noise correlations is motivated by the scaling phenomena observed in turbulence, another interesting problem in nonlinear physics.

In the case of turbulence, the fluid is stirred at long length scales, setting up a *Kolmogorov energy cascade*[32] by which energy is transferred to shorter wave-length modes, ultimately dissipating at a microscopic scale. In the intermediate (inertial) regime, the energy density follows simple power laws. We may similarly assume that the deposition noise is correlated over long distances ℓ, such that $D(\mathbf{k})$ is large only for $|\mathbf{k}| < 1/\ell$. Distances intermediate between the atomic scale and ℓ are then analogous to the inertial regime in turbulence. Adapting the arguments of Kolmogorov to the KPZ equation leads to

$$z \approx \frac{4}{3}, \qquad \alpha \approx \frac{2}{3}, \qquad \beta \approx \frac{1}{2}, \tag{81}$$

in all dimensions. (Because there is no conservation of a corresponding energy for the KPZ equation in $d \neq 1$, the arguments leading to Eq. (81) are considerably less compelling than those in turbulence. Hence, the above exponents should be regarded as a first approximation.) The exponent $\beta = 1/2$ is similar to that of random deposition. Indeed, an early experiment on recrystallization of amorphous GaAs films,[33] obtained an exponent of $\beta \approx 0.50$ for highly correlated surfaces.

3. Moving Flux Lines and Polymers

3.1. *Some properties of the KPZ equation*

First consider *deterministic growth*, such as a slow and uniform snowfall, on an initial profile which at $t = 0$ is described by $h_0(\mathbf{x})$. The nonlinear equation can in fact be *linearized* with the aid of a "Cole-Hopf" transformation,

$$W(\mathbf{x}, t) = \exp\left[\frac{\lambda}{2\nu} h(\mathbf{x}, t)\right]. \tag{82}$$

The function $W(\mathbf{x}, t)$ evolves according to the diffusion equation with *multiplicative noise*,

$$\frac{\partial W(\mathbf{x}, t)}{\partial t} = \nu \nabla^2 W + \frac{\lambda}{2\nu} W \eta(\mathbf{x}, t). \tag{83}$$

In the absence of noise, $\eta(\mathbf{x}, t) = 0$, Eq. (83) can be solved subject to the initial condition $W(\mathbf{x}, t = 0) = \exp[\lambda h_0(\mathbf{x})/2\nu]$, and leads to the growth profile,

$$h(\mathbf{x}, t) = \frac{2\nu}{\lambda} \ln\left\{\int d^d\mathbf{x}' \exp\left[-\frac{|\mathbf{x} - \mathbf{x}'|^2}{2\nu t} + \frac{\lambda}{2\nu} h(\mathbf{x}, t)\right]\right\}. \tag{84}$$

It is instructive to examine the $\nu \to 0$ limit, which is indeed appropriate to snowfalls since there is not much rearrangement after deposition. In this limit, the integral in Eq. (84) can be performed by the saddle point method. For each $\mathbf{x}$ we have to identify a point $\mathbf{x}'$ which maximizes the exponent, leading to a collection of paraboloids described by

$$h(\mathbf{x}, t) = \max\left\{h_0(\mathbf{x}') - \frac{|\mathbf{x} - \mathbf{x}'|^2}{2\lambda t}\right\}_{\mathbf{x}'}. \tag{85}$$

Such parabolic sequences are quite common in many layer by layer growth processes in nature, from biological to geological formations. The patterns for $\lambda = 1$ are identical to those obtained by the geometrical construction of Huygens, familiar from optics. The growth profile (wave front) is constructed from the outer envelop of circles of radius t drawn from all points on the initial profile. The nonlinearity in Eq. (71) thus algebraically accounts for their origin.

As growth proceeds, the surface smoothens by the *coarsening* of the parabolas. What is the typical size of the features at time t? In maximizing the exponent in Eq. (85), we have to balance a reduction $|\mathbf{x} - \mathbf{x}'|^2/2\lambda t$, by a possible gain from $h_0(\mathbf{x'})$ in selecting a point away from $\mathbf{x}$. The final scaling is controlled by the roughness of the initial profile. Let us assume that the original pattern is a *self-affine fractal* of roughness χ, i.e.

$$\overline{|h_0(\mathbf{x}) - h_0(\mathbf{x'})|} \sim |\mathbf{x} - \mathbf{x'}|^\chi \,. \tag{86}$$

Balancing the two terms in Eq. (85) gives

$$(\delta x)^\chi \sim \frac{(\delta x)^2}{t} \implies \delta x \sim t^{1/z}, \quad \text{with} \quad z + \chi = 2 \,. \tag{87}$$

For example, if the initial profile is like a random walk in $d = 1$, $\chi = 1/2$, and $z = 3/2$. This leads to the spreading of information along the profile by a process that is faster than diffusion, $\delta x \sim t^{2/3}$.

Note that the slope, $\vec{v}(\mathbf{x}, t) = -\lambda\vec{\nabla}h(x, t)$, satisfies the equation,

$$\frac{D\vec{v}(\mathbf{x}, t)}{Dt} \equiv \frac{\partial\vec{v}}{\partial t} + \vec{v}\cdot\vec{\nabla}\vec{v} = \nu\nabla^2\vec{v} - \lambda\nabla\eta \,. \tag{88}$$

This is the Navier–Stokes equation for the velocity of a fluid of viscosity ν, which is being randomly stirred by a conservative force,[34] $\vec{f} = -\lambda\nabla\eta$. The fluid is vorticity free since

$$\vec{\Omega} = \vec{\nabla}\times\vec{v} = -\lambda\nabla\times\nabla h = 0 \,. \tag{89}$$

This is the *Burgers'* equation,[35] which provides a simple example of the formation of shock waves in a fluid. The gradient of Eq. (85) in $d = 1$ gives a saw-tooth pattern of shocks which coarsen in time.

To study stochastic roughening in the presence of the nonlinear term, we first carry out a scaling analysis. Under the scaling $\mathbf{x} \to b\mathbf{x}$, $t \to b^z t$, and $h \to b^\chi h$, Eq. (71) transforms to

$$b^{\chi-z}\frac{\partial h}{\partial t} = \nu b^{\chi-2}\nabla^2 h + \frac{\lambda}{2}b^{2\chi-2}(\nabla h)^2 + \eta(b\mathbf{x}, b^z t) \,. \tag{90}$$

The correlations of the transformed noise, $\eta'(\mathbf{x}, t) = b^{z-\chi}\eta(b\mathbf{x}, b^z t)$, satisfy

$$\langle\eta'(\mathbf{x}, t)\eta'(\mathbf{x'}, t')\rangle = b^{2z-2\chi}\, 2D\, \delta^d(\mathbf{x} - \mathbf{x'})b^{-d}\delta(t - t')b^{-z}$$

$$= b^{z-d-2\chi}\, 2D\, \delta^d(\mathbf{x} - \mathbf{x'})\delta(t - t') \,. \tag{91}$$

Under such scaling the parameters of Eq. (71) are transformed to

$$\begin{cases} \nu \to b^{z-2}\nu \\ \lambda \to b^{\chi+z-2}\lambda \\ D \to b^{z-2\chi-d}D \,. \end{cases} \tag{92}$$

For $\lambda = 0$, the equation is made scale invariant upon the choice of $z_0 = 2$, and $\chi_0 = (2 - d)/2$. Close to this linear-fixed point, λ scales to $b^{z_0 + \chi_0 - 2}\lambda = b^{(2-d)/2}\lambda$, and is a relevant operator for $d < 2$. In fact a perturbative dynamic renormalization group suggests that it is *marginally relevant* at $d = 2$, and that in all dimensions a sufficiently large λ leads to new scaling behavior.

Are there any nonrenormalization conditions that help in identifying the exponents of the full nonlinear stochastic equation? Note that since Eqs. (71) and (88) are related by a simple transformation, they must have the same scaling properties. Since the Navier–Stokes equation is derivable from Newton's laws of motion for fluid particles, it has the Galilean invariance of changing to a uniformly moving coordinate frame. This symmetry is preserved under renormalization to larger scales and requires that the ratio of the two terms on the left-hand side of Eq. (88) ($\partial_t \vec{v}$ and $\vec{v} \cdot \nabla \vec{v}$) stays at unity. In terms of Eq. (71), this implies the nonrenormalization of the parameter λ, and leads to the exponent identity

$$\chi + z = 2 . \tag{93}$$

Unfortunately there is no other nonrenormalization condition except in $d = 1$. Following Eq. (48), we can write down a Fokker–Planck equation for the evolution of the configurational probability as

$$\frac{\partial P([h(\mathbf{x})], t)}{\partial t} = - \int d^d\mathbf{x} \frac{\delta}{\delta h(\mathbf{x})} \left[\left(\nu \nabla^2 h + \frac{\lambda}{2}(\nabla h)^2 \right) P - D \frac{\delta P}{\delta h(\mathbf{x})} \right] . \tag{94}$$

Since Eq. (71) was not constructed from a Hamiltonian in general, we do not know the stationary solution at long times. In $d = 1$, we make a guess and try a solution of the form

$$P_0[h(x)] \propto \exp\left[-\frac{\nu}{2D} \int dx (\partial_x h)^2 \right] . \tag{95}$$

Since

$$\frac{\delta P_0}{\delta h(x)} = -\partial_x \frac{\delta P_0}{\delta(\partial_x h)} = \frac{\nu}{D}(\partial_x^2 h)P_0 , \tag{96}$$

Equation (94) leads to

$$\frac{\partial P_0}{\partial t} = - \int dx \frac{\delta P_0}{\delta h(x)} \left(\nu \partial_x^2 h + \frac{\lambda}{2}(\partial_x h)^2 - D \frac{\nu}{D} \partial_x^2 h \right)$$

$$= -\frac{\lambda}{2} P_0 \int dx \frac{\nu}{D} (\partial_x^2 h)(\partial_x h)^2 = -\frac{\lambda\nu}{2D} P_0 \int dx \partial_x \left(\frac{(\partial_x h)^3}{3} \right) = 0 . \tag{97}$$

We have thus identified the stationary state of the one-dimensional equation. (This procedure does not work in higher dimensions as it is impossible to write the final result as a total derivative.) Surprisingly, the stationary distribution is the same as the one in equilibrium at a temperature proportional to D/ν. We can thus immediately identify the roughness exponent $\chi = 1/2$, which together with the exponent identity in Eq. (93) leads to $z = 3/2$, i.e. super-diffusive behavior.

The values of the exponents in the strongly nonlinear regime are not known exactly in higher dimensions. However, extensive numerical simulations of growth have provided fairly reliable estimates.[11] In the physically relevant case ($d = 2$) of a surface grown in three dimensions, $\chi \approx 0.39$ and $z \approx 1.61$.[36]

As an aside, we remark that some exact information is available for the *anisotropic* KPZ equation in $2 + 1$ dimensions. Using a perturbative RG approach, Wolf showed[37] that in the equation

$$\partial_t h = K\nabla^2 h + \frac{\lambda_x}{2}(\partial_x h)^2 + \frac{\lambda_y}{2}(\partial_y h)^2 + \eta(x, y, t)\,, \tag{98}$$

the nonlinearities $\{\lambda_x, \lambda_y\}$ renormalize to zero if they initially have opposite signs. This suggests logarithmic fluctuations for the resulting interface, as in the case of the linear Langevin equation. In fact, it is straightforward to demonstrate that Eq. (98) also satisfies a fluctuation dissipation condition if $\lambda_x = -\lambda_y$. When this condition is satisfied, the associated Fokker–Planck equation has a steady-state solution

$$\mathcal{P} = \exp\left(-\frac{\nu}{2D} \int dx\,dy\,(\nabla h)^2\right)\,. \tag{99}$$

This is a nonperturbative result which again indicates the logarithmic fluctuations resulting from Eq. (98). In this context, it is interesting to note that the steady-state distribution for an exactly solvable discrete model of surface growth belonging to the above universality class has also been obtained in Ref. 38. (We have shown that there are no other similarly generalized KPZ equations that satisfy a fluctuation dissipation condition in higher dimensions.[39])

3.2. *A moving flux line*

Let us now turn to the case of a line in three dimensions. Fluctuations of the line can be indicated by a two dimensional vector **r**. (The notations for the parameters in this section are chosen to conform with the literature for polymers.) Even in an isotropic medium, the drift velocity $\mathbf{v}$ breaks the isotropy in $\mathbf{r}$ by selecting a direction. A gradient expansion up to second order for the equation of motion gives[40]

$$\partial_t r_\alpha = [K_1 \delta_{\alpha\beta} + K_2 v_\alpha v_\beta] \partial_x^2 r_\beta$$

$$+ [\lambda_1(\delta_{\alpha\beta}v_\gamma + \delta_{\alpha\gamma}v_\beta) + \lambda_2 v_\alpha \delta_{\beta\gamma} + \lambda_3 v_\alpha v_\beta v_\gamma] \frac{\partial_x r_\beta \partial_x r_\gamma}{2} + \eta_\alpha\,, \quad (100)$$

with random force correlations

$$\langle \eta_\alpha(x,t)\eta_\beta(x',t')\rangle = 2[T_1 \delta_{\alpha\beta} + T_2 v_\alpha v_\beta]\delta(x-x')\delta(t-t')\,. \quad (101)$$

Higher order nonlinearities can be similarly constructed but are in fact irrelevant. In terms of components parallel and perpendicular to the velocity, the equations are

$$\begin{cases} \partial_t r_\| = K_\| \partial_x^2 r_\| + \dfrac{\lambda_\|}{2}(\partial_x r_\|)^2 + \dfrac{\lambda_\times}{2}(\partial_x r_\perp)^2 + \eta_\|(x,t)\,, \\[2mm] \partial_t r_\perp = K_\perp \partial_x^2 r_\perp + \lambda_\perp \partial_x r_\| \partial_x r_\perp + \eta_\perp(x,t)\,, \end{cases} \quad (102)$$

with

$$\begin{cases} \langle \eta_\|(x,t)\eta_\|(x',t')\rangle = 2T_\|\delta(x-x')\delta(t-t')\,, \\[2mm] \langle \eta_\perp(x,t)\eta_\perp(x',t')\rangle = 2T_\perp\delta(x-x')\delta(t-t')\,. \end{cases} \quad (103)$$

The noise-averaged correlations have the dynamic scaling form

$$\begin{cases} \langle [r_\|(x,t) - r_\|(x',t')]^2\rangle = |x-x'|^{2\zeta_\|} g_\| \left(\dfrac{|t-t'|}{|x-x'|^{z_\|}} \right), \\[4mm] \langle [r_\perp(x,t) - r_\perp(x',t')]^2\rangle = |x-x'|^{2\zeta_\perp} g_\perp \left(\dfrac{|t-t'|}{|x-x'|^{z_\perp}} \right). \end{cases} \quad (104)$$

In the absence of nonlinearities ($\lambda_\| = \lambda_\times = \lambda_\perp = 0$), Eqs. (102) can easily be solved to give $\zeta_\| = \zeta_\perp = 1/2$ and $z_\| = z_\perp = 2$. Simple dimensional counting indicates that all three nonlinear terms are relevant and may modify the exponents in Eq. (104). Studies of related stochastic equations[41,37] indicate that interesting dynamic phase diagrams may emerge from the competition between nonlinearities. Let us assume that $\lambda_\|$ is positive and finite (its sign can be changed by $r_\parallel \to -r_\parallel$), and focus on the dependence of the scaling exponents on the ratios $\lambda_\perp/\lambda_\parallel$ and $\lambda_\times/\lambda_\parallel$. (It is more convenient to set the vertical axis to $\lambda_\times K_\parallel T_\perp / \lambda_\parallel K_\perp T_\parallel$.)

The properties discussed for the KPZ equation can be extended to this higher dimensional case:

(1) *Galilean Invariance (GI)*: Consider the infinitesimal reparametrization

$$\begin{cases} x' = x + \lambda_\parallel \epsilon t, & t' = t, \\ r'_\parallel = r_\parallel + \epsilon x, & r'_\perp = r_\perp. \end{cases} \tag{105}$$

Equations (102) are invariant under this transformation provided that $\lambda_\parallel = \lambda_\perp$. Thus *along this line* there is GI, which again implies the exponent identity

$$\zeta_\parallel + z_\parallel = 2. \tag{106}$$

(2) *Fluctuation–Dissipation (FD) Condition*: The Fokker–Planck equation for the evolution of the joint probability $\mathcal{P}[r_\parallel(x), r_\perp(x)]$ has a stationary solution

$$\mathcal{P}_0 \propto \exp\left(-\int dx \left[\frac{K_\parallel}{2T_\parallel}(\partial_x r_\parallel)^2 + \frac{K_\perp}{2T_\perp}(\partial_x r_\perp)^2 \right] \right), \tag{107}$$

provided that $\lambda_\times K_\parallel T_\perp = \lambda_\perp K_\perp T_\parallel$. Thus for this special choice of parameters, if $\mathcal{P}$ converges to this solution, the long-time behavior of the correlation functions in Eq. (104) can be directly read off Eq. (107), giving $\zeta_\parallel = \zeta_\perp = 1/2$.

(3) *The Cole–Hopf (CH) Transformation*: is an important method for the exact study of solutions of the KPZ equation. Here we generalize this transformation to the complex plane by defining, *for $\lambda_\times < 0$*,

$$\Psi(x,t) = \exp\left(\frac{\lambda_\parallel r_\parallel(x,t) + i\sqrt{-\lambda_\parallel \lambda_\times}\, r_\perp(x,t)}{2K} \right). \tag{108}$$

The linear diffusion equation

$$\partial_t \Psi = K\partial_x^2 \Psi + \mu(x,t)\Psi,$$

then leads to Eqs. (102) if $K_\parallel = K_\perp = K$ and $\lambda_\parallel = \lambda_\perp$. [Here $\mathrm{Re}(\mu) = \lambda_\parallel \eta_\parallel/2K$ and $\mathrm{Im}(\mu) = \sqrt{-\lambda_\parallel \lambda_\times}\, \eta_\perp/2K$.] This transformation enables an exact solution of the *deterministic* equation, and further allows us to write the solution to the *stochastic* equation in the form of a path integral

$$\Psi(x,t) = \int_{(0,0)}^{(x,t)} \mathcal{D}x(\tau) \exp\left\{-\int_0^t d\tau \left[\frac{\dot{x}^2}{2K} + \mu(x,\tau)\right]\right\}. \qquad (109)$$

Equation (109) has been extensively studied in connection with quantum tunneling in a disordered medium,[42] with Ψ representing the wave function. In particular, results for the tunneling probability $|\Psi|^2$ suggest $z_{\parallel} = 3/2$ and $\zeta_{\parallel} = 1/2$. The transverse fluctuations correspond to the phase in the quantum problem which is not observable. Hence this mapping does not provide any information on $\zeta_{\perp}$ and $z_{\perp}$, which are in fact observable for the moving line.

At the point $\lambda_{\perp} = \lambda_{\times} = 0$, $r_{\parallel}$ and $r_{\perp}$ decouple, and $z_{\perp} = 2$ while $z_{\parallel} = 3/2$. However, in general $z_{\parallel} = z_{\perp} = z$ unless the effective $\lambda_{\perp}$ is zero. For example at the intersection of the subspaces, with GI and FD, the exponents $z_{\parallel} = z_{\perp} = 3/2$ are obtained from the exponent identities. Dynamic RG recursion relations can be computed to one-loop order,[40,43] by standard methods of momentum-shell dynamic RG.[34,31]

The renormalization of the seven parameters in Eqs. (102), generalized to n transverse directions, give the recursion relations

$$\frac{dK_{\parallel}}{d\ell} = K_{\parallel}\left[z - 2 + \frac{1}{\pi}\frac{\lambda_{\parallel}^2 T_{\parallel}}{4K_{\parallel}^3} + n\frac{1}{\pi}\frac{\lambda_{\perp}\lambda_{\times}T_{\perp}}{4K_{\parallel}K_{\perp}^2}\right],$$

$$\frac{dK_{\perp}}{d\ell} = K_{\perp}\left[z - 2 + \frac{1}{\pi}\frac{\lambda_{\perp}((\lambda_{\times}T_{\perp}/K_{\perp}) + (\lambda_{\perp}T_{\parallel}/K_{\parallel}))}{2K_{\perp}(K_{\perp} + K_{\parallel})}\right.$$
$$\left. + \frac{1}{\pi}\frac{K_{\perp} - K_{\parallel}}{K_{\perp} + K_{\parallel}}\frac{\lambda_{\perp}((\lambda_{\times}T_{\perp}/K_{\perp}) - (\lambda_{\perp}T_{\parallel}/K_{\parallel}))}{K_{\perp}(K_{\perp} + K_{\parallel})}\right],$$

$$\frac{dT_{\parallel}}{d\ell} = T_{\parallel}\left[z - 2\zeta_{\parallel} - 1 + \frac{1}{\pi}\frac{\lambda_{\parallel}^2 T_{\parallel}}{4K_{\parallel}^3}\right] + n\frac{1}{\pi}\frac{\lambda_{\times}^2 T_{\perp}^2}{4K_{\perp}^3},$$

$$\frac{dT_{\perp}}{d\ell} = T_{\perp}\left[z - 2\zeta_{\perp} - 1 + \frac{1}{\pi}\frac{\lambda_{\perp}^2 T_{\parallel}}{K_{\perp}K_{\parallel}(K_{\perp} + K_{\parallel})}\right], \qquad (110)$$

$$\frac{d\lambda_{\parallel}}{d\ell} = \lambda_{\parallel}\left[\zeta_{\parallel} + z - 2\right],$$

$$\frac{d\lambda_\perp}{d\ell} = \lambda_\perp \left[\zeta_\parallel + z - 2 - \frac{1}{\pi} \frac{\lambda_\parallel - \lambda_\perp}{(K_\perp + K_\parallel)^2}((\lambda_\times T_\perp/K_\perp) - (\lambda_\perp T_\parallel/K_\parallel)) \right],$$

$$\frac{d\lambda_\times}{d\ell} = \lambda_\times \left[2\zeta_\perp - \zeta_\parallel + z - 2 + \frac{1}{\pi} \frac{\lambda_\parallel K_\perp - \lambda_\perp K_\parallel}{K_\perp K_\parallel (K_\perp + K_\parallel)}((\lambda_\times T_\perp/K_\perp) \right.$$

$$\left. -(\lambda_\perp T_\parallel/K_\parallel)) \right].$$

The RG flows naturally satisfy the constraints imposed by the nonperturbative results: the subspace of GI is closed under RG, while the FD condition appears as a *fixed line*. The RG flows, and the corresponding exponents, are different in each quadrant of parameter space, which implies that the scaling behavior is determined by the relative signs of the three nonlinearities. This was confirmed by numerical integrations[40,43] of Eqs. (102), performed for different sets of parameters.

The analysis of analytical and numerical results can be summarized as follows:

- $\lambda_\perp \lambda_\times > 0$: In this region, the scaling behavior is understood best. The RG flows terminate on the fixed line where FD conditions apply, hence $\zeta_\parallel = \zeta_\perp = 1/2$. All along this line, the one loop RG exponent is $z = 3/2$. These results are consistent with the numerical simulations. The measured exponents rapidly converge to these values, except when $\lambda_\perp$ or $\lambda_\times$ are small.
- $\lambda_\times = 0$: In this case the equation for $r_\parallel$ is the KPZ equation (71), thus $\zeta_\parallel = 1/2$ and $z_\parallel = 3/2$. The fluctuations in $r_\parallel$ act as a strong (multiplicative and correlated) noise on $r_\perp$. The one-loop RG yields the exponents $z_\perp = 3/2$, $\zeta_\perp = 0.75$ for $\lambda_\perp > 0$, while a negative $\lambda_\perp$ scales to 0 suggesting $z_\perp > z_\parallel$. Simulations are consistent with the RG calculations for $\lambda_\perp > 0$, yielding $\zeta_\perp = 0.72$, surprisingly close to the one-loop RG value. For $\lambda_\perp < 0$, simulations indicate $z_\perp \approx 2$ and $\zeta_\perp \approx 2/3$ along with the expected values for the longitudinal exponents.
- $\lambda_\perp = 0$: The transverse fluctuations satisfy a simple diffusion equation with $\zeta_\perp = 1/2$ and $z_\perp = 2$. Through the term $\lambda_\times (\partial_x r_\perp)^2/2$, these fluctuations act as a correlated noise[31] for the longitudinal mode. A naive application of the results of this Ref. 31 gives $\zeta_\parallel = 2/3$ and $z_\parallel = 4/3$. Quite surprisingly, simulations indicate different behavior depending on the sign of $\lambda_\times$. For $\lambda_\times < 0$, $z_\parallel \approx 3/2$ and $\zeta_\parallel \approx 1/2$ whereas for $\lambda_\times > 0$, longitudinal fluctuations are much stronger, resulting in $z_\parallel \approx 1.18$ and $\zeta_\parallel \approx 0.84$. Actually, $\zeta_\parallel$ increases

steadily with system size, suggesting a breakdown of dynamic scaling, due to a change of sign in $\lambda_\perp \lambda_\times$.

- $\lambda_\perp < 0$ and $\lambda_\times > 0$: The analysis of this region is the most difficult in that the RG flows do not converge upon a finite fixed point and $\lambda_\perp \to 0$, which may signal the breakdown of dynamic scaling. Simulations indicate strong longitudinal fluctuations that lead to instabilities in the discrete integration scheme, excluding the possibility of measuring the exponents reliably.

- $\lambda_\perp > 0$ and $\lambda_\times < 0$: The *projected* RG flows in this quadrant converge to the point $\lambda_\perp / \lambda_\parallel = 1$ and $\lambda_\times T_\perp K_\parallel / \lambda_\parallel T_\parallel K_\perp = -1$. This is actually not a fixed point, as $K_\parallel$ and $K_\perp$ scale to infinity. The applicability of the CH transformation to this point implies $z_\parallel = 3/2$ and $\zeta_\parallel = 1/2$. Since $\lambda_\perp$ is finite, $z_\perp = z_\parallel = 3/2$ is expected, but this does not give any information on $\zeta_\perp$. Simulations indicate strong transverse fluctuations and suffer from difficulties similar to those in the previous region.

Equations (102) are the simplest nonlinear, local, and dissipative equations that govern the fluctuations of a moving line in a random medium. They can be easily generalized to describe the time evolution of a manifold with arbitrary internal ($\mathbf{x} \in R^d$) and external ($\mathbf{r} \in R^{n+1}$) dimensions, and to the motion of curves that are not necessarily stretched in a particular direction. Since the derivation only involves general symmetry arguments, the given results are widely applicable to a number of seemingly unrelated systems. We will discuss one application to drifting polymers in more detail in the next section, explicitly demonstrating the origin of the nonlinear terms starting from more fundamental hydrodynamic equations. A simple model of crack front propagation in three dimensions[44] also arrives at Eqs. (102), implying the self-affine structure of the crack surface after the front has passed.

3.3. *Drifting polymers*

The dynamics of polymers in fluids is of much theoretical interest and has been extensively studied.[45,46] The combination of polymer flexibility, interactions, and hydrodynamics make a first principles approach to the problem quite difficult. There are, however, a number of phenomenological studies that describe various aspects of this problem.[47] One of the simplest is the Rouse model[48]: The configuration of the polymer at time t is described by a vector $\mathbf{R}(x,t)$, where $x \in [0,N]$ is a continuous variable replacing the discrete monomer index.

Ignoring inertial effects, the relaxation of the polymer in a viscous medium is approximated by

$$\partial_t \mathbf{R}(x,t) = \mu \mathbf{F}(\mathbf{R}(x,t)) = K\partial_x^2 \mathbf{R}(x,t) + \eta(x,t)\,, \tag{111}$$

where μ is the mobility. The force $\mathbf{F}$ has a contribution from interactions with near neighbors that are treated as springs. Steric and other interactions are ignored. The effect of the medium is represented by the random velocities η with zero mean. The Rouse model is a linear Langevin equation that is easily solved. It predicts that the mean square radius of gyration, $R_g^2 = \langle |\mathbf{R} - \langle \mathbf{R} \rangle|^2 \rangle$, is proportional to the polymer size N, and the largest relaxation times scale as the fourth power of the wave number (i.e. in dynamic light scattering experiments, the half width at half maximum of the scattering amplitude scales as the fourth power of the scattering wave vector $\mathbf{q}$). These results can be summarized as $R_g \sim N^\nu$ and $\Gamma(\mathbf{q}) \sim q^z$, where ν and z are called the *swelling* and *dynamic* exponents, respectively.[49] Thus, for the Rouse Model, $\nu = 1/2$ and $z = 4$.

The Rouse model ignores hydrodynamic interactions mediated by the fluid. These effects were originally considered by Kirkwood and Risemann[50] and later on by Zimm.[51] The basic idea is that the motion of each monomer modifies the flow field at large distances. Consequently, each monomer experiences an additional velocity

$$\delta_H \partial_t \mathbf{R}(x,t) = \frac{1}{8\pi\eta_s} \int dx' \frac{\mathbf{F}(x')r_{xx'}^2 + (\mathbf{F}(x') \cdot \mathbf{r}_{xx'})\mathbf{r}_{xx'}}{|\mathbf{r}_{xx'}|^3}$$

$$\approx \int dx' \frac{\gamma}{|x - x'|^\nu}\partial_x^2 \mathbf{R}\,, \tag{112}$$

where $\mathbf{r}_{xx'} = \mathbf{R}(x) - \mathbf{R}(x')$ and the final approximation is obtained by replacing the actual distance between two monomers by their average value. The modified equation is still linear in $\mathbf{R}$ and easily solved. The main result is the speeding up of the relaxation dynamics as the exponent z changes from 4 to 3. Most experiments on polymer dynamics[52] indeed measure exponents close to 3. Rouse dynamics is still important in other circumstances, such as diffusion of a polymer in a solid matrix, stress and viscoelasticity in concentrated polymer solutions, and is also applicable to relaxation times in Monte Carlo simulations.

Since both of these models are linear, the dynamics remains invariant in the center of mass coordinates upon the application of a uniform external force.

Hence the results for a drifting polymer are identical to a stationary one. This conclusion is in fact not correct due to the hydrodynamic interactions. For example, consider a rodlike conformation of the polymer with monomer length b_0, where $\partial_x R_\alpha = b_0 t_\alpha$ everywhere on the polymer, so that the elastic (Rouse) force vanishes. If a uniform force $\mathbf{E}$ per monomer acts on this rod, the velocity of the rod can be solved using Kirkwood Theory, and the result is[45]

$$\mathbf{v} = \frac{(-\ln \kappa)}{4\pi \eta_s b_0} \mathbf{E} \cdot [\mathbf{I} + \mathbf{tt}] . \tag{113}$$

In the above equation, η_s is the solvent viscosity, $\mathbf{t}$ is the unit tangent vector, and $\kappa = 2b/b_0 N$ is the ratio of the width b to the half length $b_0 N/2$ of the polymer. A more detailed calculation of the velocity in the more general case of an arbitrarily shaped slender body by Khayat and Cox[53] shows that *nonlocal* contributions to the hydrodynamic force, which depend on the whole shape of the polymer rather than the local orientation, are $\mathcal{O}(1/(\ln \kappa)^2)$. Therefore, corrections to Eq. (113) are small when $N \gg b/b_0$.

Incorporating this tilt dependence of polymer mobility requires adding terms nonlinear in the tilt, $\partial_x \mathbf{r}$, to a *local* equation of motion. Since the overall force (or velocity) is the only vector breaking the isotropy of the fluid, the structure of these nonlinear terms must be identical to Eq. (100). Thus in terms of the fluctuations parallel and perpendicular to the average drift, we again recover the equations,

$$\begin{cases} \partial_t R_\| = U_\| + K_\| \partial_x^2 R_\| + \dfrac{\lambda_\|}{2}(\partial_x R_\|)^2 + \dfrac{\lambda_\times}{2} \sum_{i=1}^2 (\partial_x R_{\perp i})^2 + \eta_\|(x,t) , \\[2mm] \partial_t R_{\perp i} = K_\perp \partial_x^2 R_{\perp i} + \lambda_\perp \partial_x R_\| \partial_x R_{\perp i} + \eta_{\perp i}(x,t) , \end{cases} \tag{114}$$

where $\{\perp i\}$ refers to the two transverse coordinates of the monomer positions. The noise is assumed to be white and gaussian but need not be isotropic, i.e.

$$\begin{cases} \langle \eta_\|(x,t)\eta_\|(x',t') \rangle = 2T_\| \delta(x - x')\delta(t - t') , \\[2mm] \langle \eta_{\perp i}(x,t)\eta_{\perp j}(x',t') \rangle = 2T_\perp \delta_{i,j}\delta(x - x')\delta(t - t') . \end{cases} \tag{115}$$

At zero average velocity, the system becomes isotropic and the equations of motion must coincide with the Rouse model. Therefore, $\{\lambda_\|, \lambda_\times, \lambda_\perp, U, K_\| - K_\perp, T_\| - T_\perp\}$ are all proportional to E for small forces. The relevance of these nonlinear terms are determined by the dimensionless scaling variable

$$y = \left(\frac{U}{U^*}\right) N^{1/2} ,$$

where U^* is a characteristic microscopic velocity associated with monomer motion and is roughly 10–20 m/s for polystyrene in benzene. The variable y is proportional to another dimensionless parameter, the Reynolds number Re, which determines the breakdown of hydrodynamic equations and onset of turbulence. However, typically Re $\ll y$, and the hydrodynamic equations are valid for moderately large y. Equations (114) describe the static and dynamical scaling properties of the nonlinear and anisotropic regime when $U > U^* N^{-1/2}$.

Equations (114) is just a slight variation from Eqs. (102), with two transverse components instead of one. Thus, the results discussed in the previous lecture apply. A more detailed calculation of the nonlinear terms from hydrodynamics[54] shows that all three nonlinearities are positive for small driving forces. In this case, the asymptotic scaling exponents are isotropic, with $\nu = 1/2$ and $z = 3$. However, the fixed points of the RG transformation are in general anisotropic, which implies a kinetically induced form birefringence *in the absence of external velocity gradients*. This is in marked contrast with standard theories of polymer dynamics where a uniform driving force has essentially no effect on the internal modes of the polymer.

When one of the nonlinearities approaches zero, the swelling exponents may become anisotropic and the polymer elongates or compresses along the longitudinal direction. However, the experimental path in the parameter space as a function of E is not known and not all of the different scaling regimes correspond to actual physical situations. The scaling results found by the RG analysis are verified by direct integration of equations, as mentioned in the earlier sections. A more detailed discussion of the analysis and results can be found in our earlier work.[43]

In constructing Eqs. (114), we only allowed for local effects, and ignored the nonlocalities that are the hallmark of hydrodynamics. One consequence of hydrodynamic interactions is the *back-flow* velocity in Eq. (112) that can be added to the evolution Eqs. (114). Dimensional analysis gives the recursion relation

$$\frac{\partial \gamma}{\partial \ell} = \gamma[\nu z - 1 - (d - 2)\nu] + O(\gamma^2), \tag{116}$$

which implies that, at the nonlinear fixed point, this additional term is surprisingly irrelevant for $d > 3$, and $z = 3$ due to the nonlinearities. For $d < 3$, $z = d$ due to hydrodynamics, and the nonlinear terms are irrelevant. The situation in three dimensions is unclear, but a change in the exponents is unlikely. Similarly, one could consider the effect of self-avoidance by including the force generated by a softly repulsive contact potential

$$\frac{b}{2} \int dx \ dx' \ \mathcal{V}(\mathbf{r}(x) - \mathbf{r}(x')) \,. \tag{117}$$

The relevance of this term is also controlled by the scaling dimension $y_b = \nu z - 1 - (d-2)\nu$, and therefore this effect is marginal in three dimensions at the nonlinear fixed point, in contrast with both Rouse and Zimm models where self-avoidance becomes relevant below four dimensions. Unfortunately, one is ultimately forced to consider nonlocal *and* nonlinear terms based on similar grounds, and such terms are indeed relevant below four dimensions. In some cases, local or global arclength conservation may be an important consideration in writing down a dynamics for the system. However, a local description is likely to be more correct in a more complicated system with screening effects (motion in a gel that screens hydrodynamic interactions) where a first principles approach becomes even more intractable. Therefore, this model is an important starting point towards understanding the scaling behavior of polymers under a uniform drift, a problem with great technological importance.

4. Ordering Phenomena on Growing Films

In many growth processes particles are highly mobile in an active layer at the surface, but are relatively immobile once incorporated in the bulk. We study models in which atoms are allowed to interact, equilibrate, and order on the surface, but are frozen in the bulk. Order parameter correlations in the resulting bulk material are highly anisotropic, reflecting its growth history. In a flat (layer by layer) growth mode, correlations perpendicular to the growth direction are similar to a two-dimensional system in equilibrium, while parallel correlations reflect the dynamics of such a system. When the growing film is rough, various couplings between height and order parameter fluctuations are possible. Such couplings modify the dynamic scaling properties of surface roughness, and may also change the critical behavior of the order parameter. Even the deterministic growth of the surface profile can result in interesting textures for the order parameter.

4.1. *Introduction*

For many technological applications, high quality films are grown by the process of vapor deposition. The properties of such films can be quite different from the same material produced in bulk equilibrium,[55,56] reflecting their preparation history. For example, during the growth of some binary alloys, the deposited atoms are highly mobile on the surface, but relatively immobile in the bulk.[56] Consequently, the surface fluctuations occurring during the growth process are frozen into the bulk. A characteristic feature of such (metastable) phases is *anisotropic* correlations related to the growth direction which are absent in bulk equilibrium.

A number of models for composite film growth have been introduced in the past[57–63] Generally in these models, the probability that an incoming atom sticks to a given surface site depends on the state of neighboring sites in the layer below. Once a site is occupied, its state does not change any more, and thus the surface configuration becomes frozen in the bulk. Such growth rules are equivalent to (stochastic) cellular automata, where each site is updated in parallel as a function of the states of its neighbors. Subsequent states of the cellular automaton correspond to successive layers in the crystal.

It is in general not possible to calculate exact correlation functions for such (nonequilibrium) growth processes. The exception occurs in special cases where the growth rules satisfy a detailed balance condition, relating their stationary behavior to an equilibrium system of one lower dimension.[64] However, it can be shown that if d-dimensional probabilistic cellular automata with two states, and up-down symmetry, undergo a symmetry breaking, their critical behavior is identical to the corresponding Ising model in equilibrium.[64] Correlations in time are then equivalent to those generated by Glauber dynamics of the Ising system. $(d+1)$-dimensional crystals grown according to the rules of these cellular automata therefore have an order–disorder phase transition with correlations perpendicular to the growth direction characterized by the critical exponent ν, and those parallel to the growth direction by the exponent νz of the d-dimensional Ising model (z being the appropriate dynamical critical exponent).

In the next section, I will introduce a model for *layer by layer* growth of binary films. The atoms on the top layer are assumed to equilibrate completely (by surface diffusion *or* desorption–resorption mechanisms) before another layer is added.[65] Such an assumption is realistic only if the growth rate is much slower than characteristic equilibration times of the surface layer. The model satisfies detailed balance, and can therefore be analyzed with methods from equilibrium statistical physics. This discrete model is then used to justify a continuum formulation to the problem which is identical to the time dependent Landau–Ginzburg equation for model A dynamics.[3]

In general, a layer by layer growth mode is unstable, and growing surfaces are rough, as described in the previous chapters. The Kardar, Parisi, Zhang (KPZ) equation describes the dynamic fluctuations in the height of an amorphous surface.[21] The interplay between roughness and ordering phenomena is then considered by introducing simple equations that couple fluctuations in height and the order parameter. Long-range correlations occur at the critical point for the onset of ordering in the surface binary mixture. This in turn leads to greater roughness fluctuations, whose scaling can be explored perturbatively around $d = 4$ dimensions.[66]

While the order parameter for binary deposition is a scalar, we can more generally examine the case of a continuous order parameter formed on the surface layer. There are soft modes associated with such continuous symmetry breaking whose coupling to height fluctuations are explored at the end. In particular, the deterministic relaxation of the order parameter on an initially rough surface can in fact be described exactly through a generalized Cole-Hopf transformation. Interestingly, the relaxation process is super-diffusive and occurs through coarsening of domains (separated by sharp domain walls) on surface mounds.

4.2. *Layer by layer growth*

4.2.1. *Discrete model*

Binary growth is modeled by two kinds of atoms, A and B, which occupy the sites of a $(d+1)$-dimensional hypercubic lattice. Let ϵ_{AA}, ϵ_{AB}, and ϵ_{BB} denote the interaction energies between neighboring atoms of types AA, AB, and BB, respectively. When each layer has N sites, there are 2^N possible configurations for a layer. The energy cost for adding a new layer of configuration γ on top of one in configuration α is the sum of the internal energy E_γ of the new layer, and the interaction energy $V_{\alpha\gamma}$ with the previous layer. These energies are the sums of all local bonds ϵ_{ij} between nearest neighbors ij within the new layer, and between the two layers, respectively. In addition, E_γ contains a chemical potential $\mu_A N_A + \mu_B N_B$ related to the partial pressures of A and B atoms in the gas phase.

Assuming that the top layer is in thermal equilibrium, the conditional probability that it is in configuration γ, given configuration α for the layer below, is

$$W_{\gamma\alpha} = \frac{\exp[-\beta(E_\gamma + V_{\alpha\gamma})]}{\sum_\delta \exp[-\beta(E_\delta + V_{\alpha\delta})]}, \tag{118}$$

where $T = (k_B\beta)^{-1}$ is the temperature at which the crystal is grown. After adding many layers, the steady-state probability for finding a configuration γ is determined by the stationarity condition

$$P_\gamma = \sum_\alpha W_{\gamma\alpha} P_\alpha , \tag{119}$$

which has the solution

$$P_\alpha = \frac{\sum_\gamma \exp[-\beta(E_\alpha + E_\gamma + V_{\alpha\gamma})]}{\sum_{\delta,\nu} \exp[-\beta(E_\delta + E_\nu + V_{\delta\nu})]} \equiv \frac{\sum_\gamma \exp[-\beta H_{\alpha\gamma}]}{\sum_{\delta,\nu} \exp[-\beta H_{\delta\nu}]} . \tag{120}$$

The above expression is the equilibrium probability for the top layer of a two-layer system, obtained after summing over the states of the bottom layer. Transverse correlation functions (i.e. perpendicular to the growth direction) are therefore exactly the same as correlation functions in a two-layer system.

From Eqs. (118) and (120) it follows that the system satisfies detailed balance, i.e.

$$W_{\alpha\gamma} P_\gamma = W_{\gamma\alpha} P_\alpha . \tag{121}$$

Thus, beyond a transient thickness, the crystal looks the same along or against the growth direction, and the sequence of layers corresponds to time evolution of thermodynamic equilibrium states. This generalizes previous results for cellular automata, which are obtained by setting the in-plane interactions E_α to zero. As in such cellular automata, the $(d + 1)$-dimensional system has transverse properties like d-dimensional models. In particular, phase transitions occur at the same temperature as for a d-dimensional two-layer system.

Generalizing the model, by allowing several layers at the surface to equilibrate, is straightforward. To mimic the large energy of the impinging particles, as well as their modified environment, we can assign each of the top ℓ layers from the surface a different temperature, through scaled interaction energies depending on its depth. The probability that a layer with configuration γ follows one in configuration α in the bulk is obtained by considering the layer at the moment when it is the ℓth layer from the top, that is, immediately before its configuration is frozen. Denoting the configuration of the first $\ell - 1$ layers by C_γ and their energy (including the coupling to the ℓth layer, and different interaction constants in the different layers) by $E(C_\gamma)$, the conditional probabilities $W_{\gamma\alpha}$ can be written as

$$W_{\gamma\alpha} = \frac{\sum_{C_\gamma} \exp\{-\beta[E_\gamma + V_{\alpha\gamma} + E(C_\gamma)]\}}{\sum_{\delta,C_\delta} \exp\{-\beta[E_\delta + V_{\alpha\delta} + E(C_\delta)]\}} .$$

Following the approach for the case $\ell = 1$, we can show that the set of weights

$$P_\alpha = \frac{\sum_{\gamma, \mathcal{C}_\gamma, \mathcal{C}_\alpha} \exp\{-\beta[E_\alpha + E_\gamma + V_{\alpha\gamma} + E(\mathcal{C}_\gamma) + E(\mathcal{C}_\alpha)]\}}{\sum_{\delta, \nu, \mathcal{C}_\delta, \mathcal{C}_\nu} \exp\{-\beta[E_\delta + E_\nu + V_{\delta\nu} + E(\mathcal{C}_\delta) + E(\mathcal{C}_\nu)]\}},$$

describe a stationary state. It is easy to verify that this stationary solution satisfies detailed balance. The stationary state corresponds to an equilibrium Hamiltonian with 2ℓ layers, with interactions which depend on the distance from the closest surface. The top (or the bottom) layer describes the deposited surface, while the middle (ℓ or $\ell + 1$) layers describe transverse correlations in the bulk. While the correlations parallel to the growth direction are more complicated, the general conclusions for $\ell = 1$ remain valid.

4.2.2. *Continuum formulation*

In the above discrete model, we can use an Ising variable $\sigma_i = \pm 1$ to indicate if site i is occupied by atom A or B. Close to the critical point, density fluctuations occur over long distances and universal properties are better captured by considering a coarse-grained order parameter $m(\mathbf{x}, t)$. Here $\mathbf{x}$ labels the d directions transverse to growth, while t which indicates time is also proportional to the coordinate parallel to the growth direction. Hence $m(\mathbf{x}, t)$ encodes the time history of the growth process. From the exact solution of the discrete problem, we know that the behavior of these configuration is equivalent to the time evolution of a d-dimensional system at equilibrium. In the continuum limit, the latter is described by the time-dependent Landau–Ginzburg equation[3]

$$\partial_t m = K \nabla^2 m + rm - um^3 + \eta_m(\mathbf{x}, t), \tag{122}$$

where $\eta_m(\mathbf{x}, t)$ is a random noise of zero mean, whose variance is proportional to the growth temperature.

Away from the critical point at $r = r_c$, fluctuations in m decay over a transverse correlation length ξ, and a longitudinal correlation "time" ξ^z. At the critical point itself, there is no intrinsic scale, and correlations decay as

$$\langle m(\mathbf{x}, t) m(\mathbf{x}', t') \rangle = \frac{1}{|\mathbf{x} - \mathbf{x}'|^{-2\chi_m}} g_m\left(\frac{|t - t'|}{|\mathbf{x} - \mathbf{x}'|^z}\right). \tag{123}$$

In dimensions $d > 4$, criticality occurs for $r = u = 0$ (the diffusion equation), leading to $z = 2$ and $\chi_m = (2 - d)/2$. On approaching criticality, the correlation length diverges as $\xi \propto |r - r_c|^{-\nu}$, with $\nu = 1/2$. For $d \leq 4$, the nonlinear term um^3 is relevant, and the exponents can be calculated perturbatively[3] in $\varepsilon = 4 - d$.

4.3. *Rough growth*

4.3.1. *Dynamic roughening*

The layer by layer growth mode cannot be maintained indefinitely, and the surface eventually becomes rough.[67] Let us denote the height of the surface at location $\mathbf{x}$ at time t by a function $h(\mathbf{x}, t)$. As discussed in the previous sections, there is considerable evidence from simulations (and some experiments) that the resulting surfaces exhibit self-affine fluctuations, well described by the continuum KPZ equation (71). The self-affine fluctuations in the surface height will be described by dynamic scaling exponents χ_h and z, defined through

$$\langle [h(\mathbf{x}, t) - h(\mathbf{x}', t')]^2 \rangle = |\mathbf{x} - \mathbf{x}'|^{2\chi_h} g_h \left(\frac{|t - t'|}{|\mathbf{x} - \mathbf{x}'|^z} \right). \tag{124}$$

The linear equation for $\lambda = 0$ gives diffusive exponents $\chi_h = (2 - d)/2$ and $z = 2$. Any nonlinearity is relevant in $d \leq 2$, while sufficiently large λ is required in $d > 2$ to produce a rough phase ($\chi_h \geq 0$).

4.3.2. *Coupling growth and ordering*

There are few studies of the interplay between fluctuations in height and the order parameter. Some numerical simulations have incorporated both elements: as a model for diamond growth, Capraro and Bar-Yam[61] introduced a variant of ballistic deposition which exhibits sublattice ordering. Kotrla and Predota[63] have examined binary deposition in $1+1$ dimensions, resulting in domains with rough surfaces. In a recent work with Barbara Drossel,[66] we took an analytical approach to this problem. A different set of equations was given by Léonard and Desai[68] for the case of phase separation during molecular beam epitaxy. Their equations reflect the situation of MBE where particle deposition is random (in contrast to having sticking probabilities that depend on the local environment), and where the order parameter can only be built up through surface diffusion.

The starting point is the continuum Eqs. (122)–(71), describing the order parameter $m(\mathbf{x}, t)$ and height $h(\mathbf{x}, t)$ fluctuations. To these equations we added all terms consistent with the symmetries of the problem. The lowest order (potentially relevant) terms result in the following pair of coupled differential equations

$$\begin{cases} \partial_t h = \nu\nabla^2 h + \dfrac{\lambda}{2}(\nabla h)^2 + \zeta_h - \dfrac{\alpha}{2}m^2 \,, \\[2ex] \partial_t m = K\nabla^2 m + rm - um^3 + \zeta_m + a\nabla h \cdot \nabla m + bm\nabla^2 h + \dfrac{c}{2}m(\nabla h)^2 \,. \end{cases} \tag{125}$$

(Note that these equations satisfy the symmetry $m \mapsto -m$.) Fluctuations of the surface are modified by coupling to the order parameter, through the term proportional to αm^2. There are also three coupling constants a, b, and c, which modify the order parameter fluctuations due to coupling to h.

As long as the binary mixture is disordered ($r > r_c$), fluctuations in m and hence m^2, are short-ranged, and αm^2 acts as another source of white noise. The surface fluctuations should thus scale with the standard KPZ exponents. However, the range of correlations increases as $r \to r_c$ and $\xi \sim |r - r_c|^{-\nu} \to \infty$. This modifies (most likely increases) the overall amplitude of surface roughness, and height fluctuations over a scale L behave as

$$\sqrt{\langle \delta h^2(L,r)\rangle} = \xi^{\chi_h^c - \chi_h} L^{\chi_h} g(L/\xi) \,, \tag{126}$$

where χ_h^c is the roughness exponent at criticality, which is discussed next.

4.3.3. *Critical roughness*

Under a change of scale $\mathbf{x} \mapsto b\mathbf{x}$, $t \mapsto b^z t$, $h \mapsto b^{\chi_h} h$, and $m \mapsto b^{\chi_m} m$, the nonlinear coefficients in Eq. (125) scales as $x \mapsto b^{y_x} x$, with

$$y_\lambda = y_a = y_b = \chi_h + z - 2, \quad y_c = 2\chi_h + z - 2, \quad y_\alpha = 2\chi_m - \chi_h + z. \tag{127}$$

The critical point in dimensions $d \geq 4$ occurs at $r = u = 0$. The linear diffusion equations at this point result in the bare field dimensions $\chi_h^0 = \chi_m^0 = (2-d)/2$. Taking account of the nonlinearities, we observe the following behaviors:

- $d > 6$: All nonlinearities are (perturbatively) irrelevant; the surface is smooth, and the order parameter goes through a classical phase transition.
- $4 < d < 6$: The leading nonlinearity is the term αm^2 describing the correlated noise acting on the surface height, with ($z = 2$)

$$y_\alpha^0 = 4 - d - \chi_h^c = 3 - d/2 \,. \tag{128}$$

In these dimensions, the correlated noise is more relevant than the white noise from the flux variations.[31] The correct result can in fact be obtained simply by setting y_α to zero, leading to critical height fluctuations with

$$\chi_h^c = 4 - d > \frac{2-d}{2}\,.$$

Note that while the roughness exponent is larger than its bare value, it is still negative. The scaling of the order parameter is not modified, and $\chi_m = (2-d)/2$.

- $d \leq 4$: When the roughness exponent is positive, all the couplings λ, a, b, and c become relevant. Also in $d \leq 4$, the critical point of the Landau–Ginzburg model is no longer at $r = u = 0$, and a full renormalization group (RG) study is called for.[66] Ignoring the feedback from height fluctuations to the order parameter, we find to leading order an RG equation of the form

$$\frac{1}{(\alpha\lambda)}\frac{d(\alpha\lambda)}{d\ell} = \varepsilon - C(\alpha\lambda)\,,$$

where $\varepsilon = 4 - d$ and C is a *positive* constant.

There is a fixed point at $\alpha\lambda = \varepsilon/C$, with roughness exponent $\chi_h^c = 0$. In $d > 4$, this is an unstable fixed point governing a transition between flat $(\chi_h^c = 4 - d < 0)$ and rough phases (occurring for $\alpha\lambda < -(d-4)/C$). For $d < 4$, this fixed point is stable and attracts all points with $\alpha\lambda > 0$. Negative values of $\alpha\lambda$ flow to a rough phase which is not perturbatively accessible. Including all nonlinearities in the equation for m complicates the analysis, but we did not find a fixed point whose critical behavior is different from the ordinary Landau–Ginzburg model (at least to lowest order).

Since the above analytical results are inconclusive, we also undertook numerical simulations. To study the interplay between surface roughening and phase separation we simulated a brick-wall restricted solid on solid model with two species of particles.[69] Already the $(1+1)$-dimensional system shows a variety of different scaling behaviors, depending on how the two phenomena are coupled. In the most interesting case, which is related to the advection of a passive scalar in a velocity field, nontrivial scaling exponents are found.

4.4. *Continuous order*

4.4.1. *Stochastic evolution*

The situation on the ordered side of the phase transition is more complex. The analogy to the dynamics of the lower-dimensional system suggests that the leading process is the gradual coarsening of the ordered domains. Such domains would then appear as cone-shaped columns in the bulk film, a reasonably common feature of growth textures. However, more work is necessary to verify and quantify this picture.

Another interesting situation is when the symmetry breaking involves a continuous, rather than a discrete (Ising like), order parameter. For example, we may consider deposition of spins which can realign on the surface but are frozen in the bulk. More interestingly, the growth of crystals involves translational and orientational symmetry breakings in the plane. In the simplest case of a vector order parameter, we can simply generalize Eqs. (125) by replacing the scalar m with an n-component vector $\vec{m}(\mathbf{x}, t)$. While the discussion of critical roughening is not significantly modified from the Ising case ($n = 1$), new issues arise pertaining to the ordered phase.

The most common excitations of the broken symmetry phase are *soft (Goldstone) modes*, which can in principle couple to the surface roughness. The simplest example is provided by the XY model ($n = 2$), where the direction of the vector can be described by an angular field $\theta(\mathbf{x}, t)$. Including the lowest order terms which satisfy rotational symmetry leads to the coupled equations of motion

$$
\begin{cases}
\partial_t h = \nu \nabla^2 h + \dfrac{\lambda}{2}(\nabla h)^2 + \zeta_h - \dfrac{\alpha}{2}(\nabla \theta)^2 \,, \\
\partial_t \theta = K \nabla^2 \theta + \zeta_\theta + a \nabla h \cdot \nabla \theta \,.
\end{cases}
\tag{129}
$$

Interestingly, these are precisely the equations that Eq. (100) proposed in the contexts of moving flux lines and drifting polymers earlier. In particular, in $d = 1$ the KPZ exponents ($\chi_h = 1/2$ and $z = 3/2$) are recovered for the surface roughness, while the angular fluctuations remove any long-range order. Further analysis is again necessary for the case $d = 2$. Specifically, an important aspect of the field θ not present in the earlier studies is its angular nature. It could thus include vortices which are *topological defects*. Such defects typically play an important role in equilibrium two-dimensional systems, and have been recently considered in a number of related nonequilibrium situation.[70]

4.4.2. *Deterministic textures*

As discussed before, the nonlinear KPZ equation can be recast as a linear diffusion equation through the Cole-Hopf transformation. This transformation can in fact be generalized to describe the coupling of the surface height to a

44   *M. Kardar*

vector order parameter. Consider a field of unit spins, $|\vec{s}(\mathbf{x}, t)| = 1$, and set

$$\vec{W}(\mathbf{x}, t) = \exp\left[\frac{\lambda h(\mathbf{x}, t)}{2\nu}\right] \vec{s}(\mathbf{x}, t). \tag{130}$$

A diffusive equation of the field $\vec{W}(\mathbf{x}, t)$, as

$$\partial_t \vec{W} = \nu \nabla^2 \vec{W} + \frac{\lambda}{2\nu} \eta_h(\mathbf{x}, t) \vec{W}, \tag{131}$$

can be recast into the pair of coupled differential equations

$$\begin{cases} \partial_t h = \nu \nabla^2 h + \frac{\lambda}{2}(\nabla h)^2 + \left(\frac{2\nu^2}{\lambda}\right) \vec{s} \cdot \nabla^2 \vec{s} + \eta_h(\mathbf{x}, t), \\ \partial_t \vec{s} = \nu[\nabla^2 \vec{s} - (\vec{s} \cdot \nabla^2 \vec{s})\vec{s}] + \lambda \nabla h \cdot \nabla \vec{s}. \end{cases} \tag{132}$$

Note that the transverse component of $\nabla^2 \vec{s}$ contributes to $\partial_t \vec{s}$, thus ensuring that the magnitude of $\vec{s}$ is not changed in time, while the longitudinal component of this quantity couples to the surface height.

It can be checked easily that for $n = 2$, the parametrization $\vec{s} = (\cos\theta, \sin\theta)$, reduces Eqs. (132) to Eqs. (129) in the special limit of $\alpha\lambda = 4\nu^2$, $K = \nu$, $a = \lambda$, and $\eta_\theta = 0$. It is also possible to construct other Cole-Hopf transformations for cases when $\alpha\lambda < 0$.

Starting from any arbitrary initial condition at $t = 0$, the deterministic limit ($\eta_h = 0$) of Eqs. (131)–(132) is easily solved using the diffusion kernel, as

$$\vec{W}(\mathbf{x}, t) = \exp\left[\frac{\lambda h(\mathbf{x}, t)}{2\nu}\right] \vec{s}(\mathbf{x}, t)$$

$$= \int \frac{d^d \mathbf{x}'}{(4\pi\nu t)^{d/2}} \exp\left[-\frac{(\mathbf{x} - \mathbf{x}')^2}{4\nu t} + \frac{\lambda h(\mathbf{x}', 0)}{2\nu}\right] \vec{s}(\mathbf{x}', 0). \tag{133}$$

The saddle–point evaluation of the above integral (formally exact as $\nu \to 0$) captures the long-time behavior of the solution. The surface profile

$$h(\mathbf{x}, t) = \min_{\mathbf{x}'} \left[h(\mathbf{x}', 0) - \frac{(\mathbf{x} - \mathbf{x}')^2}{2\lambda t}\right], \tag{134}$$

consists of a set of parabolic mounds centered at locations $\mathbf{x}' = \mathbf{x}_0(\mathbf{x}, t)$ corresponding to high points of the initial surface. Note that the evolution of the surface profile in this limit is independent of $\vec{s}$. The evolution of spins on the other hand is completely controlled by the surface height, and given by

$$\vec{s}(\mathbf{x}, t) = \vec{s}(\mathbf{x}_0(\mathbf{x}, t), 0)\,, \tag{135}$$

that is, each of the surface mounds carries the spin of its initial high point! Such behavior is quite different from the diffusive evolution of spins in the absence of coupling to the surface profile. For self-affine initial surface profiles, the relaxation of the height and spins is now both diffusive. Furthermore, the spin textures produced by this process are domains separated by sharp domain walls, very different from the soft modes and vortices that characterize diffusive relaxation. Similar extensions of the Cole-Hopf transformation to matrix order parameters are also possible, and could for example describe relaxation of crystalline substrates.

References

1. *Selected Papers on Noise and Stochastic Processes*, ed. Nelson Wax (Dover, New York, 1954).
2. S. Dattagupta, *Relaxation Phenomena in Condensed Matter Physics* (Academic Press, London, 1987).
3. P. C. Hohenberg and B. I. Halperin, *Rev. Mod. Phys.* **49**, 435 (1977).
4. S.-K. Ma, *Modern Theory of Critical Phenomena* (Benjamin-Cummings, Reading, MA, 1976).
5. D. Forster, *Hydrodynamic Fluctuations, Broken Symmetries, and Correlation Functions* (Benjamin-Cummings, Reading, MA, 1975).
6. M. Kardar, in *Disorder and Fracture*, eds. J. C. Charmet, S. Roux, and E. Guyon (Plenum, New York, 1990), p. 3.
7. T. Hwa and M. Kardar, *Phys. Rev.* **A45**, 7002 (1992).
8. *Dynamics of Fractal Surfaces*, eds. F. Family and T. Vicsek (World Scientific, Singapore, 1991).
9. T. Halpin–Healy and Y.-C. Zhang, *Phys. Rep.* **254**, 215 (1995).
10. A.-L. Barabasi and H. E. Stanley, *Fractal concepts in surface growth* (CUP, Cambridge, 1995).
11. F. Family and T. Vicsek, *J. Phys.* **A18**, L75 (1985).
12. For a review see, P. Meakin, *Phys. Rep.* **235**, 191 (1993).
13. J. D. Weeks, G. H. Gilmer, and K. A. Jackson, *J. Chem. Phys.* **65**, 712 (1976).
14. F. Family, *J. Phys.* **A19**, L441 (1986).
15. M. J. Vold, *J. Coll. Sci.* **14**, 168 (1959).
16. J. M. Kim and J. M. Kosterlitz, *Phys. Rev. Lett.* **62**, 2289 (1989).
17. See e.g., D. E. Wolf, in *Scale Invariance, Interfaces, and Nonequilibrium Dynamics*, eds. A. McKane, M. Droz, J. Vannimenus, and D. Wolf (Plenum Press, New York, 1995), p. 215.

18. M. Kardar, in *Disorder and Fracture*, eds. J. C. Charmet, S. Roux, and E. Guyon (Plenum, New York, 1990), p. 3.
19. M. Kardar, *Tr. J. of Phys.* **18**, 221 (1994).
20. S. F. Edwards and D. R. Wilkinson, *Proc. R. Soc. Lond.* **A381**, 17 (1982).
21. M. Kardar, G. Parisi, and Y.-C. Zhang, *Phys. Rev. Lett.* **56**, 889 (1986).
22. J. Villain, *J. Phys.* **I1**, 19 (1991).
23. T. Sun, H. Guo, and M. Grant, *Phys. Rev.* **A40**, 6763 (1989).
24. Z. W. Lai and S. Das Sarma, *Phys. Rev. Lett.* **66**, 2348 (1991).
25. M. Siegert and M. Plischke, *Phys. Rev.* **E50**, 917 (1994).
26. J. Krug, M. Plischke, and M. Siegert, *Phys. Rev. Lett.* **70**, 3271 (1993).
27. R. L. Schwoebel, *J. Appl. Phys.* **40**, 614 (1969).
28. W. K. Burton, N. Cabrera, and F. C. Frank, *Phys. Trans. R. Soc. Lond.* **A243**, 299 (1951).
29. M. D. Johnson, C. Orme, A. W. Hunt, D. Graff, J. Sudijono, L. M. Sander, and B. G. Orr, *Phys. Rev. Lett.* **72**, 116 (1994).
30. J. Krim and G. Palasantzas, *Int. J. Mod. Phys.* **B9**, 599 (1995).
31. E. Medina, T. Hwa, M. Kardar, and Y.-C. Zhang, *Phys. Rev.* **A39**, 3053 (1989).
32. A. N. Kolmogorov, *C. R. Acad. Sci. USSR* **30**, 301 (1941); *ibid.* **32**, 16 (1941).
33. C. Licoppe, Y. I. Nissim, and C. d'Anterroches, *Phys. Rev.* **B37**, 1287 (1988).
34. D. Forster, D. R. Nelson, and M. J. Stephen, *Phys. Rev.* **A16**, 732 (1977).
35. J. M. Burgers, *The Nonlinear Diffusion Equation* (Riedel, Boston, 1974).
36. B. M. Forrest and L.-H. Tang, *Phys. Rev. Lett.* **64**, 1405 (1990)
37. D. Wolf, *Phys. Rev. Lett.* **67**, 1783 (1991).
38. M. Prahofer and H. Spohn, *J. Stat. Phys.*, in press (1997).
39. R. da Silveira and M. Kardar, in progress (1999).
40. D. Ertas and M. Kardar, *Phys. Rev. Lett.* **69**, 929 (1992).
41. T. Hwa, *Phys. Rev. Lett.* **69**, 1552 (1992).
42. E. Medina, M. Kardar, Y. Shapir, and X.-R. Wang, *Phys. Rev. Lett.* **62**, 941 (1989); E. Medina and M. Kardar, *Phys. Rev.* **B46**, 9984 (1992).
43. D. Ertas and M. Kardar, *Phys. Rev.* **E48**, 1228 (1993).
44. J. P. Bouchaud, E. Bouchaud, G. Lapasset, and J. Planes, *Phys. Rev. Lett.* **71**, 2240 (1993).
45. M. Doi and S. F. Edwards, *Theory of Polymer Dynamics* (Oxford University Press, 1986).
46. P. G. de Gennes, *Scaling Concepts in Polymer Physics* (Cornell University Press, 1979).
47. R. B. Bird, *Dynamics of Polymeric Physics*, Vols. 1 and 2 (Wiley, New York, 1987).
48. P. E. Rouse, *J. Chem. Phys.* **21**, 1272 (1953).
49. We have changed the notation to confer with the traditions of polymer science. ν is ζ and z is z/ζ in terms of the notation used previously.
50. J. Kirkwood and J. Risemann, *J. Chem. Phys.* **16**, 565 (1948).
51. B. H. Zimm, *J. Chem. Phys.* **24**, 269 (1956).
52. See, for example, M. Adam and M. Delsanti, *Macromolecules* **10**, 1229 (1977).

53. R. E. Khayat and R. G. Cox, *J. Fluid. Mech.* **209**, 435 (1989).

54. See Appendices A and B of our longer paper.[43]

55. S. Froyen and A. Zunger, *Phys. Rev. Lett.* **66**, 2132 (1991).

56. P. W. Rooney, A. L. Shapiro, M. Q. Tran, and F. Hellman, *Phys. Rev. Lett.* **75**, 1843 (1995).

57. T. R. Welberry and R. Galbraith, *J. Appl. Crystallogr.* **6**, 87 (1973).

58. I. G. Enting, *J. Phys. C: Solid State Phys.* **10**, 1379 (1977).

59. K. Kim and E. A. Stern, *Phys. Rev.* **B32**, 1019 (1985).

60. L. C. Davis and H. Holloway, *Phys. Rev.* **B35**, 2767 (1987).

61. Y. Bar-Yam, D. Kandel, and E. Domany, *Phys. Rev.* **B41**, 12 869 (1990).

62. P. W. Rooney and F. Hellman, *Phys. Rev.* **B48**, 3079 (1993).

63. M. Kotrla and M. Predota, *EuroPhys. Lett.* **39**, 251 (1997).

64. G. Grinstein, C. Jayaprakash, and Y. He, *Phys. Rev. Lett.* **55**, 2527 (1985).

65. B. Drossel and M. Kardar, *Phys. Rev.* **E55**, 5026 (1997).

66. B. Drossel and M. Kardar, work in progress; see also M. Kardar, *Physica* **A263**, 345 (1999).

67. T. Hwa, M. Kardar, and M. Paczuski, *Phys. Rev. Lett.* **66**, 441 (1991); and references therein.

68. F. Léonard and R. C. Desai, *Phys. Rev.* **B55**, 9990 (1997).

69. B. Drossel and M. Kardar, *Phys. Rev. Lett.* **85**, 614 (2000).

70. I. S. Aranson, H. chate, and L.-H. Tang, *Phys. Rev. Lett.* **80**, 2646 (1998); see also I. S. Aranson, S. Scheidl, and V. M. Vinokur, *Phys. Rev.* **B58**, 14541 (1998).

Annual Reviews of Computational Physics VIII (pp. 49–82)
Edited by Dietrich Stauffer

KINETICS OF EPITAXIAL THIN FILM GROWTH

FEREYDOON FAMILY

Department of Physics, Emory University, Atlanta GA 30322, USA

Molecular-Beam-Epitaxy is one of the most effective techniques for growing a wide variety of high purity materials. These lecture notes present an introductory review of some of the central ideas in the kinetics of submonolayer and multilayer epitaxial growth. The concepts of a critical island size, dynamical scaling of the island-size distribution, and the barrier to interlayer diffusion (Ehrlich–Schwoebel barrier) are introduced. The results of kinetic Monte Carlo simulations of a realistic model of submonolayer epitaxial growth as well as an analytical expression for the scaled island-size distribution are presented and compared with rate-equation analyses and recent experiments. The results provide a quantitative explanation for the variation of the submonolayer island density, critical island size, island-size distribution and morphology as a function of temperature and deposition rate found in recent experiments. A realistic model for multilayer homoepitaxial growth on fcc and bcc lattices which takes into account the correct crystal structure is presented. The effects of instabilities which lead to mound formation and coarsening are discussed and a unified picture of the effects of attractive and repulsive interactions at ascending and descending steps on surface morphology and island nucleation is presented. An accurate prediction of the mound angle is obtained analytically and by kinetic Monte Carlo simulations, and is compared with experiments. The general dependence of the mound angle, and mound coarsening behavior on temperature, deposition rate, and strength of the step barrier is also presented and compared with recent experiments.

1. Introduction

Molecular-Beam-Epitaxy (MBE) is one of the most effective techniques for growing high purity materials including a variety of semiconductors and magnetic materials for applications in electronic and optoelectronic devices.[1,2] In this method a constant flux of atoms impinge under ultrahigh vaccuum conditions on a substrate held at a fixed temperature to grow a high quality crystalline material. The long-standing scientific challenge in this area has been to model epitaxial growth conditions and understand what are the fundamental processes that control the evolution of epitaxial structure and morphology.

This review is divided into two parts. The first part discusses the submonolayer regime with special emphasis on the scaling of the island density and the island-size distribution. The second part is devoted to a detailed discussion of the evolution and morphology of the surface during multilayer growth. In particular, the crucial role that the crystalline microstructure plays in the development of a relatistic model of epitaxial growth is discussed.

2. Submonolayer Regime: Island-Size Distribution

2.1. *Introduction*

The fundamental physical processes in the growth of thin films by molecular-beam-epitaxy involve nucleation, aggregation and coalescence of islands on a two-dimensional substrate.[1-3] In the submonolayer and early multilayer regime this leads to the formation of a distribution of islands of various sizes and morphologies which grow and coalesce with time. The island-size distribution exhibits a general dynamic scaling behavior, similar to that observed in aggregation[4] and vapor-deposited thin-film growth,[5,6] but the island morphology depends on the microscopic details of the growth process such as the deposition rate, temperature, island-relaxation mechanisms and the structure of the substrate.

The scaling behavior in the submonolayer is important for a variety of reasons. For example, the experimental and theoretical study of the scaling behavior of the island density and distribution in the submonolayer regime may be used to determine a variety of important microscopic parameters in epitaxial growth. In particular, a study of the scaling of the island density as a function of deposition rate and temperature has enabled the determination of such quantities as the activation energy for monomer diffusion, as well as the critical island size for island nucleation in a variety of systems.[7-19] The *critical island size i* is a particularly important parameter in the study of the nucleation and growth of islands and is defined in analogy with nucleation theory.[20,21] It is assumed that there exists a critical size i, such that islands of size larger than i are more likely to grow than to shrink, while islands smaller than i are more likely to break up.

In addition, the submonolayer regime has important consequences for multilayer growth. For example, due to the existence of the Ehrlich–Schwoebel (ES) barrier,[22] i.e. an additional barrier beyond that for normal diffusion for atoms to hop down from the edge of an island, the quality of multilayer film growth may be strongly affected by the island distribution and morphology in the submonolayer.[12,23] In particular, the effects of the barrier, combined with the island-size distribution and morphology, may determine whether growth is "layer-by-layer" or three-dimensional. As an example, the re-entrant RHEED oscillations which occur with decreasing temperature in Pt/Pt(111) growth have been shown[24,25] to be due to a change in island morphology with decreasing temperature combined with the existence of an ES barrier. Similarly, the addition of impurities on a surface has recently been shown[12] to lead to improved layer-by-layer growth, since the increased island density due to heterogeneous nucleation reduces island size thus promoting interlayer diffusion.

2.2. *Rate equation theory*

The fundamental quantity in the kinetic description of island growth is the island-size distribution function $N_s(t)$, which gives the density of islands of size s (where s is the number of atoms or particles in the island) at time t. The traditional method for studying the kinetics of cluster growth processes is based on the theoretical approach developed by Smoluchowski[26] who wrote down an equation for the evolution of the cluster size distribution using a mean-field argument which neglects fluctuations and geometry. For the case in which only monomers diffuse, one can write down a simple set of equations[20,21] governing the density of monomers N_1 and the density $N_s(t)$ of islands of size s at time t. Including dissociation but ignoring both coalescence and the contribution to island growth due to arriving adatoms landing directly on top of an island, one has generally,

$$\frac{dN_1}{dt} = F - 2K_1 D N_1^2 - N_1 D \sum_{s \geq 2} K_s N_s + \sum_{s > 1} \gamma_s N_s \,, \tag{1a}$$

$$\frac{dN_s}{dt} = N_1 D (K_{s-1} N_{s-1} - K_s N_s) - \gamma_s N_s + \gamma_{s+1} N_{s+1}, \qquad (s > 1) \tag{1b}$$

where F is the deposition rate and K_s (γ_s) governs the rate of attachment (detachment) of adatoms on the substrate from islands of size s. The first three terms on the right of Eq. (1a) correspond to the processes of monomer deposition, monomer capture by the formation of dimers, and monomer attachment to islands respectively, while the fourth term corresponds to the rate at which monomers detach from existing islands. Similarly, the first term in parentheses on the right of Eq. (1b) corresponds to the rate at which islands of size

$s - 1$ are converted to islands of size s by addition of monomers, while the second term corresponds to the creation of islands of size $s + 1$ by addition of monomers to islands of size s. The solution of the rate-equations involves simply the knowledge of the detachment rates γ_s for $s < i$ as well as the capture probabilities K_s.

We now consider the classical rate-equation theory for the simplest possible model — the point-island model[27,28] — in which the capture probability K_s can be taken to be independent of island size.

In the point-island model, the islands have zero spatial extent[27,28] and therefore the capture probability K_s is taken to be independent of island size s i.e. $K_s \sim s^p$ with $p = 0$.[29] Assuming the existence of a critical stable island size i such that $\gamma_s = 0$ for $s > i$, and summing Eq. (1b) for all $s \geq i + 1$ to obtain the stable island density $N_x = \sum_{s>i} N_s$ one obtains (taking $K_s = a$)

$$\frac{dN_1}{d\theta} = 1 - 2aRN_1^2 - aRN_1N + \frac{1}{F} \sum_{2 \leq s \leq i} \gamma_s N_s \,, \tag{2a}$$

$$\frac{dN_x}{d\theta} = aRN_1N_i \,, \tag{2b}$$

where $R = D/F$ and we have replaced the time derivative by a coverage derivative. In order to solve these equations, one needs an expression for N_i. Typically one invokes the quasi-equilibrium Walton relation[30] that $N_i \sim A_i N_1^i$ where $A_i = e^{E_i/k_BT}$ is a temperature-dependent prefactor and E_i is the binding energy of the critical cluster of size i. This implies

$$\frac{dN_1}{d\theta} = 1 - aRN_1^2 - aRN_1N + \frac{1}{F} \sum_{2 \leq s \leq i} \gamma_s N_s \,, \tag{3a}$$

$$\frac{dN_x}{d\theta} = aRN_1^{i+1}A_i \,. \tag{3b}$$

In the late-time (asymptotic) limit, we assume that for $s \leq i$, $N_s \ll N \simeq N_x$ and $\frac{dN_1}{d\theta} \simeq 0 \simeq 1 - aRN_1N$ to obtain the point-island rate-equation prediction,[21]

$$N_1 = (1/B_i)R^{-2/(i+2)}\theta^{-1/(i+2)} \,, \tag{4a}$$

$$N = (B_i/a)R^{-\chi_i}\theta^{1/(i+2)} \,, \tag{4b}$$

where $\chi_i = i/(i + 2)$ and $B_i = [(i + 2)a^2 A_i]^{1/i+2}$. Substituting for A_i and B_i yields the temperature dependence of the island density and monomer density at fixed coverage,[21]

$$N \sim e^{[E_i/(i+2)+\chi_i E_a]/k_B T} , \qquad (5a)$$

$$N_1 \sim e^{[(-E_i+2E_a)/(i+2)]/k_B T} , \qquad (5b)$$

where E_i is the critical cluster binding energy and E_a is the activation energy for monomer diffusion.

The above results imply that for fixed coverage one has $N \sim r_1^{-\phi_i} R^{-\chi_i}$ where r_1 is the reduced one-bond detachment rate and $\phi_i = (i-1)/(i+2)$ for $i = 1, 2$, and 3. If we now assume[31] that the crossover scaling function for the island density N (at fixed coverage) from $i = 1$ to $i = k$ has the general form,

$$N \sim R^{-1/3} f_{1k}(r_1^{x_{1k}} R) , \qquad (6)$$

where x_{1k} is a crossover exponent[31] and where $f(u) \sim$ const for $u \ll 1$ and $f(u) \sim u^{-2(k-1)/(3(k+2))}$ for $u \gg 1$, then the rate-equation prediction (5) implies that the value of the crossover exponent is given by $x_{12} = x_{13} = 3/2$.

2.3. *Scaling of the island-size distribution*

Since the coverage $\theta = Ft$ is often more convenient for comparison with experiments, we express the time dependence of the island-size distribution and its moments in terms of θ. Defining the total island density N (excluding monomers) and the coverage θ by

$$N = \sum_{s \geq 2} N_s(\theta) \quad \text{and} \quad \theta = \sum_{s \geq 1} sN_s , \qquad (7)$$

then the average island size S can be written in terms of the zeroth and first moments as[27,32]

$$S = \frac{\sum_{s \geq 2} sN_s(\theta)}{\sum_{s \geq 2} N_s(\theta)} = \frac{(\theta - N_1)}{N} , \qquad (8)$$

where N_1 is the monomer density.

According to the dynamic scaling assumption[4] there exists only one characteristic size in the problem which is the mean island size $S(\theta)$ defined in Eq. (8). This implies that $N_s(\theta)$ scales with $S(\theta)$ and one may write generally $N_s(\theta) = A(S, \theta) f_i(s/S)$ where $f_i(u)$ is the scaling function for the island-size distribution corresponding to the case in which the value of the critical island size is equal to i. Using the definition of θ and the scaling form for the island-size distribution one can write, $\theta = \sum_{s\geq 1} sN_s(\theta) = A(S,\theta)S^2 \int_0^\infty f_i(u)u\,du$ which implies $A(S,\theta) \sim \theta/S^2$. Taking $A(S,\theta) = \theta/S^2$, one may write the general scaling form,[4–6,27–32]

$$N_s(\theta) = \theta S^{-2} f_i(s/S), \qquad (s \geq 2) \tag{9}$$

where the scaling function $f_i(u)$ satisfies

$$\int_0^\infty f_i(u)\, u\, du = 1, \tag{10}$$

and $f_i(u) \sim u^{\omega_i}$ for $u \ll 1$. At late time, one expects the walker density N_1 to be much smaller than the coverage θ so that in the scaling regime $N_1 \ll \theta$ and $S \simeq \theta/N$. This implies the scaling relation,

$$N \simeq \theta/S, \tag{11}$$

which leads to the additional sum-rule,

$$\int_0^\infty f_i(u)du = 1. \tag{12}$$

In simulations[32] of irreversible island growth ($i = 1$) we have observed good scaling for the island-size distribution of the form of Eq. (9) over an extended region of coverage in the precoalescence regime (i.e. $\theta \leq 0.4$). This corresponds to a regime in which the island-density N remains constant, while the average island size is proportional to the coverage. In the asymptotic limit of large D/F, this regime becomes wider and extends to lower and lower coverage while the scaling function $f_i(u)$ becomes independent of both D/F and coverage. In addition, with increasing critical island size i, this regime appears to occur at lower coverage and to become more pronounced.

We note that the scaling form (9) is valid in the aggregation regime before a single cluster spans the system. At higher coverages clusters begin to coalesce. In fact, there exists a critical coverage θ_c at which a single cluster extends across the entire system.[32] This is the percolation transition point and the size distribution at this coverage is expected to obey Stauffer scaling.[33]

2.4. Analytic form for the scaling function $f_i(u)$

The simple rate-equation theory does a fairly good job of predicting the scaling of the island density for $i = 1 - 3$ as a function of D/F, but poorer agreement is found for the island-size distribution. Substituting the scaling form for the island-size distribution (9) into Eq. (2a) with $s > i$ ($\gamma_s = 0$ and $K_s = a$) and using the fact that for moderately large coverage, $S = \theta/N \sim R^{i/(i+2)}\theta^{(i+1)/(i+2)}$, and $aRN_1 N \simeq 1$, one obtains the point-island rate-equation prediction for the asymptotic scaled island-size distribution function.[31,32]

$$f_i(u) = \frac{1}{i+2}\left(1 - \frac{i+1}{i+2}\,u\right)^{-\frac{i}{i+1}} \qquad 0 \le u \le \frac{i+2}{i+1} \tag{13a}$$

$$f_i(u) = 0 \qquad u > \frac{i+2}{i+1} \tag{13b}$$

The divergence of $f_i(u)$ for finite u is a particularly unphysical feature of this result. This is due to the neglect of the effects of island size in the point-island model as well as the neglect of correlations.

Since an accurate calculation of the scaled island-size distribution for realistic models appears to be fairly difficult, we have derived an approximate analytical expression[31] for $f_i(u)$ which appears to agree quite well with our asymptotic simulation results and with experimental data in the precoalescence regime. Our analytical expression for $f_i(u)$ is based in part on the observation[32] that for the case of fractal islands with $i = 1$, the island-size distribution scaling function $f_1(u)$ has approximately linear behavior for small u (with $f_1(0)$ approaching zero with increasing D/F). For $i > 1$ we expect smaller islands to have a lower density and $f_i(u)$ to go to zero faster than the first power. We have therefore conjectured[31] that for general i, the island-size distribution behaves as u^i for small u. We also expect $f_i(u)$ to have a peak at $u = 1$ corresponding to the average island size and to decrease rapidly for large u with an exponential cutoff. These assumptions lead to the following approximate general scaling form for $i \ge 1$:

$$f_i(u) = C_i u^i e^{-ia_i u^{1/a_i}}, \tag{14a}$$

where the constants C_i and a_i satisfy the expressions

$$\frac{\Gamma[(i+2)a_i]}{\Gamma[(i+1)a_i]} = (ia_i)^{a_i}, \qquad C_i = \frac{(ia_i)^{(i+1)a_i}}{a_i\Gamma[(i+1)a_i]}, \tag{14b}$$

obtained from the two sum-rules (10) and (12). Figure 1 shows a plot of this analytical form for $f_i(u)$ for $i = 1$ to $i = 4$. As one can see, the peak of the island-size distribution increases and becomes more sharply peaked with increasing i.

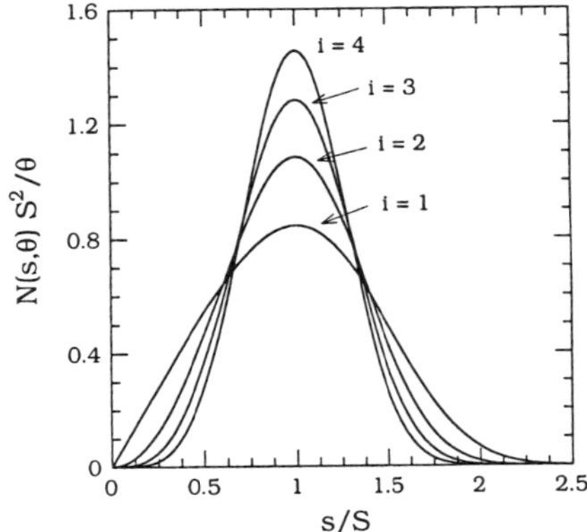

Fig. 1. Analytic form (14) for the scaled island-size distribution for $i = 1 - 4$.

2.5. *Kinetic Monte Carlo simulations of submonolayer growth*

In order to investigate the scaling behavior of the submonolayer island density and size-distribution as a function of critical island size i, we have studied a simplified model in which atoms are randomly deposited at a rate F per site per unit time onto a regular lattice (square or triangular) while monomers, including those deposited on top of existing islands, are assumed to diffuse at the rate $D = D_0 e^{-E_a/k_B T}$ in units of nearest-neighbor hops per unit time. In our model a monomer becomes stable when it has at least z nearest-neighbors, so that the critical island size i depends on both the underlying lattice and z (see Fig. 2). For atoms which have $0 < n < z$ nearest-neighbor surface atoms, we assume that the activation energy for diffusion is given by E_n so that the relative diffusion rate is given by $r_n = D_n/D = e^{-\Delta E_n/k_B T}$ where $\Delta E_n = E_n - E_a$.

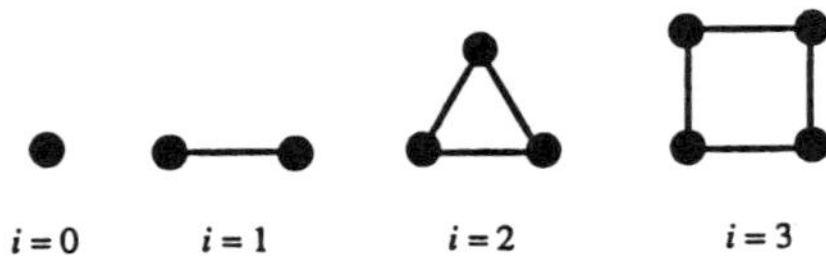

Fig. 2. Diagram showing stable island configurations for different critical island sizes $i = 0$ to 3. If we assume that an atom becomes stable when it has at least z in-plane nearest-neighbors, then the case $z = 1$ corresponds to $i = 1$ on any lattice, while $z = 2$ corresponds to $i = 2$ and 3 on triangular and square lattices, respectively. Also shown is the case $i = 0$ which corresponds to freezing of monomers.

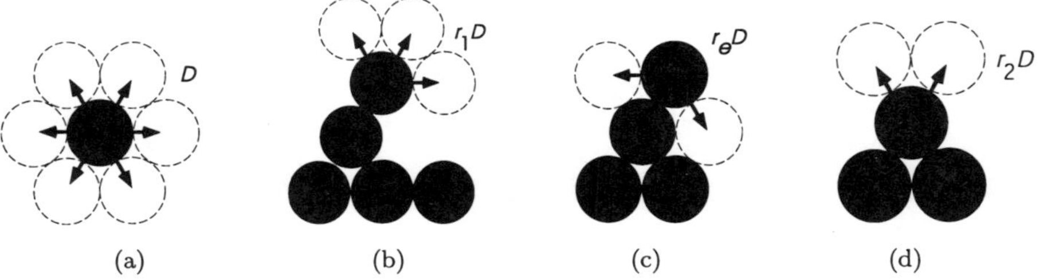

Fig. 3. Diagram showing possible motions of adatom on a triangular lattice with (a) zero nearest-neighbors or bonds, (b) one nearest-neighbor bond, (c) enhanced one-bond edge diffusion, and (d) two nearest-neighbor bonds.

In order to study the effects of edge-diffusion on island morphology we have also included an additional parameter $r_e = e^{-\Delta E_e/k_B T}$ corresponding to diffusion of atoms with one nearest-neighbor (i.e. "1-bond diffusion") along the edge of an island. Figure 3 shows the possible adatom motions corresponding to diffusion with one and two nearest-neighbors and to enhanced edge-diffusion on a triangular lattice. By varying r_n, r_e, and z, we have studied the island-size distribution, density, and morphology for $i = 0, 1, 2$ and 3.

2.6. *Island-size distribution for $i = 1$*

The case $i = 1$ corresponds to irreversible nearest-neighbor attachment (i.e. $z = 1$, see Fig. 2). The scaling of the island density and island-size distribution for this case has been extensively studied in a variety of different models.[31,32] Figure 4 shows a log–log plot of the peak island density as a function of D/F for the case of irreversible nearest-neighbor attachment on a square lattice. In the limit of large D/F the peak island density scales as $N \sim (D/F)^{-\chi_1}$ with $\chi_1 \simeq 1/3$ in good agreement with the standard rate-equation prediction $\chi_i = i/(i+2)$.

Figure 5(a) shows the scaled island-size distribution obtained from simulations with $i = 1$ on a square lattice,[31] along with similarly scaled data from simulations on a triangular lattice. Experimental results[14] for the island-size distribution in the case of Fe/Fe(100) deposition in the temperature range $20-207°C$ (for which the critical island size is believed to be equal to 1) are also shown. Also included are simulation results for deposition on a square-lattice with enhanced edge diffusion ($r_e \neq 0$) leading to compact islands. Figure 6 shows the range of morphologies from fractal to compact islands obtained in our simulations. As can be seen in Fig. 5(a), in the asymptotic limit of large

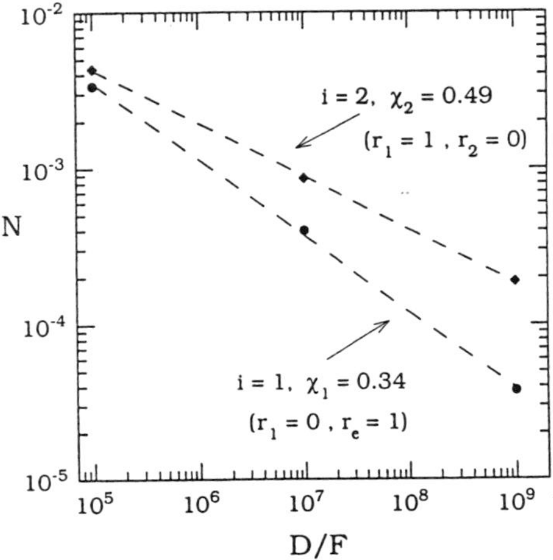

Fig. 4. Island density N as a function of D/F for $i = 1$ and $i = 2$.

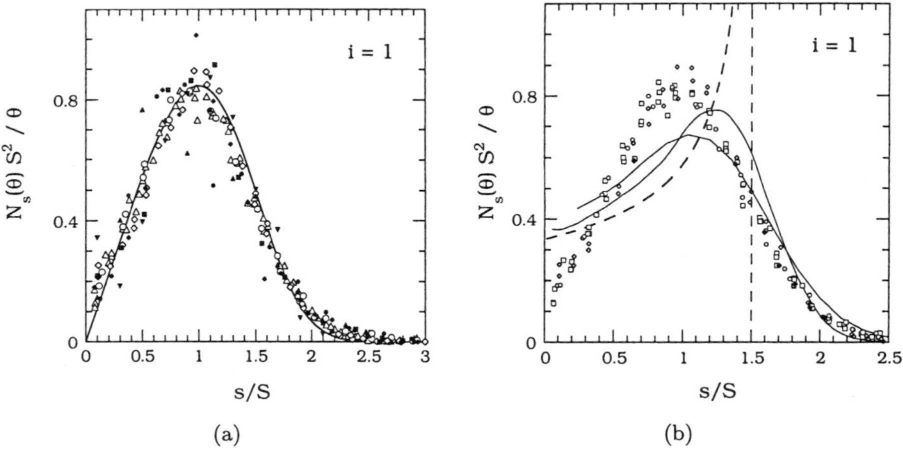

Fig. 5. (a) Island-size distribution scaling function $f_i(u)$ for $i = 1$. Open triangles correspond to simulation results on square-lattice with $R = 10^9$, $r_1 = r_e = 0$ and $\theta = 0.1 - 0.4$. Open diamonds correspond to $r_e = 10^{-4}$ and $\theta = 0.12 - 0.3$. Open circles correspond to triangular lattice simulations with $R = 10^8$, $r_1 = r_e = 0$ and $\theta = 0.2$ and $\theta = 0.3$. Completely filled symbols correspond to the experimental results[14] at $T = 20°$C (circles), $T = 108°$C (squares), $T = 132°$C (diamonds), $T = 163°$C (triangles), and $T = 207°$C (inverted triangles). Solid line is Eq. (14) for $i = 1$. (b) Comparison of simulation results (open symbols) for $f_1(u)$ with point-island model prediction. Solid curves are point-island model simulation results for $\theta = 0.1$ and $D/F = 10^6$ (lower peak) and $D/F = 10^8$ (upper peak). Dashed curve is asymptotic point-island result Eq. (13) for $i = 1$.

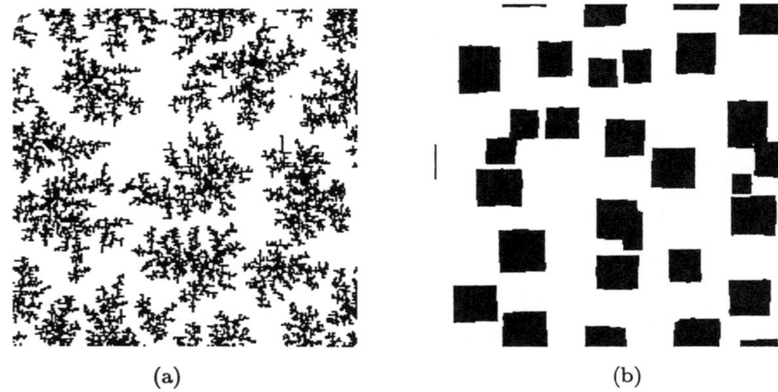

(a)   (b)

Fig. 6. Pictures of the island morphology for the case $i = 1$, with $D/F = 10^9$ at coverage $\theta = 0.3$. (a) Fractal islands formed for $r_e = 0$ ($L = 200$) and (b) Square islands formed with $r_e = 10^{-2}$ ($L = 300$).

D/F, the island-size distribution scaling function is essentially independent of the lattice. In addition, our simulation results agree quite well with the experimental results at low temperature thus confirming that the critical island size at room temperature for Fe/Fe(100) is indeed equal to 1. Also shown (solid line) is our analytical form (14) for $f_1(u)$ which agrees quite well with our simulations as well as the experimental results.

For contrast, we show in Fig. 5(b) a comparison between our analytical form (14) for $i = 1$ and results obtained from point-island model simulations. In the point-island model simulations any particle which lands on a site occupied by another particle or particles is immediately "frozen." As can be seen, the point-island model results differ considerably from the more realistic Monte Carlo simulations presented earlier. In particular the peak of the scaled island-size distribution for the point-island model appears to be shifting to the right with increasing D/F and approaching the prediction of Eq. (13) (dashed line) corresponding to a peak (divergence) at $u = 3/2$ rather than a peak at $u = 1$.

2.7. *Island-size distribution and morphology for $i = 2$*

The case $i = 2$ was simulated[31,39] on a triangular lattice with the restriction that any atom with two or more nearest-neighbors is irreversibly "frozen" (i.e. $z = 2$, see Fig. 2) so that a trimer is the smallest stable island. As has been pointed out,[39] this leads to fractal islands on a triangular lattice. Figure 7(a) shows our simulation results for the island-size distribution scaling function

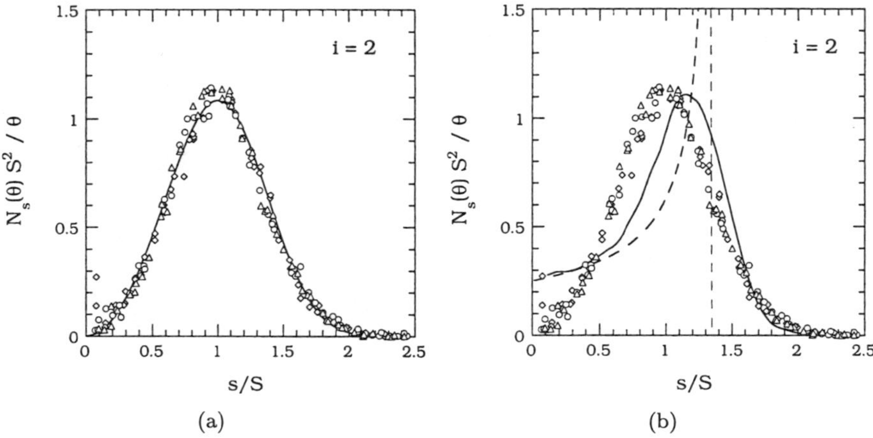

Fig. 7. (a) Island-size distribution scaling function $f_i(u)$ for $i = 2$. Simulation results are for $R = 10^7$, $r_e = 0$ with $r_1 = 0.05$ (circles) and $r_1 = 1$ (triangles) and for $R = 10^8$, $r_e = 0$ with $r_1 = 0.003$ (diamonds) on a triangular lattice with $\theta = 0.1 - 0.4$. Solid line is Eq. (14) for $i = 2$. (b) Comparison of simulation results (open symbols) for $f_2(u)$ with point-island model predictions (solid and dashed curves). Solid curve is point-island model simulation result for $\theta = 0.1$ with $D/F = 10^6$ while dashed curve shows asymptotic point-island rate-equation result (13) for $i = 2$.

for $i = 2$ with $D/F = 10^7 - 10^8$, $\theta = 0.1 - 0.4$ and $r_1 = 0.003 - 1.0$. The simulation data over a wide range of parameters — coverage, D/F, and one-bond detachment probability r_1 — all collapse onto the same curve. Also shown is our approximate analytic form for $f_2(u)$ which shows excellent agreement with the simulation results. We note that in recent experiments on Rh/Rh(111) deposition at high temperature[19] (believed to correspond to $i = 2$) excellent agreement was found between the experimental-scaled island distribution and our simulation and analytical results.[31]

For contrast, we show in Fig. 7(b) a comparison between our more realistic simulation results and those obtained from point-island model simulations. As can be seen, the point-island model gives very poor agreement with our simulation results for $i = 2$. In particular, the peak of the scaled island-size distribution for the point-island model is shifted away from $u = 1$, while at small u the distribution is finite.

We have also studied[31–40] the scaling of the island density N as a function of D/F for $i = 2$. For large D/F and r_1 not too small we find $\chi_2 \simeq 0.5$ (see Fig. 4) in agreement with the rate-equation prediction $\chi_i = i/(i+2)$. However for very small r_1 and finite D/F we find $N \sim R^{-1/3}$ corresponding to $i = 1$. Figure 8

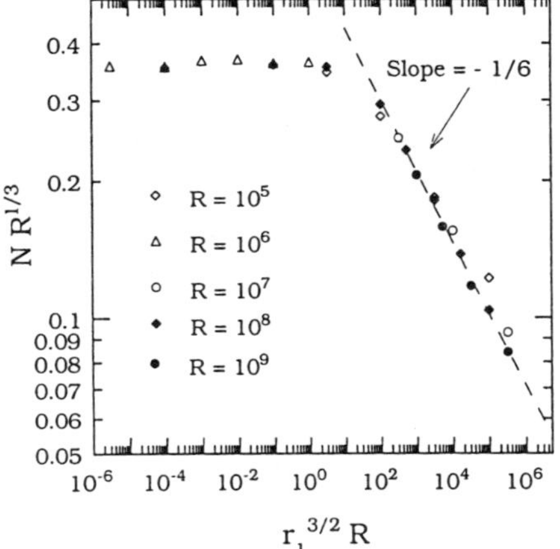

Fig. 8. Plot of crossover scaling function $f_{12}(u) = N(D/F)^{1/3}$ with $u = r_1^{3/2}(D/F)$ as in Eq. (6).

shows our results for the crossover scaling function $f_{12}(r_1^{3/2}R) = NR^{1/3}$ (where N is the island density at fixed coverage $\theta = 0.1$) for the case without edge-diffusion ($D_e = 0$) using the rate-equation prediction $x_{12} = 3/2$. As can be seen, there is good agreement with the crossover scaling form (12) using the exponents $\phi_2 = 1/4$, $x_{12} = 3/2$ predicted by Eqs. (10) and (11). In particular, for small values of $u = r_1^{3/2}R$ one has $f_{12}(u) \sim$ const corresponding to $i = 1$ behavior, while for very large values of u one has $f_{12}(u) \sim u^{-1/6}$, corresponding to $i = 2$.

In order to understand the unusually large arm thickness observed in the fractal islands formed in Au/Ru(0001)[9] we have also studied the morphology of fractal islands on a triangular lattice. Although the case $i = 1$, $z = 1$ (without edge diffusion) gives fractal islands[32,28] (see Fig. 6(a)) the small arm thickness and the large magnitude of the island density in this case do not agree with experiments. This indicates that the critical island size is larger than one. We note that, even in the presence of a significant amount of one-bond detachment, the islands will still be fractal on a triangular lattice.[39] However, the existence of a large amount of two-bond detachment will lead to significantly more compact islands.

Figure 9 shows the island morphology obtained on a triangular lattice with $r_1 = 10^{-4}$ and $D/F = 10^{11}$, at an island density close to but somewhat higher than that in the experiments, at coverage $\theta = 0.37$ and $\theta = 0.69$. A

(a)   (b)

Fig. 9. Morphology of islands formed on a triangular lattice with $D/F = 10^{11}$, $r_1 = 10^{-4}$, $r_2 = 10^{-8}$, and $r_e = 10^{-2}$ (system size $L = 600$). (a) $\theta = 0.37$ and (b) $\theta = 0.69$.

small amount of two-bond detachment ($r_2 = 10^{-8}$) as well as enhanced one-bond edge-diffusion ($r_e = 10^{-2}$) have also been included. As can be seen, the island morphology and arm thickness are quite similar to what is observed in experiment. Thus, this model may provide a partial explanation of the large arm thickness observed in the fractal islands formed in Au/Ru(0001). We note that the flux dependence of the island density observed experimentally for Au/Ru(0001) indicates that the critical island size is actually larger than 2. This may be due to the existence of enhanced two-bond detachment from small islands or possibly small cluster mobility.

2.8. Island-size distribution for $i = 3$

It has been suggested[14] that at elevated temperatures, the critical cluster size for Fe/Fe (100) deposition is equal to 3. In this case the minimal stable configuration is a tetramer, and in order to determine the island-size distribution, we have studied a model similar to that for $i = 2$ (i.e. $z = 2$, see Fig. 2) but on a square lattice. In contrast to the triangular lattice, the existence of significant one-bond detachment on a square-lattice leads to compact square islands.[31]

Figure 10 shows our results for the island-size distribution scaling function for a variety of values of $R = D/F$ and r_1 both with and without edge-diffusion. The simulation results cover a wide range of coverages (including $\theta \simeq 0.07$ for which the island-size distribution was measured for Fe/Fe(100)) and correspond approximately to the experimental island density observed at $T = 301°C$ and $T = 356°C$ in Fe/Fe(100). Also shown are experimental results for the scaled island-size distribution for Fe/Fe(100) deposition at these temperatures as well as our analytical expression (14) for $f_3(u)$. As can be seen, there is very good

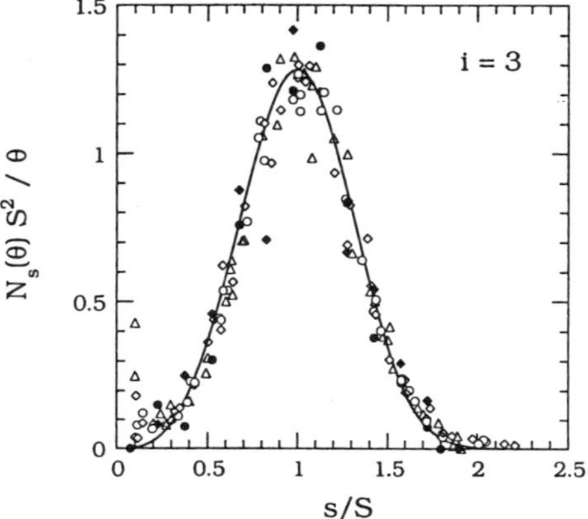

Fig. 10. Island-size distribution scaling function $f_i(u)$ for $i = 3$. Simulation results on a square lattice (open symbols) for $\theta = 0.06 - 0.3$ are for $R = 10^{11}$, $r_1 = 2.5 \times 10^{-6}$, and $r_e = 0$ (circles) and for $R = 2.0 \times 10^{11}$ and $r_1 = 10^{-6}$ (diamonds) both with and without edge diffusion ($r_e = 0$ and $r_e = 10^{-2}$). Open triangles are for $\theta = 0.04 - 0.12$ with $R = 5 \times 10^9$, $r_1 = 10^{-4}$ with $r_e = 0$ and $r_e = 10^{-2}$. Filled symbols are experimental results for Fe/Fe(100) at $T = 301°$C (diamonds) and $T = 356°$C (circles). Solid line is Eq. (14) for $i = 3$.

agreement between our simulation results and the analytical form (14) as well as with the experiments. This confirms that $i = 3$ is the critical island size at elevated temperatures for Fe/Fe(100) deposition.

We have also studied[39,40] the scaling of the island density N with D/F for the case $i = 3$ corresponding to large r_1 and find reasonable agreement with the rate equation prediction as shown by the inset of Fig. 11. We note, however, that the result $\phi_3 \simeq 0.34$ does not agree with the standard rate-equation prediction ($\phi_3 = 0.4$). Taking into account the measured value of χ_3 gives a value for the crossover scaling exponent x_{13} ($x_{13} \simeq 1.33$) that is significantly below the expected value of $3/2$ predicted by Eq. (5).

Figure 11 shows the island density crossover scaling function $f_{13}(r_1^{x_{13}} R) = NR^{1/3}$ (for the case of no edge diffusion) using the value of the crossover scaling exponent obtained from our estimate of ϕ_3 in the inset ($x_{13} = 1.33$). As can be seen, there is reasonably good scaling while the slope of the crossover scaling function for large u is consistent with the measured value of χ_3.

2.9. *Island-size distribution for $i = 0$*

The case of "zero" critical island size ($i = 0$) corresponds to the spontaneous "freezing" of single monomers which may nucleate stable islands, so that the

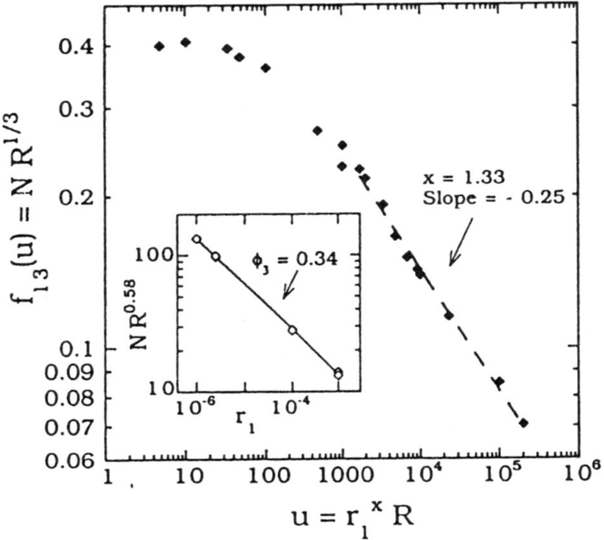

Fig. 11. Island density crossover-scaling function $f_{13}(u) = NR^{1/3}$ where $u = r_1^{x_{13}}R$ and crossover exponent $x_{13} = 1.33$ obtained from simulations of restricted pair-bond model on square lattice. Inset shows scaling behavior of island density N as function of detachment rate r_1 in the $i = 3$ regime.

smallest stable "cluster" contains one island. This case has recently been studied experimentally by Chambliss and Johnson.[16] In this experiment, isolated Fe atoms which have been deposited on a Cu(100) surface spontaneously embed into the surface and form stable islands leading to a critical island size $i = 0$. The case of zero critical island size also corresponds to deposition in the presence of impurities or surfactants which lead to heterogenous nucleation.

To simulate this case,[31,38] we have studied a model in which, in addition to the usual diffusion with hopping rate D, monomers spontaneously "freeze" and form stable islands at a rate given by $R_s = r_s D$, where $r_s = e^{-\Delta E_s/k_B T}$ corresponds to the extra activation energy beyond that for normal diffusion. In addition, any monomers which become nearest-neighbors of an embedded island are immediately added irreversibly to the embedded island. As before, adatoms which land on top of an existing island continue to diffuse with diffusion rate D, until they fall off the edge of the island, at which point they attach irreversibly to the island.

Figure 12 shows our simulation results for the island-size distribution scaling function for $r_s = 10^{-3} - 10^{-5}$ and $\theta = 0.06 - 0.3$, as well as various values

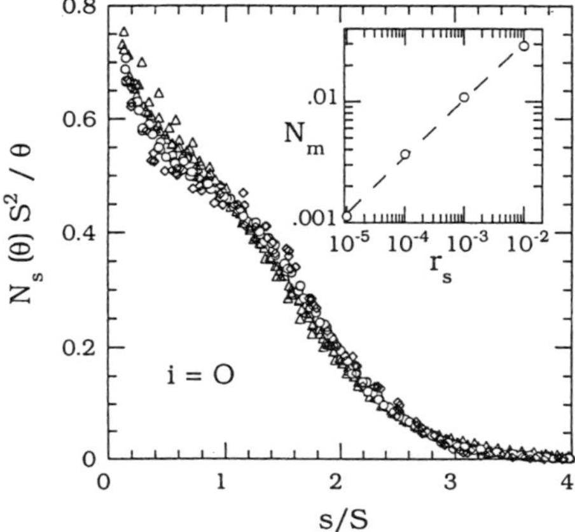

Fig. 12. Island-size distribution scaling function $f_0(u)$ form simulations on a square lattice for $i = 0$ with $r_s = 10^{-3}$ (triangles), 10^{-4} (circles) and 10^{-5} (diamonds) with $R \geq 10^9$. Inset shows log–log plot of island density N versus relative embedding probability r_s. Dashed line fit has slope equal to 0.48 in agreement with the rate-equation prediction $\phi_s \simeq 1/2$.

of $D/F > 10^9$. As can be seen, the island-size distribution scaling function f_0 is essentially independent of the embedding probability ratio r_s as well as D/F over all coverage for $D/F \geq 10^9$. In addition, the scaling function looks quite similar to that obtained in experiment.

3. Multilayer Growth

3.1. *Introduction*

One of the goals of developing growth models is to simulate the evolution of surface morphology. Recently, a morphological instability[41,42] leading to the formation of large "mounds," ripples, and/or facetted structures which coarsen with time has been observed in a variety of experiments on homoepitaxial growth. For example, the formation of mounds with a selected "magic" mound angle of approximately $13°$ was observed in Fe/Fe(100) deposition at room temperature,[43] while in Fe/Fe(001)/MgO(001) deposition at $T \simeq 180°C$,[44] pyramidal structures with [012] facets were observed. Similarly, experiments on Cu/Cu(100) deposition indicate the formation of mounds with low-angle [113] facets at $T = 160$ K and [115] facets at 200 K.[45] Low-angle mound formation has also been observed in GaAs/GaAs(100),[42,46] Ge/Ge(100),[47] Ag/Ag(100),[48] and Rh/Rh(111).[19]

From a microscopic point of view, the presence of an instability to mound formation in homoepitaxial growth may be explained by the existence of a positive Ehrlich–Schwoebel step barrier E_B[22] (defined as the difference between the energy barrier for an adatom to hop from a terrace to the layer below and the energy barrier for diffusion on a flat terrace) which prevents adatoms landing on a large island or terrace from diffusing to the layer below and enhances the nucleation of new islands on the upper terrace. This leads to the formation of large "wedding-cake" structures which coarsen with time. From the standpoint of recent continuum theories,[41,49] the reflection of diffusing adatoms from descending steps leads to a net positive (uphill) surface current. The presence of such an uphill current implies a negative surface tension which destabilizes the surface. However, for large enough slopes (e.g., at angles corresponding to high-symmetry surfaces[50]) the surface current will go to zero, thus stabilizing the surface. The value of the selected mound angle then corresponds to the slope for which the surface current is zero.

Recently, it has been shown[51–58] that for deposition on metal fcc(100) and bcc(100) surfaces, the crystal structure plays a crucial role. The presence of a selected mound angle in these systems may be explained as due to the competition between the effects of crystal geometry (which leads to a negative current due to downward funneling[56] of atoms deposited near step edges) and an upward, positive surface current (due for example to the reflection of adatoms from descending steps). This leads to a selected mound angle which is significantly smaller than predicted by crystal symmetry but in good agreement with experiment. In particular, in recent kinetic Monte Carlo simulations of Fe/Fe(100) growth at room temperature,[52,58] we have obtained good agreement with experiment for the selected mound angle and mound coarsening behavior, using reasonable estimates for the value of the interlayer step barrier for Fe/Fe(100). We have also carried out a general analytical calculation of the surface current as a function of step barrier and temperature[58,57] on fcc(100) and bcc(100) surfaces which was found to give reasonable agreement with our simulation results[57] as well as a reasonable prediction for the selected mound angle for Fe/Fe(100) at room temperature. The asymptotic surface roughening and coarsening behavior and its dependence on temperature and strength of the step barrier was also investigated.[58] These results appear to confirm the scenario of a kinetically determined mound angle based on a competition between upward and downward currents.

We have also carried out a general analysis of the effects of step-adatom interactions on epitaxial growth.[55] In particular, we have recently shown that, even in the absence of a step-barrier, the presence of short-range adatom attraction to ascending steps (as has recently been observed in Ir/Ir(111)[59]) leads to mound formation in epitaxial growth. This indicates that in addition to the Erhlich–Schwoebel barrier at descending steps, the effects of attraction (or repulsion) at ascending steps may also play an important role in epitaxial growth. A general calculation of the surface current as a function of attraction (repulsion) from ascending and descending steps on an fcc/bcc(100) surface leads to a unified picture.

3.2. *Kinetic Monte Carlo simulations of epitaxial growth*

In our model atoms are randomly deposited onto a square lattice corresponding to the bcc(100) substrate (see Fig. 13). Due to the bcc(100) geometry, the initial surface is "corrugated" in a checkerboard pattern, so that initially half of the surface sites (pluses) are in the top layer while the other half (open circles) are in the layer below forming four-fold hollow sites. Due to the epitaxial nature of the growth, deposited atoms are incorporated into the surface only at the four-fold hollow sites corresponding to sites for which the four nearest-neighbor atoms are only one layer below. This implies that if the deposited atom lands directly on a four-fold hollow site then it becomes part of the surface. However, if one or more of the nearest-neighbor sites has height lower than one less than the height of the deposited atom then the freshly deposited atom "cascades" randomly (downward funneling[56]) to one of these lower sites. This process is then repeated at the new site, until a four-fold hollow is found, although typically, no more than two cascade processes ever takes place.

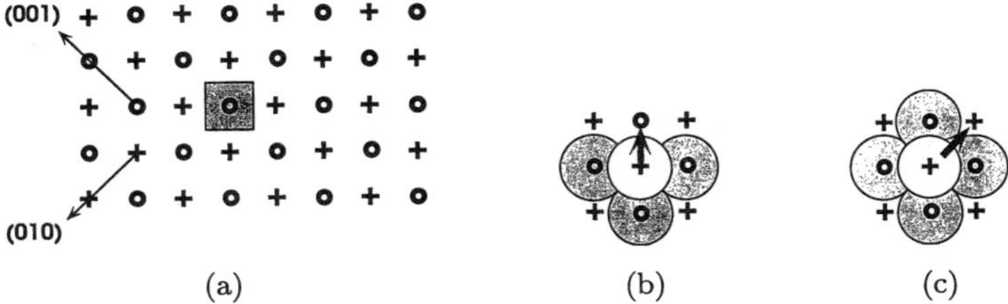

(a)   (b)   (c)

Fig. 13. (a) Square lattice of deposition sites used in simulation of growth on a bcc (100) substrate. Initially, the sites labeled + correspond to the top layer and form four-fold hollows while the sites labeled with open circles are one layer below. The shaded square shows the catchment area for the deposition site in the center. (b) Cascade to an epitaxial site after deposition. (c) Diffusion across a bridge site to a next nearest-neighbor site.

Once atoms have reached the nearest four-fold hollow site, they are then allowed to diffuse. In our model, diffusion takes place along the next nearest-neighbor direction and is determined by the number of next nearest-neighbor (in-plane) bonds of each atom as well as by whether or not the atoms jump down a step. In particular, the hopping rate for atoms on a flat surface with no next nearest-neighbor bonds is given by $D = D_0 e^{-E_a/k_B T}$ where E_a is the activation energy for monomer diffusion. At room temperature ($T = 20°\text{C}$) there is effectively no detachment of adatoms with one or more lateral bonds (in agreement with experimental results for Fe/Fe(100)[14]) so that the hopping rate of adatoms with one or more bonds away from a cluster is effectively zero. However, in order to take into account the observed square morphology of islands at room temperature, edge-diffusion of one-bonded atoms along the edge of a cluster is included at a rate given by $D_e = D e^{-E_e/k_B T}$. In order to include the effect of a step barrier, for monomers which diffuse to a site which is not a four-fold hollow site (which would correspond to going down a step) an extra step-barrier energy E_B is assumed so that the diffusion rate is given by $D_{\text{step}} = D e^{-E_B/k_B T}$.

Simulations were carried out using parameters appropriate for recent experiments on Fe/Fe(001) deposition at room temperature ($E_a = 0.45$ eV, $D_0 = 1.8 \times 10^{11}$ sec^{-1}, $E_e = 0.1 - 0.125$ eV, $E_B = 0.05 - 0.07$ eV) as well as for different values of the step-barrier E_B. Simulations were carried out using both a "fast" deposition rate ($F = 0.51$ ML/sec) corresponding to a previous experiment[43] on Fe/Fe(100) as well as a "slow" deposition rate ($F = 0.0257$ ML/sec) close to that used in an experiment by He *et al.*[60] In order to quantify our results, the circularly averaged height–height correlation function $G(r) = \langle \tilde{h}(0) \tilde{h}(r) \rangle_C$ was calculated, where $h(\mathbf{r})$ is the height in layers at site $\mathbf{r}$, $\tilde{h}(\mathbf{r}) = h(\mathbf{r}) - \langle h \rangle$ and $\langle h \rangle$ is the average layer height. The average feature separation or mound size $2r_c$ was also calculated from the position r_c of the first zero crossing of $G(r)$. In addition, the surface roughness was monitored by calculating the root-mean-square surface width $w = [G(0)]^{1/2}$. To study the time evolution of the mound slope, the ratio w/r_c (which is proportional to the average mound angle) was also calculated.

Figure 14 shows a typical example of the surface morphology in the late stages of growth after 100 monolayers have been deposited. This result was obtained with a slow deposition rate, but the surface morphology for fast deposition is qualitatively similar[58] showing the formation of mound structures with a selected angle.

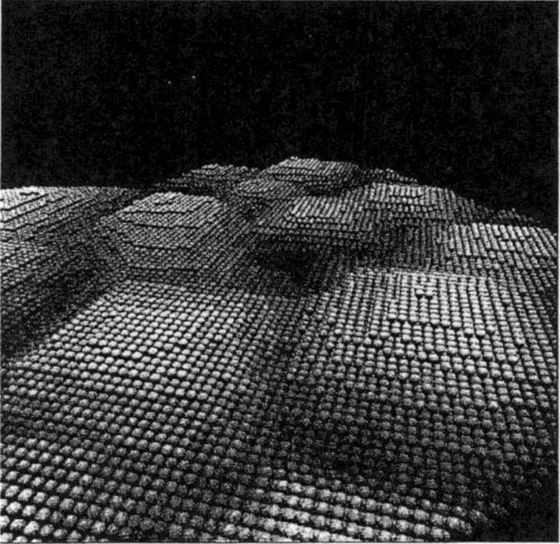

Fig. 14. Surface morphology obtained from bcc lattice simulations at $T = 20°$C ($F = 0.0257$ ML/sec) after 100 layers have been deposited.

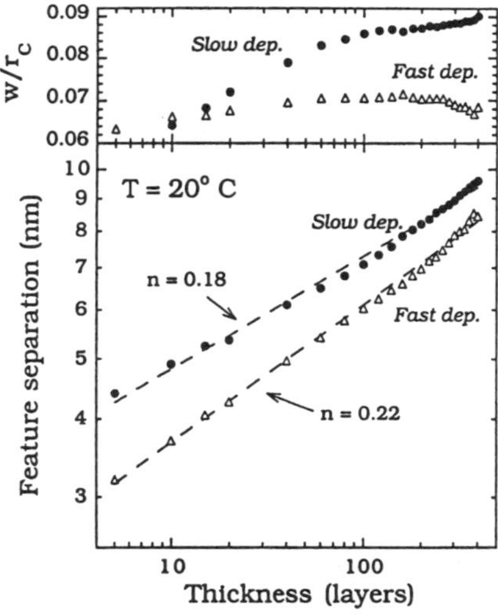

Fig. 15. Calculated feature separation ($2r_c$) and mound angle ratio w/r_c versus film thickness at $T = 20°$C for slow and fast deposition rates.

3.3. *Mound formation and coarsening exponent*

Figure 15 shows the characteristic ratio w/r_c for both fast and slow deposition rates, as a function of film thickness. As can be seen, beyond the first few layers, the mound angle is essentially constant. The calculated value for w/r_c

for the case of fast deposition is very close to the experimental value of $w/r_c \simeq$ 0.06 − 0.07.[43] In addition, the weak dependence of w/r_c on deposition rate is also consistent with the experimental results in which slow deposition rates were used up to 20 ML while fast deposition rates were used up to a thickness of several hundred layers.[43]

We also show in Fig. 15 the feature separation as a function of thickness obtained in our simulation. Power-law fits to the data indicate that the feature separation grows as $\langle h \rangle^n$ indicating the existence of a coarsening exponent n which is found to be relatively insensitive to the deposition rate, with $n \simeq 0.19 \pm 0.02$ for the slow deposition rate, and $n \simeq 0.21 \pm 0.02$ for the fast deposition rate. These values are consistent with but slightly above the experimental estimate $n \simeq 0.16 \pm 0.04$.[43] Thus, our use of the correct crystal geometry appears to explain both the observed characteristic mound angle, as well as the observed mound coarsening exponent n. We note however, that fits to the later-time data give somewhat higher values ($n \simeq 0.22$ and $n \simeq 0.25$ for slow and fast deposition rates respectively) close to the value $n \simeq 1/4$ previously found in numerical simulations of a continuum model with angle-selection.[50]

In order to investigate the asymptotic coarsening exponent in our model, we have carried out simulations at room temperature and slow deposition rate, using a very large step barrier ($E_B = 0.6$ eV). Figure 16 shows a plot of the scaled surface skewness $S_3 = \langle (h - \langle h \rangle)^3 \rangle / w^3$ at room temperature for different values of the step barrier E_B. As can be seen, for the case of a moderate

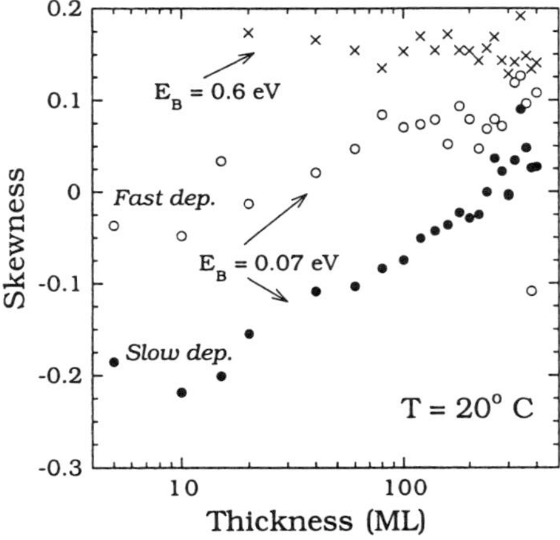

Fig. 16.   Scaled surface skewness as function of step barrier at $T = 20°$C.

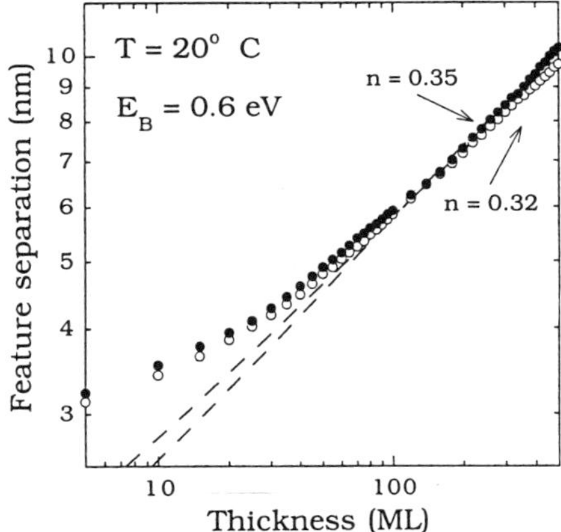

Fig. 17. Calculated feature separation $(2r_c)$ versus film thickness at $T = 20°\text{C}$ for the case of a large step barrier E_B.

step barrier $(E_B = 0.07 \text{ eV})$, the scaled skewness is still increasing with time, indicating that the asymptotic regime has not yet been reached for this case. However, for the case of a large step-barrier the skewness saturates rapidly indicating that we are in the asymptotic, scaling regime.

Figure 17 shows the feature separation $2r_c$ as a function of coverage for the case of a large step barrier both with (filled symbols) and without (open symbols) transient kinetics at a step edge. The value of the late-time coarsening exponent ($n = 0.35 \pm 0.02$ with transient kinetics and $n = 0.32 \pm 0.01$ without transient kinetics) is significantly larger than previously obtained[58] with a moderate step-barrier and is close to the value ($n \simeq 0.33$) obtained experimentally for Rh/Rh(111).[19] Similar results have also been obtained at both higher temperature ($T = 80°\text{C}$) and for a smaller but still significantly large step-barrier ($E_B = 0.15 \text{ eV}$). This indicates that for the case of a strong step-barrier (and without detachment from steps) the asymptotic coarsening exponent is close to 1/3. In general, however, one expects that n may depend on the detailed mechanisms of mound coalescence and relaxation.

3.4. *Surface kinetic roughening*

We have also studied kinetic roughening of the surface in our simulations.[61] Figure 18 shows the corresponding results for the surface width as a function of film thickness. The effective kinetic roughening exponent β defined by $w \sim \langle h \rangle^\beta$ was found to be 0.24 ± 0.02 for slow deposition, and 0.22 ± 0.02 for

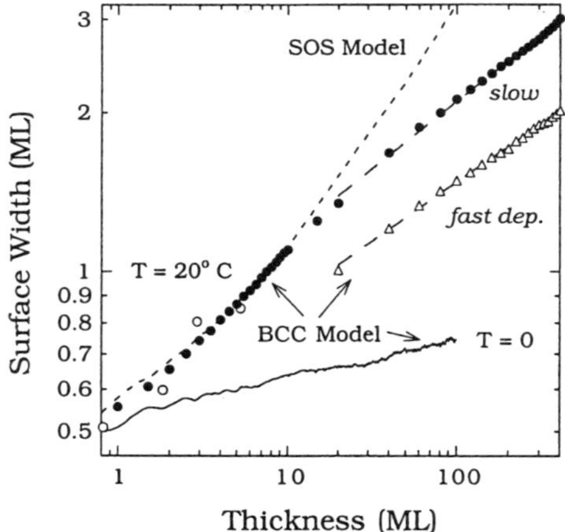

Fig. 18. Surface width w for bcc growth model at $T = 20°$C (slow and fast deposition) as well as corresponding results for simple-cubic-lattice (SOS) model (slow deposition) and bcc model at $T = 0°$C. Power-law fits to late-time data (dashed lines) at $T = 20°$C give exponents, $\beta = 0.45 \pm 0.01$ (SOS model), $\beta = 0.24 \pm 0.01$ (bcc model, slow deposition), and $\beta = 0.22 \pm 0.02$ (bcc model, fast deposition). Open circles are experimental results.[62]

fast deposition, in good agreement with the experimental estimate[60] of $\beta = 0.22 \pm 0.02$. We note that similar results for the roughness exponent β have been obtained by Bartelt and Evans.[51] Also shown in Fig. 18 is a comparison with the results for the corresponding simple-cubic-lattice model with transient kinetics. As can be seen, while the surface width for the sc lattice model agrees with experiment for the first few layers deposited, it increases much more rapidly as more layers are deposited and leads to a significantly larger value for β at room temperature. For this model the mound angle increases without bound due to the existence of large cliffs while the feature separation increases much more slowly than in the experiment.

3.5. *Surface current and angle selection*

One explanation for the angle selection observed in our simulations is that the initial cascade upon deposition to a four-fold hollow site leads to a downward current which counterbalances the uphill current[42] due to the Ehrlich–Schwoebel step barrier E_B. For sufficiently large local slope this leads to a stabilizing *negative* current. In order to test this idea quantitatively we have calculated the surface mass current $J(m)$ as a function of imposed slope m (tilting the surface around the (001) direction) using our model as shown in

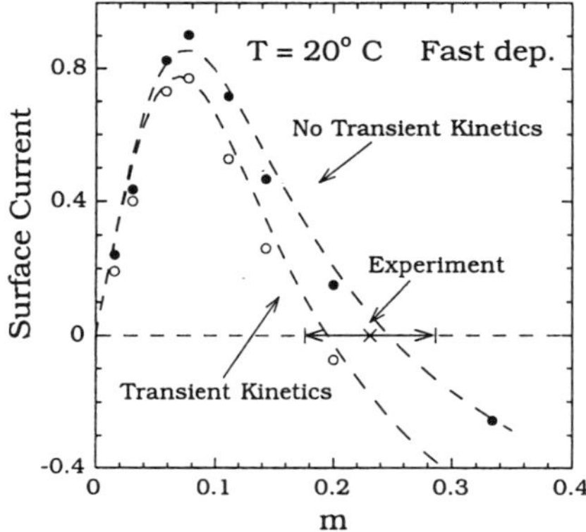

Fig. 19. Surface mass current $J/2$ as a function of imposed slope m using the parameters given in Figs. 14–16, both with and without transient kinetics. Dashed lines are theoretical fits of the form given in the text with $\sigma \simeq 5.0$. Cross with arrows indicates experimental estimate[43] of the mound angle for room-temperature Fe/Fe(100) deposition.

Fig. 19. As can be seen, the current is negative for sufficiently large slopes both with and without transient kinetics. The value of the slope ($m = m_0$) for which the current is zero corresponds to the selected mound angle. The range of values of m_0 $(0.20-0.25)$ found in our simulations is in very good agreement with the experimental result[43] ($\theta = 13 \pm 3°$, $m = 0.18 - 0.29$) for Fe/Fe(100) at room temperature.

The value of m_0 obtained in our simulations and in the experiments[43] may also be estimated from an analysis of the upward and downward currents which properly takes into account the crystal geometry.[58] Assuming a periodic array of steps running along the [001] direction and taking into account cascade events to four-fold hollow sites due to deposition at steps and assuming for simplicity a perfectly reflecting step-barrier, one obtains for the surface current,

$$J(m) = J_+(m) - J_-(m) = (1 - 2|m|)^2/2|m| - 2|m|. \tag{15}$$

Setting the surface current to zero leads to the estimate $m_0 \simeq 0.25$ or $\theta \simeq 14°$, which is in good agreement with our simulations and with experiment. Similar considerations for the case in which a small amount of transient kinetics have been included lead to $m_0 = 0.2$ or $\theta \simeq 11°$ again in good agreement with our simulations and experiment. We note that for small angles one expects that the uphill current will be proportional to the slope.[42] Accordingly, also shown in Fig. 19 is a theoretical interpolation of the form, $J(m) = \sigma^2 m/(1 +$

$\sigma^2 m/J_+(m)) - J_-(m)$ where σ corresponds to the diffusion length for atoms deposited on a flat terrace. As can be seen, this form agrees quite well with the simulation results.

3.6. *Critical temperature for mound formation*

While these results demonstrate that mound formation occurs for the case of a moderately large step barrier, it is interesting to ask what happens for smaller step barrier. Two general questions emerge: (1) What is the dependence of the mound angle on temperature or more precisely on the ratio E_B/k_BT? (2) Is there a critical temperature above which mound formation no longer occurs? In order to answer these questions, we have calculated the surface current as a function of mound angle for the case of a finite barrier.[57]

We consider a quasi-one-dimensional model[58] consisting of a regular-stepped bcc(100) or fcc(100) surface with infinitely long straight steps along the [001] direction with slope m and terrace length $l = 1/m$ (in units of $1/2$ the next nearest-neighbor distance) where $l = 2j + 1$ and j is the number of exposed rows in each (100) terrace (see Fig. 22). The notation used here refers to a bcc(100) geometry (as for Fe(100)) and in this case the tilted surface considered here corresponds to a [10l] facet. However, our calculation also applies to deposition on a fcc(100) surface. For this case the stepped surface corresponds to a [11l] facet, while the actual value of the selected slope is $\sqrt{2}$ times larger due to the difference in the relative spacing between layers.

In the usual downward funneling models[51−56] one assumes that particles that land at the edge of a step will go with equal probability to the nearest four-fold hollow sites on the upper or lower terraces. Here, we consider the general case in which a particle landing at the edge of a step will "cascade" to the four-fold hollow site on the upper terrace with probability α and to the lower terrace attachment site with probability $1 - \alpha$. Similarly, downward funneling implies that a particle landing just beyond a step will cascade with probability 1 to the attachment site 0 on the lower terrace. Again we consider the more general case in which such a particle "cascades" to the lower terrace attachment site with probability $1 - \beta$ and to the upper terrace with probability β. For comparison with Fe/Fe(100) deposition at room temperature, it appears that the simplest assumption $\alpha = 1/2$, $\beta = 0$ (corresponding to downward funneling) is most appropriate.

We have calculated the surface current by mapping our model to a one-dimensional random walk between two absorbing barriers, with biased diffusion at a step edge.[57] We obtain the following general expression for the surface current J,[57]

$$J = \frac{(2R - 3) + m(9 + 2R(2(\alpha + \beta) - 5))}{2(1 + (2R - 3)m)},$$ (16)

where $R = (\nu_0/\nu_B)e^{E_B/k_BT}$ is the asymmetry ratio for diffusion at a step-edge and where E_B is the step barrier and ν_0/ν_B is the ratio of frequency factors for hopping away from the step edge versus hopping over the step. We expect this expression to be valid for m not too large (i.e. $m \leq 1/3$). From Eq. (16), we obtain the selected mound slope as a function of R,

$$m_0 = \frac{2R - 3}{4R(1 - \alpha - \beta) + 3(2R - 3)}, \qquad (R \geq 3/2).$$ (17)

Figure 20 shows the predicted mound slope using Eq. (17) as a function of $r = \ln R = \ln(\nu_0/\nu_B) + E_B/k_BT$ for the case $\alpha = 1/2$ and $\beta = 0$ (corresponding to the usual downward funneling[56]) along with simulation results. As can be seen, there is very good agreement between the simulation results and Eq. (17). Equation (17) also implies that there is a critical value of R ($R_c = 3/2$) required for mound formation. This implies that if $E_B > 0$, then if $\nu_0/\nu_B < 3/2$ there exists a critical temperature for mound formation. Figure 21 shows a plot of the critical temperature for mound formation T_c as a function of the ratio ν_0/ν_B for the case $E_B > 0$.

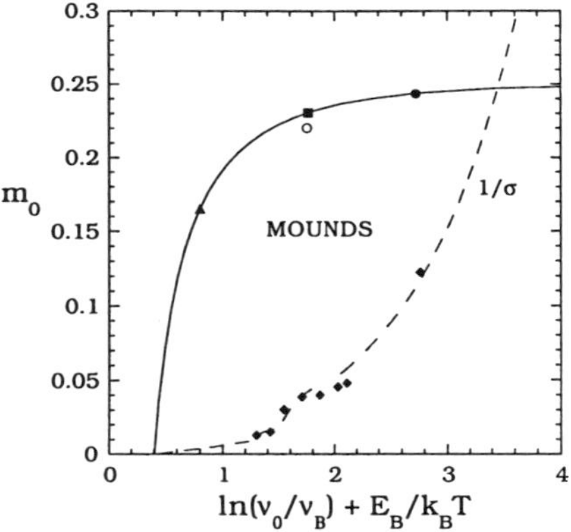

Fig. 20. Selected mound slope m_0 as a function of $r = \ln(\nu_0/\nu_B) + E_B/k_BT$ using Eq. (17) for the case $\alpha = 1/2$, $\beta = 0$ (solid line). Square is experimental estimate[43] while other symbols correspond to simulation results. Diamond-shaped symbols correspond to inverse diffusion length $1/\sigma$ for Fe/Fe(100) obtained from experimental results for the island density.

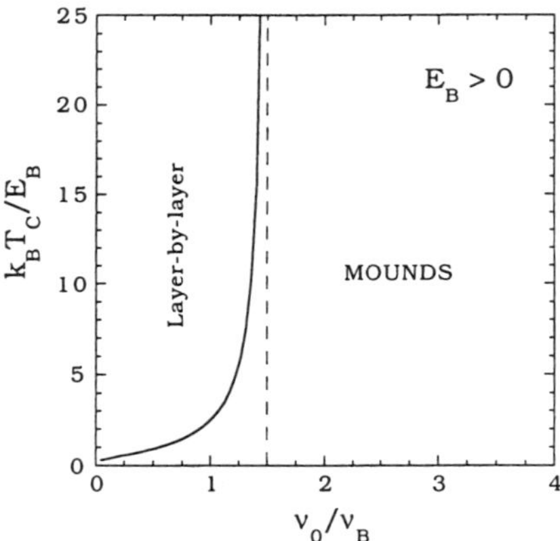

Fig. 21. Critical temperature T_c for mound formation as a function of the prefactor ratio ν_0/ν_B for the case of a positive step barrier.

3.7. *Effects of step-adatom attraction on epitaxial growth*

In detailed FIM studies of adatom diffusion on the Ir(111) surface, Wang and Ehrlich[59] have found that there exists a short-range attractive interaction between an adatom diffusing on a terrace and a cluster. Although it was pointed out that this effect increases the capture radius for a cluster, the consequences of this attraction on epitaxial growth have not been investigated.

In order to study the effects of step-adatom attraction on a specific model, we consider the stability of a bcc(100) (or equivalently fcc(100)) surface, in the presence of step-adatom attraction.[59] As for the case of a step-barrier, we consider a quasi-one-dimensional model consisting of a regular stepped bcc(100) surface (corresponding to a (1 0 l) facet) with infinitely long straight steps along the [001] direction (see Fig. 22) with terrace length $l = 1/m$ (in units of 1/2 the next nearest-neighbor distance) where $l = 2j + 1$ and j is the number of exposed rows in each (100) terrace and m is the slope of the surface. We also assume irreversible attachment at ascending steps (site 0).

In order to include the effects of short-range step-adatom attraction in our calculation we assume that an adatom one step away from an ascending step (site 2 in Fig. 22) experiences a diffusion bias in which the ratio of a hop to the left (towards the step) versus a hop to the right is given by R', where $R' > 1$ due to step-adatom attraction. Similarly, we consider the possibility of a step barrier by including a diffusion bias R on an adatom at site $l - 3$ where R is the ratio of a hop to the left (away from the step) versus a hop to the

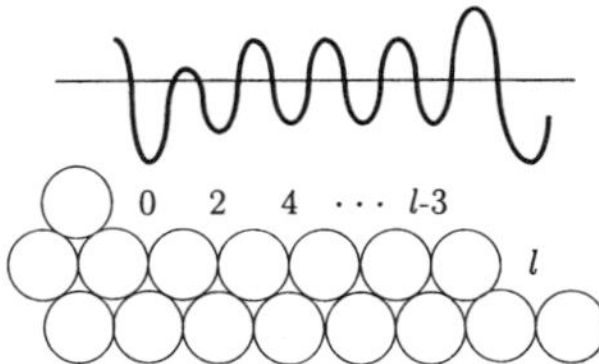

Fig. 22. Diagram showing stepped bcc/fcc(100) surface with slope m (terrace width $l = 1/m$) and straight step edges along the (001) axis (side view). Even sites correspond to four-fold hollow sites on terrace. Also shown is schematic of potential surface showing decreased potential barrier due to step-adatom attraction near the ascending step along with a possible step barrier at the descending step.

right. Accordingly, our model can be mapped to a one-dimensional random walk between two absorbing barriers (sites 0 and l in Fig. 22) with biased diffusion at site 2 due to step-adatom attraction and at site $l - 3$ due to a step barrier. As before, freshly deposited atoms at a step edge (sites $l - 2$ and $l - 1$ in Fig. 22) are assumed to cascade with probability α and β to the upper step and with probability $1 - \alpha$ and $1 - \beta$ respectively to the lower step attachment site 0, where $\alpha = 1/2$, $\beta = 0$ corresponds to the usual downward funneling.

In this case we obtain for the surface diffusion current J,

$$J = \frac{(2R - 1)R' - 2 + m(4 + 5R' - 2RR'(5 - 2(\alpha + \beta)))]}{2(R' + m[(2R - 5)R' + 2])}, \quad (m \leq 1/7). \quad (18)$$

Consideration of the sign of the current in Eq. (18) for small m as a function of R and R' leads to a unified picture of the effects of attraction and repulsion at ascending and descending steps.[55] For example, Eq. (18) implies that for $R > 1/2$ there exists a critical value of the step-adatom attraction $R'_c = 2/(2R - 1)$ such that for $R' > R'_c$ the surface current is positive for small m leading to a mound instability. We note that R'_c is independent of the parameters α and β which control the funnelling near a step edge since the effects of these parameters vanish in the limit of small slopes. In particular, Eq. (18) implies that even without a step barrier ($R = 1$) step-adatom attraction with $R' > 2$ will lead to a mound instability. From Eq. (18) the selected mound angle m_0 can also be calculated by finding the value of the slope for which the current is zero. In particular, in the absence of a step barrier ($R = 1$) but in the presence of strong short-range step-adatom attraction ($R' = \infty$), Eq. (18) implies that $m_0 = 1/(5 - 2(\alpha + \beta))$, which is the same as previously found for the case of a very large step-barrier without step attraction.[58]

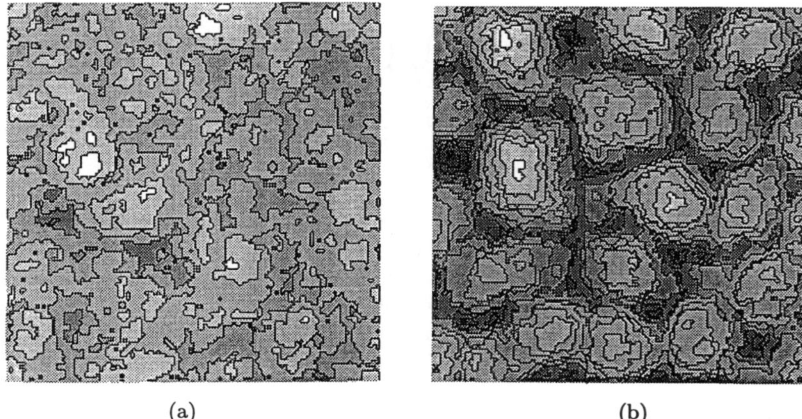

(a)   (b)

Fig. 23. Gray-scale plots (128×128) of surface morphology obtained after 100 ML deposition without a step barrier ($R = 1, E_B = 0$) for the case of *fast* deposition. Pictures correspond to a portion of a 256×256 system. (a) No step-adatom attraction ($R' = 1$) and (b) Step-adatom attraction with $R' = 10$.

In order to verify the presence of a mound instability due to step-adatom attraction, we have carried out simulations for growth on a fcc(100) (or equivalently bcc(100)) surface without a step barrier but with short-range step-adatom attraction. Figure 23 shows a comparison of the surface morphology both with and without step-adatom attraction obtained after 100 layers have been deposited. As expected, without step-adatom attraction the surface has the usual self-affine-fractal morphology indicating no typical length scale or feature size.[63] However, in the presence of step-adatom attraction large mounds with a characteristic length scale are clearly visible, indicating the presence of an instability. As evolution proceeds, these mounds continue to coarsen and increase in size as in the case of the Ehrlich–Schwoebel instability.[42] Analysis of similar images indicates that at late times the average mound slope is close to 1/4 in agreement with the prediction of Eq. (18).

Figure 24 shows a more quantitative comparison in which the aspect ratio w/r_c is plotted as a function of film thickness. A finite saturation value of w/r_c is indicative of mound formation while a small or decreasing value of w/r_c is indicative of a self-affine surface. As can be seen, for the case of deposition *without* attraction to ascending steps the value of w/r_c is very small indicating layer-by-layer growth. In contrast, inclusion of only a moderate amount of step-adatom attraction leads to a much larger mound angle ratio which

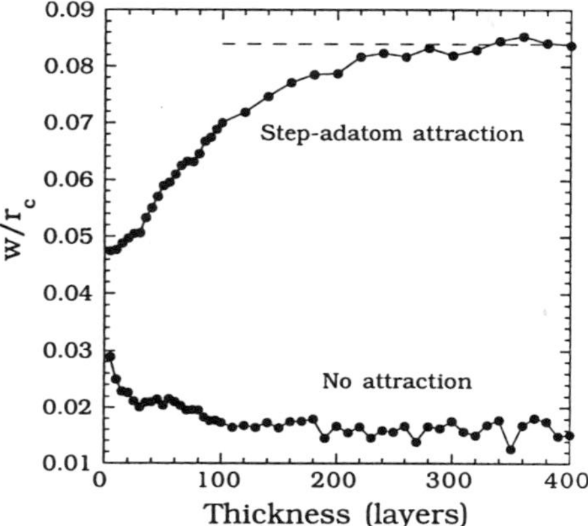

Fig. 24. Mound angle ratio w/r_c as a function of film thickness for the case of *fast* deposition with step-adatom attraction ($R' = 10$) and without step-adatom attraction ($R' = 1$).

increases with film thickness and appears to saturate at large thickness.[55] The corresponding value of the mound coarsening exponent is $n \simeq 0.19$, which is consistent with that obtained in a variety of experiments and in previous simulations with a moderate step-barrier.[55]

4. Conclusions

It is clear that over the past few years great progress has been made in our understanding of epitaxial growth using simple realistic models as well as scaling ideas and rate-equation analyses. This has led to a good quantitative explanation of recent experiments in both submonolayer and multilayer growth. Future studies are expected to focus on a more detailed understanding of the effects of microscopic processes and interactions in both submonolayer and multilayer growth. These efforts should lead to the development of more detailed models which would allow a more quantitative comparison between specific experiments and simulations and theory.

Acknowledgments

I would like to thank the organizers of the Zanjan Summer School on *Scaling and Disorderd Systems*, Professors M. R. H. Khajehpour (IASBS), M. R. Kolahchi (IASBS) and M. Sahimi (USC) for inviting me to give these lectures and for their kind hospitality in Zanjan. It was a truly enjoyable experience for me to lecture to a group of outstanding students and to visit the Institute for Advanced Studies in Basic Sciences in Zanjan for the first time. I would also like to thank Dr. Jacques G. Amar for his collaborations on many of the topics presented in these lectures. Finally, I would like to give a special thanks to Professor Dietrich Stauffer without whose constant reminders and friendly persuasions this review would not have been prepared and Dr. Mihail Popescu for a careful reading of the manuscript. The research presented in these lectures were supported in part by NSF and ONR.

References

1. J. Y. Tsao, *Materials Fundamentals of Molecular Beam Epitaxy* (World Scientific, Singapore, 1993).
2. J. W. Matthews, *Epitaxial Growth* (Academic, New York, 1975).
3. B. Lewis and J. C. Anderson, *Nucleation and Growth of Thin Films* (Academic, New York, 1978).
4. T. Vicsek and F. Family, *Phys. Rev. Lett.* **52**, 1669 (1984).
5. F. Family and P. Meakin, *Phys. Rev. Lett.* **61**, 428 (1988).
6. F. Family and P. Meakin, *Phys. Rev.* **A40**, 1998 (1989) 3836.
7. Y. W. Mo, J. Kleiner, M. B. Webb, and M. G. Lagally, *Phys. Rev. Lett.* **66** (1991).
8. H. J. Ernst, F. Fabre, and J. Lapujoulade, *Phys. Rev.* **B46**, 1929 (1992).
9. R. Q. Hwang, J. Schroder, C. Gunther, and R. J. Behm, *Phys. Rev. Lett.* **67**, 3279 (1991); R. Q. Hwang and R. J. Behm, *J. Vac. Sci. Technol.* **B10**, 256 (1992).
10. W. Li, G. Vidali, and O. Biham, *Phys. Rev.* **B48**, 8336 (1993).
11. E. Kopatzki, S. Gunther, W. Nichtl-Pecher, and R. J. Behm, *Surf. Sci.* **284**, 154 (1993).
12. G. Rosenfeld, R. Servaty, C. Teichert, B. Poelsema, and G. Comsa, *Phys. Rev. Lett.* **71**, 895 (1993).
13. J. A. Stroscio, D. T. Pierce, and R. A. Dragoset, *Phys. Rev. Lett.* **70**, 3615 (1993).
14. J. A. Stroscio and D. T. Pierce, *Phys. Rev.* **B49**, 8522 (1994).
15. J.-K. Zuo and J. F. Wendelken, *Phys. Rev. Lett.* **66**, 2227 (1991); J.-K. Zuo, J. F. Wendelken, H. Durr, and C.-L. Liu, *Phys. Rev. Lett.* **72**, 3064 (1994).
16. D. D. Chambliss and R. J. Wilson, *J. Vac. Sci. Technol.* **B9**, 928 (1991); D. D. Chambliss and K. E. Johnson, *Phys. Rev.* **B50**, 5012 (1994).
17. K. Bromann, H. Brune, H. Roder, and K. Kern, *Phys. Rev. Lett.* **75**, 677 (1995).
18. Q. Jiang and G. C. Wang, *Surf. Sci.* **324**, 357 (1995).
19. F. Tsui, J. Wellman, C. Uher, and R. Clarke, *Phys. Rev. Lett.* **76**, 3164 (1996).
20. S. Stoyanov and D. Kashchiev, in *Current Topics in Materials Science*, Vol. 7, eds. E. Kaldis (North-Holland, Amsterdam, 1981), pp. 69–141.
21. J. A. Venables, G. D. Spiller, and M. Hanbucken, *Rep. Prog. Phys.* **47**, 399 (1984).
22. G. Ehrlich and F. Hudda, *J. Chem. Phys.* **44**, 1039 (1966); R. L. Schwoebel, *J. Appl. Phys.* **40**, 614 (1969).

23. J. Tersoff, A. W. Denier van der Gon, and R. M. Tromp, *Phys. Rev. Lett.* **72**, 266 (1994).

24. R. Kunkel, B. Poelsema, L. K. Verheij, and G. Comsa, *Phys. Rev. Lett.* **65**, 733 (1990).

25. P. Smilauer and D. D. Vvedensky, *Phys. Rev.* **B48**, 17603 (1993).

26. M. von Smoluchowski, *Z. Phys. Chem.* **17**, 557 (1916); M. von Smoluchowski, *Z. Phys. Chem.* **92**, 129 (1917).

27. M. C. Bartelt and J. W. Evans, *Phys. Rev.* **B46**, 12675 (1992); *Surf. Sci.* **298**, 421 (1993).

28. L.-H. Tang, *J. de Physique* **I3**, 935 (1993).

29. J. A. Blackman and A. Wilding, *Europh. Lett.* **16**, 115 (1991).

30. D. Walton, *J. Chem. Phys.* **37**, 2182 (1962); D. Walton, T. N. Rhodin and R. W. Rollins, *J. Chem. Phys.* **38**, 2698 (1963).

31. J. G. Amar and F. Family, *Phys. Rev. Lett.* **74**, 2066 (1995).

32. J. G. Amar, F. Family, and P.-M. Lam, *Phys. Rev.* **B50**, 8781 (1994), and in *Mechanisms of Thin Film Evolution*, MRS Symposia Proceedings No. 317 (Materials Research Society, Pittsburgh, 1994), p. 167.

33. D. Stauffer, *Introduction to Percolation Theory* (Taylor and Francis, London, 1985).

34. G. S. Bales and D. C. Chrzan, *Phys. Rev.* **B50**, 6057 (1994).

35. J. W. Evans and M. C. Bartelt, *J. Vac. Sci. Technol.* **A12**, 1800 (1994).

36. C. Ratsch, A. Zangwill, P. Smilauer, and D. D. Vvedensky, *Phys. Rev. Lett.* **72**, 3194 (1994); C. Ratsch, P. Smilauer, A. Zangwill, and D. D. Vvedensky, *Surf. Sci.* **328**, L599 (1995).

37. M. Schroeder and D. E. Wolf, *Phys. Rev. Lett.* **74**, 2062 (1995).

38. F. Family and J. G. Amar, *Mat. Sci. and Eng. B* (Solid State Materials), **30**, 149 (1995).

39. J. G. Amar and F. Family, *Thin Solid Films* **272**, 208 (1996).

40. J. G. Amar and F. Family, *Surf. Sci.* **382**, 170 (1997).

41. J. Villain, *J. Phys. I* (France) **1**, 19 (1991).

42. M. D. Johnson, C. Orme, A. W. Hunt, D. Graff, J. Sudijono, L. M. Sander, and B. G. Orr, *Phys. Rev. Lett.* **72** 116 (1994).

43. J. A. Stroscio, D. T. Pierce, M. Stiles, A. Zangwill, and L. M. Sander, *Phys. Rev. Lett.* **75**, 4246 (1995).

44. K. Thürmer, R. Koch, M. Weber, and K. H. Rieder, *Phys. Rev. Lett.* **75**, 1767 (1995).

45. H.-J. Ernst, F. Fabre, R. Folkerts, and J. Lapujoulade, *Phys. Rev. Lett.* **72**, 112 (1994).

46. G. W. Smith, A. J. Pidduck, C. R. Whitehouse, J. L. Glasper, and J. Spowart, *J. Cryst. Growth* **127**, 966 (1993).

47. J. E. Van Nostrand, S. Jay Chey, M.-A. Hasan, D. G. Cahill, and J. E. Greene, *Phys. Rev. Lett.* **74**, 1127 (1995).

48. W. C. Elliott, P. F. Miceli, T. Tse, and P. W. Stephens, *Phys. Rev.* **B54**, 17938 (1996).

49. J. Krug, M. Plischke, and M. Siegert, *Phys. Rev. Lett.* **70**, 3271 (1993).

50. M. Siegert and M. Plischke, *Phys. Rev. Lett.* **73**, 1517 (1994).

51. M. C. Bartelt and J. W. Evans, *Phys. Rev. Lett.* **75**, 4250 (1995).

52. F. Family and J. G. Amar, in *Evolution of Epitaxial Structure and Morphology*, eds. R. Clarke *et al.*, MRS Proceedings Vol. 399 (Boston, 1996), p. 67; J. G. Amar and F. Family, *ibid*, p. 95.

53. M. C. Bartelt and J. W. Evans, in *Evolution of Epitaxial Structure and Morphology*, eds. R. Clarke *et al.*, MRS Proceedings Vol. 399 (Boston, 1996).

54. J. G. Amar and F. Family, *Surf. Sci.* **365**, 177 (1996).

55. J. G. Amar and F. Family, *Phys. Rev. Lett.* **77**, 4584 (1996).

56. J. W. Evans, D. E. Sanders, P. A. Thiel, and A. E. DePristo, *Phys. Rev.* **B41**, 5410 (1990); H. C. Kang and J. W. Evans, *Surf. Sci.* **271**, 321 (1992).

57. J. G. Amar and F. Family, *Phys. Rev.* **B54**, 14071 (1996).

58. J. G. Amar and F. Family, *Phys. Rev.* **B54**, 14742 (1996)

59. S. C. Wang and G. Ehrlich, *Phys. Rev. Lett.* **70**, 41 (1993).

60. Y. L. He, H. N. Yang, T. M. Lu, and G. C. Wang, *Phys. Rev. Lett.* **69**, 3770 (1992).

61. F. Family and T. Vicsek, *Dynamics of Fractal Surfaces* (World Scientific, Singapore, 1992).

62. J. A. Stroscio (private communication).

63. S. F. Edwards and D. R. Wilkinson, *Proc. R. Soc. Lond.* **A381**, 17 (1982).

64. P. Smilauer and D. D. Vvedensky, *Phys. Rev.* **B52**, 14263 (1995).

65. M. Siegert and M. Plischke, *Phys. Rev.* **E53**, 307 (1996).

66. L. Golubovic, *Phys. Rev. Lett.* **78**, 90 (1987).

Annual Reviews of Computational Physics VIII (pp. 83–111)
Edited by Dietrich Stauffer

WAVELET TRANSFORMATIONS AND DATA PROCESSING: APPLICATION TO CHARACTERIZATION AND SIMULATION OF LARGE-SCALE POROUS MEDIA

MUHAMMAD SAHIMI

*Department of Chemical Engineering, University of Southern California,
Los Angeles, CA 90089-1211, USA*

We describe wavelet transformations (WTs) and their basic properties, and discuss their application to data processing. These transformations provide a powerful method for data and image processing by (1) providing a flexible spatial-scale (and time-scale) window which narrows when focusing on small-scale features and widens on large-scale features, and (2) providing the capability for analyzing special characteristics of the data or image around specific points. We then discuss application of WTs to developing a unified and highly efficient method for characterization and simulation of porous media problems, from pore to field scale. The method uses WTs for data mining and characterization, scale-up of the fine-scale model of porous media, and interpretation of their transient flow properties. The data that can be treated by WTs include the direct data, such as various well logs and the permeability distributions, and the indirect data, such as seismic signals, and other characteristics of porous media. Wavelet transformations denoise the data, uncover their special features, and discover the structure of their distribution. They also provide an efficient method for processing seismic data, which are typically in huge amounts, and help interpret them in a meaningful manner. From these analyses one determines the spatial distribution of the fractures and construct a fine-grid model for the porous media. Wavelet transformations then coarsen the fine-grid description of the porous media, such that finer resolution is maintained in regions of high flow rates, whereas coarser property description is applied to the rest of the media. Wavelet transformations can also analyze time-dependent flow data for the porous media, such as long-term production data for oil reservoirs, as well as time-dependent seismic data (i.e. repeated measurements over regular time intervals). Therefore, one has, for the first time, a unified approach to characterization, modeling, and scale-up of models of porous media and extraction of information from their flow properties, from pore to field scale. The computational cost of the method can be *orders of magnitude* less than those of the most efficient methods currently available.

1. Introduction

One of the most important aspects of investigating and understanding a natural or man-made phenomenon is to make rational and accurate analysis of the data that are already available for the phenomenon. The available data can be *numerical*, giving quantitative information on how one or several variables of a phenomenon depend on and vary with the independent parameters of the system. Alternatively, the data may provide only *qualitative* information on the properties of the system. The third type of data may involve *images* of the system at different time and/or length scales, hence providing some crucial clues about what the model should predict about the state of the system at those time and/or length scales.

However, the data to be analyzed are often incomplete and, more importantly, *noisy*. For example, in analyzing a complex image of a system — the problem which is popularly known as *image processing* — one must extract certain types of basic patterns from the image which often contains noise. The origin of the noise may be the uncertainties in the measurements and/or the resolution of the instruments used for obtaining the data and constructing the image. If the noise conditions are favorable, that is, if the data and/or image involve high data-to-noise ratios (DNR), and if the noise is a Gaussian variable with independent and identically distributed samples (or pixels in an image), then the classical techniques for processing and analyzing the data and/or the image are applicable, and the problem is well-understood; see, for example, Refs. 1–3. On the other hand, if there is even a small deviation from a Gaussian distribution, then processing of the data and/or the image will severely deteriorate. In fact, the distribution of the image pixels that are contaminated by noise is seldom known in image processing. A more severe problem arises when the variability of DNR in various parts of the system is large. In this case, simple processing of the data (image) and removal of the noise will not be effective. In addition, in analyzing the image of a system one may encounter unwanted background patterns, which are in fact structured noise that will further degrade the performance of the image processing. In all cases, if the structure of the noise distribution cannot be fully understood, and if it cannot be separated from the actual data, any conclusion based on the analysis of the image would be subject to large uncertainties. This problem becomes particularly acute if one is to construct a three-dimensional (3D) model of the system from lower-dimensional (1D and/or 2D) data, because then one must also deal with the *uniqueness problem*: Is the 3D model, constructed based on the 1D and/or 2D data, unique, or are there *several* different models that have the *same* 1D and 2D properties as the given data?

Practical examples involving image processing are abundant and include analysis of pictures, interpretation of medical images, underwater and earth sounding, trajectory detection, and many more. For example, a typical problem in astronomy is the extraction of streaks corresponding to the trajectories of meteorites. A major concern in remote sensing is to decipher from satellite images the network of roads and the separation among fields in agriculture. Practical examples involving quantitative (numerical) data include processing of, e.g., radar signals, analyzing seismic data for natural rock arising from tectonic motions which are crucial to understanding earthquakes, and investigating various geological data for field-scale porous media, such as oil, gas and geothermal reservoirs and groundwater aquifers, and many more. The last problem is the focus of this paper, although the techniques that we discuss here are equally applicable to all the problems that have been mentioned so far.

2. Characterization of Field-Scale Porous Media

Field-scale porous media (FSPM), such as oil, gas and geothermal reservoirs, groundwater aquifers, and landfills, are highly heterogeneous at many length scales,[4,5] with their heterogeneities manifesting themselves at three different scales: (1) *microscopic*, which is at the level of pores and grains; (2) *macroscopic* — at the level of core plugs, and (3) *megascopic*, which includes the entire reservoir or aquifer. Modeling flow and transport (of oil, gas, vapor, groundwater, or contaminants) in such porous media depends critically on understanding and characterizing their structure and, in particular, the distributions of their heterogeneities. To this end, characterization of laboratory-scale (macroscopic) porous media has been done in considerable detail and reasonable understanding of them has been obtained (although, even at this scale, if a sample of a porous medium contains fractures, it would still be a formidable task to completely characterize the morphology of its pore space). However, the same is not true about FSPM whose characterization is plagued by noisy field data and wide variations in them. In addition, as discussed above, proper interpretation of the data and separation of the noise from the true property values are still very challenging problems. Even if such problems are all solved, one must tackle the fundamental problem of scale-up, namely, development of a proper model of FSPM in which all the important information, from microscopic to megascopic scales, are preserved and at the same time the model is *computationally tractable*, a difficult task given that the disparity between the relevant length scales in FSPM is enormous (about *ten* orders of magnitude), and that the time scales used in the simulation of flow and transport in FSPM are of the order of *years.*

In the early 1980s it was realized that deficiencies in predicting the performance of enhanced oil recovery processes frequently have their root in the inadequate description of the oil reservoirs. The list of the problems encountered was too long: oil-displacing agents, such as water, CO_2, natural gas, or a surfactant, reached the producing wells too early, and the oil recovery factor was disappointing and sometimes even uneconomical; new wells drilled between the existing wells did not have the expected characteristics obtained from deterministic interpolation of the existing wells' properties, and many other problems. These difficulties motivated the development of modern methods for reservoir characterization, one main focus of this article.

The same type of difficulties have been encountered in a different but related problem, namely, the spread of contaminants in groundwater aquifers. A prime example is provided by the recent discovery of underground Pu-containing plume at a significant concentration at a distance of about 1.3 km from the source at the Nevada Test Site[6] after only about 30 years since the initiating event. The Nevada Test Site is located in Yucca Mountain, which is being evaluated by the United States Department of Energy as a potential underground repository for high-level radioactive wastes. It is a highly fractured and heterogeneous rock. It had been predicted, based on conventional characterization, modeling, and simulation of flow and transport in the rock, that it would take *tens of thousands of years* for significant concentrations of Pu to spread over distances of order of one kilometer or more from the test site! This discovery has prompted serious questions on the possible causes of this anomalous contaminant transport behavior. These difficulties, together with other field observations of surprisingly fast contaminant transport, suggest that the complex structure of the underground media is the main culprit which, if not properly accounted for, will lead to unpleasant surprises in trying to predict subsurface transport of contaminants. Thus, it is important to examine the statistical characteristics of the heterogeneous structure of such FSPM and their potential impact on long-range contaminant transport.

In order to characterize FSPM, we divide their important characteristics into two groups. In one group we have what we call the *direct* data, such as those that are collected along the wells in an oil reservoir. An important such characteristic is the various logs, such as the porosity logs, that are routinely measured during well drilling. With the advent of sophisticated instrumentation, it is becoming possible to even measure the permeability *in situ* using, e.g., nuclear magnetic resonance. In the second group we have what we call the *indirect* data, the most important of which are 3D seismic recordings (and, more recently, 4D seismic data, i.e. repeated measurements of the 3D data over a period of time[7,8]), which have been increasingly used for characterization of reservoirs. Such indirect data do not provide any information on the permeability or porosity distribution of the reservoir. Nevertheless, given the extensive coverage of the 3D data, with proper calibration at the well locations (see below), they provide us with the opportunity with a more accurate characterization. For example, they provide valuable information about the spatial distribution of the fractures and faults. Fractures are of course critical to flow and transport in FSPM, as they provide high permeability paths for fluid flow in the porous media. Furthermore, using "time-lapse" 3D seismic, in conjunction with the production data, we may be able to update the reservoir's model for an effective *dynamic characterization.*[9]

A critical problem is the proper treatment of such characteristic data, ranging from denoising them, uncovering their special features and discovering their distribution functions, to interpreting them. If this is done properly, then the data and their characteristics can be incorporated in a realistic simulator. However, this problem is fraught with complexities, and although it has been studied for a long time, a definitive solution has not emerged yet. One important obstacle to finding the solution is perhaps the fact that different types of data have been analyzed with different methods, often yielding conflicting results.

Compounding these difficulties are two important results that have emerged over the past decade. The first is that it has been shown that the porosity logs and permeability distribution of many FSPM follow fractal stochastic processes. In particular, ample data[4,5,10] suggest that for many FSPM the porosity logs in the direction perpendicular to the bedding may obey the statistics of fractional Gaussian noise (fGn), while those parallel to the bedding may follow a fractional Brownian motion (fBm) which is a stochastic process $B_H(\mathbf{r})$, the power spectrum $S(\boldsymbol{\omega})$ of which (i.e. the Fourier transform of its variance) is given by

$$S(\boldsymbol{\omega}) = \frac{a}{\left(\sum_{i=1}^{d} \omega_i^2\right)^{H+d/2}}, \tag{1}$$

where H is called the Hurst exponent, d is the dimensionality of the system (or of the data set), a is a constant, and $\boldsymbol{\omega} = (\omega_1, \ldots, \omega_d)$. Fractional Brownian motion is not differentiable, but by smoothing it over an interval one can obtain its numerical "derivative," which is the fGn. For example, since the well logs are usually 1D data, then the power spectrum of fGn in 1D is given by

$$S(\omega) = \frac{b}{\omega^{2H-1}}, \tag{2}$$

where b is a constant [the corresponding 1D power spectrum for a fBm is given by, $S(\omega) = a/\omega^{2H+1}$]. The most remarkable property of the fBm and fGn is that they generate correlations whose extent is as large as the linear size of the system. Moreover, the type of the correlations depends on H. If $H > 1/2$, then the data display *persistence*, that is, a trend (e.g., a high or low value) at x is likely to be followed by a similar trend at $x + \Delta x$. If $H < 1/2$, then the data contain *antipersistence*, that is, a trend at x is not likely to be followed by a similar trend at $x + \Delta x$. For $H = 1/2$ the increments in the fBm-type data are uncorrelated.

The second important emerging result is about the structure of the fracture network of FSPM. Many FSPM contain fractures on a large number of different length scales which are a consequence of the fracturing process and the interaction of the stress field in rock at the time of fracturing with its fluid content. There is increasing evidence, ranging from Monte Carlo simulations of fracture mechanics of rock[11] to field observations[12,13] that, (1) the fracture network is very irregular at *all* length scales, with various properties of the fractures, such as their orientations, lengths, and apertures following certain statistical distributions, and (2) in many FSPM the fracture network is a fractal object, that is, the number of fractures N_ℓ of length ℓ is given by

$$N_\ell \sim \ell^{-D_f}, \tag{3}$$

where D_f is the fractal dimension of the network. Typical values of D_f are between 2.2 and 2.7. The most important implication of this result is that, the connectivity of the fracture network is incomplete and bears no resemblance to the perfectly ordered network of fractures that are used in the traditional simulations of FSPM. In addition, due to its fractal nature, the spatial distribution of the fractures is highly correlated. Flow and transport in such a network is completely different from those in a simple-cubic network of fractures which is used in the traditional models. Thus, if in addition to being noisy, the data also exhibit fractal characteristics (i.e. long-range correlations), their proper treatment and incorporation into a realistic simulator is a highly difficult problem, since the long-range correlations must be preserved.

The above difficulties are not restricted to the fractal distribution of the type discussed above. In some FSPM, the distributions of the heterogeneities contain long tails. For this type of FSPM low probability events (such as having regions of very high porosity or permeability), though unlikely, cannot be neglected since they have a disproportionate effect on the entire system at large length scales and times. This is in contrast to Gaussian characteristics in which low probability events can be discarded because they do not impact on the overall or long-term behavior of the entire system. There have been extensive field observations that the distributions of the field porosity, permeability, and fractures can exhibit these types of structural characteristics. These regions encompass FSPM at various depths, including the near-surface regions, such as the vadose zone, which extends from the soil surface down to the top of the water table at a typical distance of about a hundred meters. The lateral distance of interest may extend to a kilometer or more from the likely contaminant source. There are a variety of lithologic units within such regions, and there is also a high degree of heterogeneity within this lithology, such as sills, clastic dykes, fractures and faults, and porosity at various scales. Such extreme types of heterogeneities cannot be characterized by conventional methods of data analysis, and one must develop new techniques for this task.

Assuming that the data are denoised, treated and interpreted properly, one generates, through the use of modern geostatistical methods (see below), a highly detailed 3D representation of FSPM, which we refer to it as the *geological model*. However, the geological model typically contains several million grid blocks or cells (or nodes). Although such detailed models are important to management of FSPM, as fine-scale details in the formation properties can dominate the behavior of FSPM, their use in the simulation of flow and transport in FSPM implies solving several million discretized equations thousands of times, as one solves a time-dependent problem, which is currently not feasible.

To tackle this problem, methods have been developed for scale up of heterogeneous FSPM. The main goal is to coarsen the highly detailed geological model to levels of detail suitable for simulation, while maintaining the integrity of the model in terms of avoiding the loss of important information, so that the simulation results with the coarsened model are still accurate and representative of the true behavior of FSPM. Ideally, the scale-up method must have three attributes: (1) since during the scale-up process some information about the porous media's permeability distribution is lost, one has to somehow compensate for this loss, so that the coarsened model is still a realistic representation of the porous media, (2) the method must be general enough to be applicable to fractured FSPM, and (3) the scale-up method must be computationally efficient, in the sense that the total computational cost (developing the coarsened model *plus* the cost of transport and flow simulations) must be much less than the computational cost of the same simulations but with the fine-grid geological model. Most of the scale-up methods that have been proposed so far[14-28] are either too simplistic and unrealistic, or not computationally efficient, or not applicable to fractured FSPM. They range from simple averages of the permeability distribution to detailed fine-grid numerical flow simulation. These methods average out the effects of extreme values, such as those associated with thin communicating layers, large flow barriers, and partially communicating faults. Most importantly, practically all the current scale-up methods are not applicable to fractured FSPM since, because of the large differences between the fracture and matrix permeabilities, the interface between the two regions typically gives rise to singularities in the up-scaling scheme that cannot be resolved or removed. Hence, despite significant progress, there still remain several important unsolved problems, even with the most efficient scale-up methods currently available.

Even if the geological model of FSPM is scaled up properly, we still have to deal with another important problem which arises when one attempts to interpret transient flow data, such as long-term pressure transients and production data of, e.g., gas or oil reservoirs that provide important clues about their structure as they produce, and also to update their model. This is a complex problem, and although many methods have been developed, none is completely satisfactory because they mostly involve in one way or another some sort of empiricism or ad hoc assumption.

In this paper we discuss a unified approach to characterization of FSPM, construction and up-scaling of a realistic model for them, time-updating of the model, and proper interpretation of the results of simulation of flow and transport in FSPM. The approach that we discuss is based on the use of wavelet transformations at each of these steps. In what follows we first introduce wavelet transformations and discuss their important properties that are relevant to our discussion. We then discuss how the direct (well) data can be denoised and analyzed by the wavelets, after which we outline the treatment of the seismic data, the interpretation of transient flow data, and time-updating of the model of FSPM. As already mentioned, although we focus on modeling of FSPM, the basic properties of the wavelet transformations and most of the data treatment methods that we discuss here are quite general and applicable to many other problems involving data processing.

3. Continuous and Discrete Wavelet Transformations

In its primitive form, a wavelet transformation (WT) was developed in the early 1980s for analyzing seismic data.[29–31] However, significant theoretical work took place in the late 1980s and early 1990s,[32–40] providing rigorous foundations for the many applications that WTs have found. Wavelet transformations can be defined for functions that depend solely on a time variable t, as well as for those that depend on a spatial variable $\mathbf{x}$. The main focus of this paper is on the latter type of functions, but our discussion is equally applicable to the former ones when we discuss 1D WTs. The *continuous* WT of a function $f(\mathbf{x})$, denoted by $\hat{f}(a, \mathbf{b})$, is defined by

$$\hat{f}(a, \mathbf{b}) = \int_{-\infty}^{\infty} f(\mathbf{x})\psi_{a,b}(\mathbf{x})d\mathbf{x}, \tag{4}$$

where

$$\psi_{a,b}(\mathbf{x}) = \frac{1}{\sqrt{a}}\psi\left(\frac{\mathbf{x} - \mathbf{b}}{a}\right). \tag{5}$$

Here $a > 0$ is a dilation or rescaling parameter, $\mathbf{b}$ represents the translation of the wavelet, and $\psi(\mathbf{x})$ is called the *mother wavelet*. We should keep in mind that $\psi_{a,b}(\mathbf{x})$ has the same shape for all values of a. It is important to note that the wavelet function $\psi(\mathbf{x})$ is *not* unique: Depending on the intended applications, one may use a variety of wavelets. However, its choice is also not arbitrary. For example, in 1D $\psi(x)$ is a function with the property that

$$\int |\psi(x)|^2 dx = 1, \tag{6}$$

so that, (1) it has a compact support (or sufficiently fast decay) to achieve localization in space, and (2) it has zero mean, i.e.

$$\int_{-\infty}^{\infty} \psi(x)dx = 0, \tag{7}$$

although higher moments of $\psi(x)$ can also be zero. Equation (7) implies that $\psi(x)$ is wave-like, while Eq. (6) ensures that $\psi(x)$ is not a sustaining wave, hence the name wavelets (i.e. small waves). The compactness of the support of $\psi(x)$ (which is also a requirement for 2D and 3D wavelets) means that computations with WTs can be parallelized, since the wavelets are nonzero only over finite intervals. The inverse WT is given by

$$f(\mathbf{x}) = \frac{1}{C_\psi} \int_{-\infty}^{\infty} \int_{0}^{\infty} \frac{1}{a^{d+1}} \hat{f}(a, \mathbf{b}) \psi_{a,b}(\mathbf{x}) \, da \, d\mathbf{b} \,, \tag{8}$$

where C_ψ is a constant that depends on the wavelet used in the WT, and d is the dimensionality of the system.

Varying a has the effect of dilating ($a > 1$) or contracting ($a < 1$) $f(\mathbf{x})$. Hence, as a increases, the wavelet becomes more spread out and takes into account only the large-scale behavior of $f(\mathbf{x})$ [or the long-time behavior, if a function $f(t)$ is considered], and vice versa. On the other hand, varying $\mathbf{b}$ means that one analyzes $f(\mathbf{x})$ around different points $\mathbf{b}$. These two fundamental properties of WTs, namely, (1) their ability for providing a flexible spatial-scale [or time-scale in case of $f(t)$] window that narrows when focusing on small-scale features and widens on large-scale features, and (2) their capability for analyzing special characteristics of a function around various points, make them an ideal tool for data (and also image) processing. One can interpret the WT as a microscope where the magnification is given by $1/a$ and the optics by the choice of the wavelet $\psi(\mathbf{x})$. A typical example of a 2D wavelet is given by

$$\psi(x, y, \sigma) = [2 - (x^2/\sigma^2 + y^2)] \exp[-(x^2/\sigma^2 + y^2)/2] \,, \tag{9}$$

which, in the limit $\sigma = 1$, is called the *Mexican Hat* and is shown in Fig. 1.

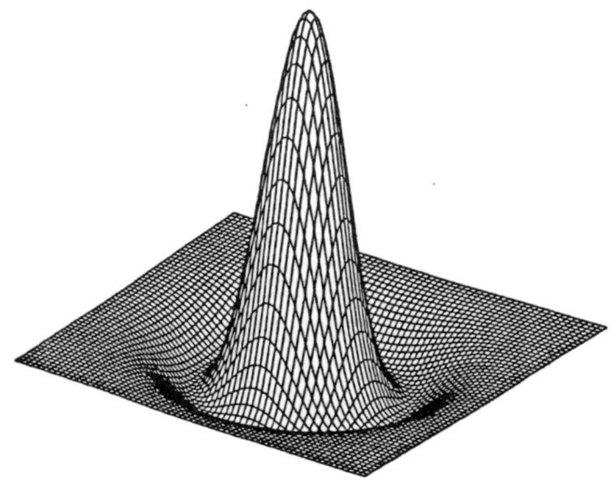

Fig. 1. The 2D Mexican Hat wavelet.

So far we have discussed continuous WTs. However, in order to use a WT for a discrete set of data, one must discretize WTs. To see how a discrete WT is developed, consider as an example, the 1D case. We may choose $a = a_0^j$, where j is an integer and $a_0 > 1$ is a fixed dilation. Moreover, we can also take b to be $b = kb_0a_0^j$, where $b_0 > 0$ depends on the choice of $\psi(x)$ and k is an integer. Therefore, a_0^j plays the role of a magnifier, and thus the WT studies the system at a particular location with the given magnification and then moves on to another location. Hence, if we define

$$\psi_{j,k}(x) = \frac{1}{\sqrt{a_0^j}}\psi\left(\frac{x - kb_0a_0^j}{a_0^j}\right) = a_0^{-j/2}\psi(a_0^{-j}x - kb_0)\,, \tag{10}$$

then the resulting WT is given by

$$\hat{f}(j,k) = a_0^{-j/2}\int_{-\infty}^{\infty} f(x)\psi(a_0^{-j}x - kb_0)dx\,. \tag{11}$$

The conditions for choosing a_0 and b_0 are discussed by Daubechies.[36] They are fairly broad and thus a_0 and b_0 admit very flexible ranges. A popular choice is $a_0 = 2$ and $b_0 = 1$ in which case the resulting 1D WT of a discrete data array $f(x)$,

$$D_j(k) = 2^{-j/2}\int_{-\infty}^{\infty} f(x)\psi(2^{-j}x - k)dx\,, \tag{12}$$

is usually called the *wavelet-detail coefficient* of the data, where $k = 1, 2, \ldots, n$ with n being the size of the data array, and the js being integer. The resulting set of wavelets $\{\psi_{j,k}\}$ for all j and k form an orthonormal basis. A remarkable property of these functions is that they are also orthonormal to their translates and dilates. Any square integrable function $f(x)$ can be approximated, up to an arbitrary high degree of precision, by

$$f(x) = \sum_{j=-\infty}^{\infty}\sum_{k=-\infty}^{\infty} D_j(k)\psi_{j,k}(x)\,. \tag{13}$$

This representation of $f(x)$ also implies that $D_j(k)$ measures the contribution to $f(x)$ of scale 2^j at location 2^jk. Thus, such series representation of $f(x)$ is akin to a Fourier series, except that series (13) is doubled indexed (indicating scale and location), and that the basis functions have localization property.

The *Haar* wavelet is the simplest of all orthogonal wavelets and is given by

$$\psi(x) = \begin{cases} 1 & 0 \le x \le 1/2 \\ -1 & 1/2 \le x < 1 \\ 0 & \text{otherwise}. \end{cases} \tag{14}$$

Compactly supported orthonormal wavelets, that is, those that are nonzero over only small intervals of x, are formed by dilation and translation of $\psi(x)$ and another function $\phi(x)$, called the *scaling function*, through the following equations:

$$\psi_{j,k}(x) = 2^{-j/2}\psi(2^{-j}x - k), \tag{15}$$

and

$$\phi_{j,k}(x) = 2^{-j/2}\phi(2^{-j}x - k). \tag{16}$$

Daubechies orthonormal and compactly supported wavelets of order M[33,34,36] possess the property that their first M moments are zero, and their $\psi(x)$ and $\phi(x)$ are related to those at the finer length scales by

$$\phi(x) = \sqrt{2} \sum_{k=0}^{L-1} h_k \phi(2x - k), \tag{17}$$

$$\psi(x) = \sqrt{2} \sum_{k=0}^{L-1} g_k \phi(2x - k), \tag{18}$$

where $L = 2M$. h_k and g_k are called *filter coefficients* and are related by

$$g_k = (-1)^k h_{L-k-1}, \qquad k = 0, 1, \ldots, L - 1. \tag{19}$$

The simplest ($M = 1$) of the Daubechies basis is the Haar wavelet (see above) for which $\phi(x) = 1$ for $0 \le x < 1$, and $\phi(x) = 0$ otherwise (i.e. $h_0 = h_1 = g_0 = -g_1 = \sqrt{2}/2$). Another 1D example is shown in Fig. 2.

Two-dimensional wavelets can be constructed in separable form by tensor products of 1D wavelets, in which case there is one scaling function

$$\Phi_{j,k_1,k_2}(x, y) = \phi_{j,k_1}(x)\phi_{j,k_2}(y), \tag{20}$$

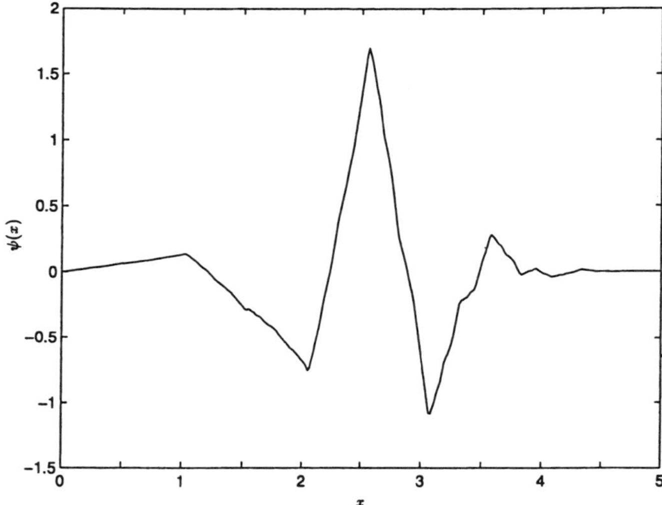

Fig. 2. The 1D Daubechies wavelet function $\psi(x)$.

and three wavelets,

$$\Psi^{(1)}_{j,k_1,k_2}(x,y) = \phi_{j,k_1}(x)\psi_{j,k_2}(y)\,, \tag{21}$$

$$\Psi^{(2)}_{j,k_1,k_2}(x,y) = \psi_{j,k_1}(x)\phi_{j,k_2}(y)\,, \tag{22}$$

$$\Psi^{(3)}_{j,k_1,k_2}(x,y) = \psi_{j,k_1}(x)\psi_{j,k_2}(y)\,. \tag{23}$$

The extension to 3D is straightforward.

4. Applications of Wavelet Transformations

Having introduced WTs and discussed their main properties, we can now describe their applications to data processing, and in particular characterization of FSPM.

4.1. *Data denoising and identification of scales of variations*

As discussed above, an important aspect of image and data processing is the ability to separate the noisy components from the actual part of a data set. For example, in characterization of FSPM one is interested to separate the noise in the seismic recordings from the significant part of the data, so that their interpretation can be done accurately. The same problem arises in the analysis of the well logs. In other contexts, one has the same problem in isolation of coherent structures of turbulent flows, and in isolation of noise from a true signal. There are several ways of separating noise from the actual data. For example, similar to the power spectrum (the Fourier transformation of the variance), one can define a *wavelet spectrum*[41–43]

$$E(a) = \int_{-\infty}^{\infty} \hat{f}(a, \mathbf{b}) d\mathbf{b} , \tag{24}$$

which for orthonormal wavelets is given by

$$E(j) = \sum_{k} |D_j(k)|^2 . \tag{25}$$

This quantity has been very successful in identifying the dominant scales or modes of variations in turbulent flows. What the wavelet spectrum does in effect is that, it condenses the phase plane (also called the scalogram) into a single function of scale. The reason that the wavelet spectrum provides a better measure of variance in the data than the power spectrum is that, in WT the coefficients are influenced by *local* events, whereas in the power spectrum the Fourier coefficients are affected by the data (or the function) *over their entire domain*. Local maxima in the wavelet spectrum provide clues about the scales at which important features provide significant contributions. The contributions can be made in one of two ways: they can be made by one feature with a large contribution, or they can be made by a series of smaller features.

Another and more general method of denoising data is by *threshold partitioning*. In this method, one calculates the wavelet-detail coefficients of the data set, sets a threshold ϵ (as a fraction of the maximum value of the coefficients) and eliminates (i.e. sets them to zero) all the coefficients that are less than ϵ. Then, the denoised data are reconstructed again with the zero and nonzero detail coefficients, that is, the inverse of the discrete WT of the denoised data is computed. If the appropriate choice of the threshold is made, then the method is very powerful; see Ref. 44 for rules regarding the choice of the threshold.

4.2. *Construction and analysis of fractal data*

Direct data for FSPM, such as the well logs (e.g., porosity, resistivity, and gamma ray logs), are typically very complex. Their degree of complexity is even

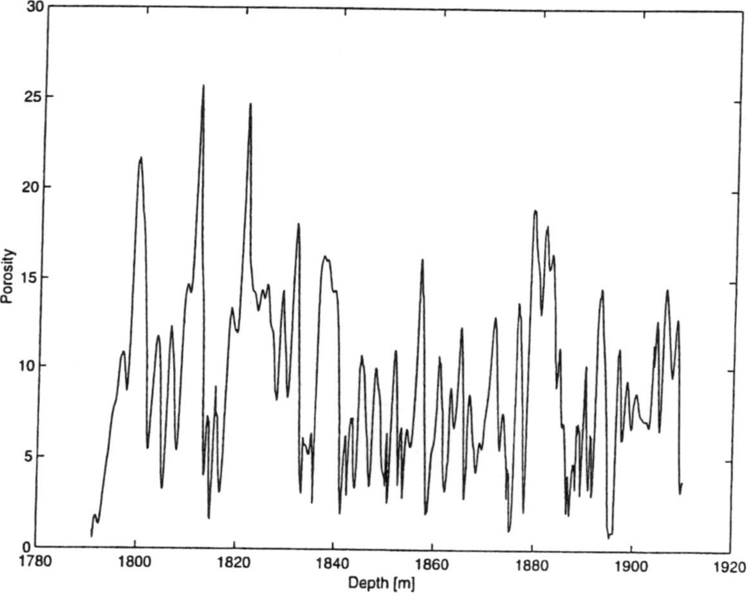

Fig. 3. A vertical porosity log along an oil well in an Iranian oil field. The depths are in meters, while the porosities are in percentages.

more if they follow a fractal stochastic process. For example, Fig. 3 presents a vertical porosity log that was collected along an oil well in Iran. The porosity was measured every 20 cm, and over 3000 data points were collected. Given such complex data, the question is: how can one accurately analyze them? In particular, how can one investigate whether such data follow fractal statistics? In the case of fBm- or fGn-type data, we have shown[45] that a highly efficient and accurate analysis can be carried out by using WTs. In this method one calculates the wavelet-detail coefficients of the data defined by Eq. (12) by fixing j, varying k, and computing $D_j(k)$. For each j one determines n such numbers and calculates their variance $\sigma^2(j)$. Then, it can be shown that, regardless of the type of the wavelet function ψ, one has

$$\log_2[\sigma^2(j)] = (2H + 1)j + \text{constant}. \tag{26}$$

Thus, plotting $\log_2[\sigma^2(j)]$ versus j yields H. Figure 4 presents such an analysis for a synthetic data generated by a fBm with $H = 0.2$. Figure 5 shows the wavelet analysis of the porosity log of Fig. 3.

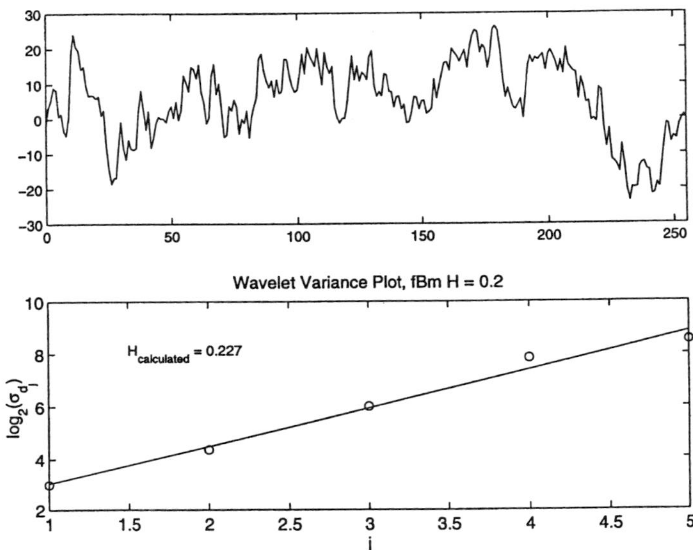

Fig. 4. A synthetic porosity log generated by a 1D fractional Brownian motion and its analysis by the wavelet method. The input Hurst exponent H was 0.2, while the wavelet method found it to be about 0.21.

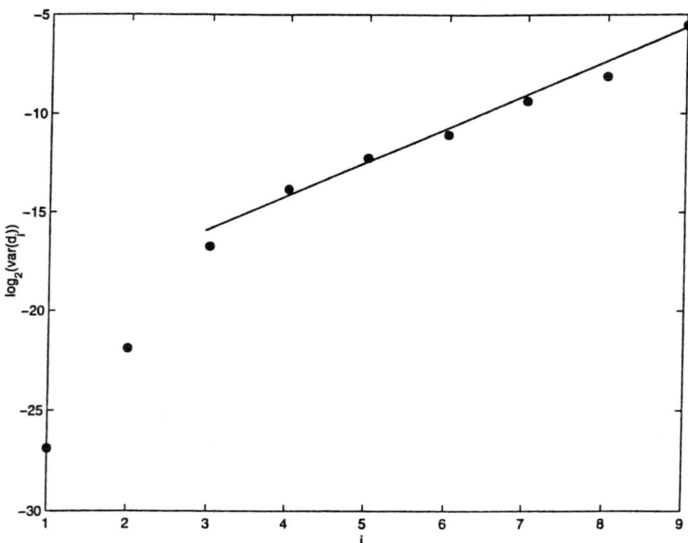

Fig. 5. Wavelet analysis of the porosity log of Fig. 3.

More generally, Mallat[34] has shown that for stochastic processes whose power spectrum is of power-law type (e.g., fBm and fGn), the wavelet spectrum $E(j)$ satisfies the following equation,

$$E(j) = 2^{2H} E(j+1) \,, \tag{27}$$

so that the wavelet spectra at different scales have a linear relationship on a logarithmic scale. Note that the spectrum $E(j)$ can be interpreted as a sort of the "energy" of the system, and therefore Mallat's relation has a particularly attractive physical interpretation for self-similar stochastic processes. Moreover, Flandrin[46] has shown that, while fBm itself is *not* a stationary process (note that the increments $B_H(\mathbf{x} + \Delta\mathbf{x}) - B_H(\mathbf{x})$ of a fBm *are* stationary), its WT is stationary. Finally, Wornell[47] has shown that one may use orthonormal wavelets to construct self-similar stochastic processes $X(x)$ by writing

$$X(x) = \sum_j \sum_k d_j(k)\psi_{j,k}(x) \,, \tag{28}$$

where the coefficients $d_j(k)$ are uncorrelated for any distinct pair j and k and have a power-law variance. Although certain conditions regarding the wavelets used in Eq. (28) must be met, the class of wavelets that do satisfy these conditions is fairly broad and include the Haar wavelets and the compactly supported wavelets constructed by Daubechies.[32,35,36]

Summarizing this section, not only can WTs analyze fractal data sets accurately and efficiently, they can also be used for constructing *synthetic* fractal data [Eq. (28)], which are of great use in construction of the geological model of FSPM, since if the heterogeneities of FSPM do follow fractal distributions, then one must use fractal interpolation between those regions of FSPM for which data are available in order to generate realistic estimates of the properties for those regions for which no data are available, e.g., the interwell zones in an oil or gas reservoir (see below).

4.3. *Identification of the spatial distribution of the fractures*

The next step in characterization of fractured FSPM is to identify the spatial distribution of the fractures. This is usually done through the interpretation of the seismic data which provide information about the distribution of the heterogeneities at the largest length scales, i.e. fractures and faults. However, processing and interpreting seismic data are subject to great uncertainties, even when WTs are used (see below), as one must use empirical and semiempirical correlations for example, for the wave speeds through FSPM, in order to obtain information about FSPM. Moreover, up until recently identification of the spatial distribution of the fractures and faults based on the porosity or resistivity log was thought to be not possible, as there was no way to interpret any particular patterns in the logs as indicating the intersections of the fractures and/or faults with the wells. However, we have recently developed[48] a method for precisely pinpointing the locations of the intersections of the fractures with the wells. This information can then be used to calibrate the seismic data in order to make their interpretation accurate and reliable. As such, the method is extremely useful for characterization of fractured FSPM.

Thus, in what follows we first describe briefly how WTs can be used to locate the intersections of the fractures with the wells, after which we discuss application of WTs for processing of seismic data.

4.3.1. *Wavelet identification of fractures from direct data*

Suppose that a well which is drilled in a porous medium does not pass through any fractured zone. Then, the wavelet-detail coefficients of the direct data that are collected along the well, e.g., a porosity or resistivity log, versus the depth will be a smooth plot, indicating no particular features. However, if the well does intersect the fractures, then the corresponding plot of the wavelet-detail coefficients of the well data versus the depth will again be smooth, *except where the fractures are located*, at which it exhibits local maxima. This is due to the fact that local maxima in the spectrum of the wavelet-detail coefficients provide information about the scales at which important features (e.g., fractures) or coherent events provide a significant contribution. Thus, using this fact one can identify the spatial distribution of the fractures around the wells. An example is shown in Fig. 6, where we show the plot of the wavelet-detail coefficients of the porosity log shown in Fig. 3. The local maxima indicate the presense of the fractures. Merging this information with seismic data in order to calibrate them provides a highly accurate map of the fracture network.[48]

4.3.2. *Wavelet treatment of seismic data*

Wavelets can also be used for denoising and interpreting the indirect information, and in particular seismic data. As mentioned above, development of WTs was actually motivated by its application to processing and interpreting seismic data. The analysis of such data and their proper interpretation have,

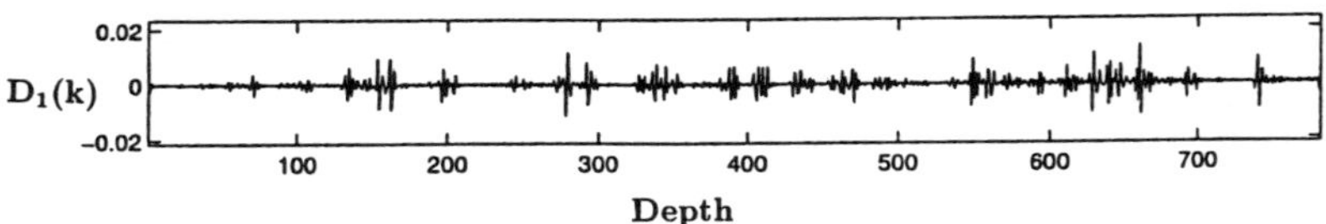

Fig. 6. Wavelet-detail coefficients of the porosity log of Fig. 3. Observe the smoothness of the curve over much of the depths' range, but also the occasional local maxima, indicating the presence of fractures at those depths.

of course, a long history. However, it became increasingly clear that the conventional methods for the analysis of such data cannot provide deep insight into their structure and what they imply for FSPM. Instead, the most significant advances in the processing and analysis of seismic data should be provided by what we call *unconventional* methods for describing wave propagation in heterogeneous FSPM, and in particular fractal distributions and wavelet transformations. The problem of processing and analyzing seismic data poses a significant computational challenge, because due to large volumes of the data, especially when repeated measurements are made (say, once every few months), an effective means to process and analyze the data is critical. Wavelet transformations have already been suggested for compression, transmission and decompression of seismic data.[49–52] These analyses attempt to deduce information about the structure of the fractures by carrying out a wavelet analysis of the seismic time series, similar to what we discussed above for the direct type of data. As discussed above, in order to make the analysis reliable, one can calibrate the information extracted from the seismic data with the spatial distribution of the fractures around the wells, obtained by the wavelet analysis of the direct data, that is to make sure that the seismic data also predict the same distribution around the wells. In this manner, one can make the information provided by the seismic data to be very reliable.

Wavelets have also been used for detecting the most important features of fracture pattern of a given sample of rock.[53–55] For example, Ouillon *et al.*[54] have developed a method that allows transformation to large-scale maps from smaller scales ones and quantification of the multiscale behavior of fracturing anisotropy. Their technique is based on finding at each point of the fracture map an optimum wavelet that can reveal the local structure, and hence helps one to detect the most important fractures in a complex network of fractures which is very useful for simulation of flow and transport in fractured FSPM. This research field is only in its initial steps, and we expect to see much progress over the next several years.

5. Wavelet Scale-up of the Geological Model of Field-Scale Porous Media

Once all the direct and indirect data have been analyzed, the fine-grid geological model of FSPM can be constructed. This is done through the use of modern geostatistical approaches. One preserves the original data in the geological model, and for those zones of FSPM for which no data are available geostatistical approaches provide estimates of the properties of interest. There are several methods for doing so, most of which have been discussed in detail by Jensen et $al.$[56] Typically, this is done by an accurate technique called the $random$ $residual$ $additions$ method, by which one adds stochastic uncertainties to the deterministic interpolated values of the properties of interest for the interwell zones. The statistical distributions of the uncertainties are obtained from the analysis of the direct data discussed above. If extensive data are available, one may also use simulated annealing method whose application to porous media problems is discussed in great details by Sahimi.[4,5] The fine-grid geological model of FSPM must then be scaled up, as it contains millions of grid points. We now describe a new scale up method based on the WTs.[57,58]

5.1. Wavelet scale-up of nonfractured porous media

Suppose that a porous medium is represented by its permeability distribution $f[K(\mathbf{x})]$, where $K(\mathbf{x})$ is the permeability at $\mathbf{x}$. Wavelet transformation of $K(\mathbf{x})$ is representation of $K(\mathbf{x})$ in the wavelet space at different length scales and positions, using the wavelets defined above. Thus, WT of $K(\mathbf{x})$ is defined by

$$\hat{K}(a, \mathbf{b}) = \int_{-\infty}^{\infty} K(\mathbf{x})\psi_{ab}(\mathbf{x})d\mathbf{x}. \tag{29}$$

In effect, by wavelet transforming of $K(\mathbf{x})$, one analyzes it at increasingly coarser length scales, and thus compressing information about $f[K(\mathbf{x})]$ at such length scales. *Therefore, given $f[K(\mathbf{x})]$, one can systematically average out unneeded information about f at finer length scales where detailed information is not needed, and retain only the most important data.* As discussed above, the ability to contract data is a fundamental property of the wavelets.

We assume that $f[K(\mathbf{x})]$ is broad and correlated, and is obtained either from extensive data, or by a combination of limited field data and the geostatistical methods (other characteristics of FSPM, such as their porosity distribution, can be scaled up by some sort of volume averaging). We represent the fine-grid geological model of the reservoir by a 2D or 3D grid, to whose nodes or blocks we assign a permeability K selected from $f[K(\mathbf{x})]$. We then apply a one-level discrete WT to $K(\mathbf{x})$. This means that we attempt to coarsen the medium by a factor of 2, but the coarsening is not done uniformly. Associated with WT at every block whose center is at (k_1, k_2) in 2D or at (k_1, k_2, k_3) in 3D are four wavelet coefficients (8 in 3D) defined by

$$S_j(k_1, k_2) = \int_{\mathbf{D}} K(x, y)\Phi_{j,k_1,k_2}(x, y)dxdy , \qquad (30)$$

$$D_j^{(d)}(k_1, k_2) = \int_{\mathbf{D}} K(x, y)\Psi_{j,k_1,k_2}^{(d)}(x, y)dxdy , \qquad (31)$$

where j is the level of coarsening ($j = 1$ for the first level), and $\mathbf{D}$ is the domain of the problem. $S_j(k_1, k_2)$, called the *scale coefficients*, contain information about K at (k_1, k_2) [at (k_1, k_2, k_3) in 3D] in the coarser mesh, while the wavelet-detail coefficients $D_j^{(d)}(k_1, k_2)$ measure the *contrast* between K at (k_1, k_2) in the coarser scale and those of its neighbors in the previous finer scale, with $d = 1, 2$, and 3 (in 3D, $d = 1, \ldots, 7$) measuring the contrasts between the nodes or blocks in the y, x, and the diagonal directions, respectively. In practice, the wavelet coefficients at any level j are calculated from the scale coefficients at the previous finer level $j - 1$ in 2 steps (3 steps in 3D). First, a 1D Mallat algorithm is applied to the scale coefficients associated with the blocks in a given direction, e.g., the x-direction. The Mallat algorithm is defined[33,34] by

$$S_j(k_1) = \sum_{l=0}^{L-1} h_l S_{j-1}(l + 2k_1) \qquad (32)$$

$$D_j^{(d)}(k_1) = \sum_{l=0}^{L-1} g_l S_{j-1}(l + 2k_1) , \qquad (33)$$

where h_l and g_l are the filter coefficients described above, and $S_{j-1}(l + 2k_1)$ is the scale coefficient at level $j - 1$. The scale coefficients at the $j = 0$ level are taken to be the values of $K(\mathbf{x})$ at the grid block. One then applies the 1D Mallat algorithm to the calculated $S_j(k_1)$ and $D_j^{(d)}(k_1)$ in y-direction. Each application results in two coefficients, and thus the four wavelet coefficients (in 2D) are obtained.

   We now define two thresholds ϵ_1 and ϵ_2 for S_j and $D_j^{(d)}$ (in terms of fractions of their corresponding largest values), where ϵ_1 is a measure of the permeability at a given node or block, and ϵ_2 measures the contrast in the permeability values between neighboring blocks. We then check the scale coefficient at each block. If it is higher than ϵ_1, then we do nothing and move on to the next block. However, if it is smaller than ϵ_1, we check the detail coefficients and set to zero all the $D_j^{(d)}(k_1, k_2)$ that are smaller than ϵ_2. Setting $D_j^{(d)}(k_1, k_2) = 0$ means that the neighbor of (k_1, k_2) corresponding to the direction (d), which in the fine scale is just one block (or one diagonal block) away from (k_1, k_2), is removed, that is, the two blocks join and form a larger block. Therefore, depending on the structure of $f[K(\mathbf{x})]$, a number of blocks (or nodes) in the original fine-scale

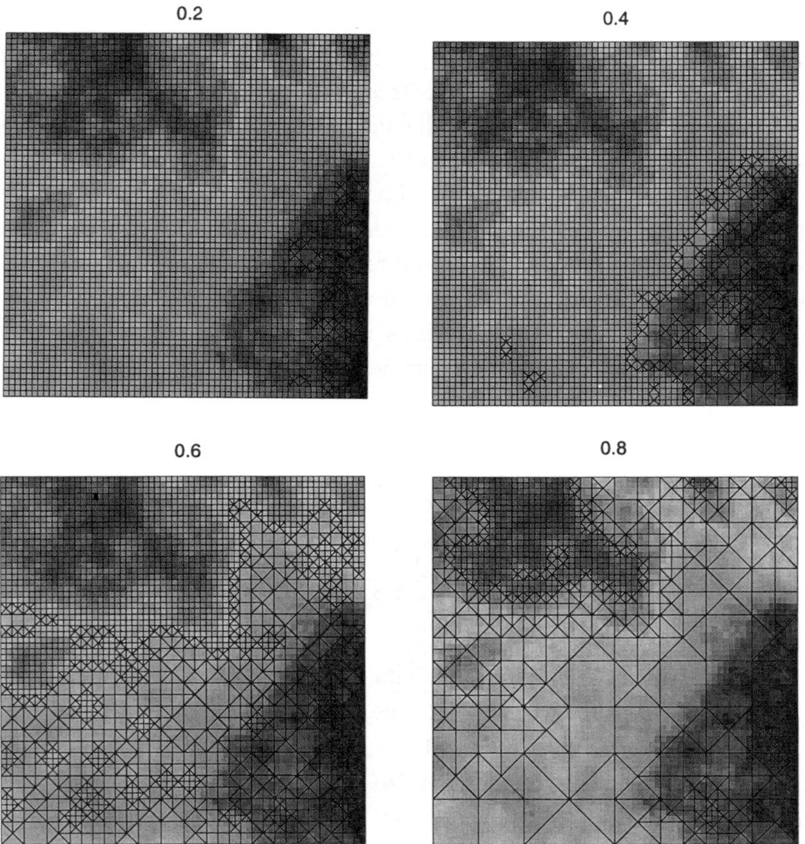

Fig. 7. Four different grids coarsened by the wavelet method, starting from the same original model. Numbers indicate the threshold $\epsilon_1 = \epsilon_2$. The gray map shows the permeability distribution.

model are coarsened. If $f[K(\mathbf{x})]$ is relatively narrow, the blocks are coarsened more or less uniformly throughout the model, whereas with a broad and correlated $f[K(\mathbf{x})]$ the coarsened blocks are scattered throughout. Figure 7 shows four different coarsened models obtained by this method where, as an example, the fine-scale permeability distribution was generated by an fBm described above, and we took $\epsilon_1 = \epsilon_2$. We then obtain a coarser model by applying again the discrete WT to the scale coefficients obtained at the previous level [which contain information about $K(\mathbf{x})$ at the previous level], and calculate a new set of four detail coefficients (8 in 3D) for each block of the coarsened mesh. The new detail coefficients are again set to zero if they are smaller than ϵ_2, and the corresponding blocks in the present grid structure are coarsened. In effect, at coarsening level j, each block is compared with those that are at a distance 2^{j-1} from it, where the distance is measured in units of the blocks' length in the fine-grid model. This process is repeated again until no significant number of the blocks are coarsened (removed). Typically, after three or four levels, the network can no longer be effectively coarsened, thus yielding very fast the final coarsened model for the specified ϵ_1 and ϵ_2. The values of the thresholds are fixed by the level of the detail that we would like to include in the model, and the amount of computational time that we can afford.

Alternatively, and perhaps more realistically, instead of using the permeability distribution, one may first solve for the distribution of the local flow rates (between grid points) in the geological model (i.e. carry out single-phase flow computations) by a method similar to a random resistor network. Then, the WT is applied to the distribution of the local flow rates, and the corresponding wavelet-detail coefficients are computed. The rest of the procedure is the same as above, namely, a threshold ϵ is set and those coefficients that are less than ϵ are set to zero, and so on. The advantage of this version of the method is that, it coarsens the model based on the local flow rates in the fine-grid geological model, and thus it automatically identifies the regions of high flow rates that are crucial to modeling of flow and transport in FSPM at large length scales. For example, the zones around the wells must be modeled carefully. Since such zones usually represent high-flow rate areas, the WT method based on the distribution of the local flow rates will be highly accurate and realistic.

An important issue is assignment of the effective permeabilities of the coarsened blocks. Once the size and shape of each coarsened block is determined, its effective permeability can be computed by any one of several available methods, such as the effective-medium approximation,[4,5] the renormalization group method,[15,19,21] or by direct single-phase flow simulations.

The wavelet method is also capable of generating a completely irregular grid, beginning with a standard rectangular (square or simple-cubic) one. In order to do this, we first proceed as before and set the wavelet-detail coefficients to zero where appropriate (see above). We then create a binary map in which the locations with nonzero wavelet coefficients are taken to be unity. The resulting binary map also represents the spatial distribution of the low and high permeability regions and the contrast between them. The irregular coarsened model is then created such that the regions with high density of unities (the high permeability zones) in the binary map are resolved, that is, they contain more blocks or nodes. To make the discussion concrete, we consider a 2D grid; its extension to 3D will be clear. The coarsening process and generation of the irregular network is achieved by a recursive *quadtree* (*octree* in 3D) decomposition on a rectangular grid. The entire domain is divided into four equal rectangular domains called *cells*. Each cell is then divided into smaller cells if it contains unit elements in the binary map. The resulting set of cells from the quadtree decomposition is then exposed to further refinement such that the neighboring cells are at most one level (scale) different.

The resulting structure created by this algorithm will contain three types of nodes: (1) Those that have four neighbors (6 in 3D). The governing equations for such nodes are discretized using standard nonuniform finite-difference schemes; (2) In the second group are the boundary nodes, for which the boundary conditions are discretized in the usual way; (3) The last type is the set of internal nodes that are at the interface between coarsened and the fine-grid cells. Such nodes have only three neighbors (see Fig. 8), and therefore the governing equations, written for such points, cannot be discretized in the usual way. Although one may use interpolation schemes among neighboring nodes, the interpolation can produce large errors in the solution of the governing equations. To overcome this problem, we use the following scheme. For a node of this type we add an extra node to the system, at the center of the neighboring coarsened block, and hence increase the number of the neighbors to four. Therefore, the governing equations at such internal nodes are written in the usual manner, except that the coordinates (x, y) are first rotated 45° to new coordinates (ξ, η), which are related to (x, y) by the following equations:

$$\xi = \frac{\sqrt{2}}{2}(x + y), \qquad \eta = \frac{\sqrt{2}}{2}(x - y). \tag{34}$$

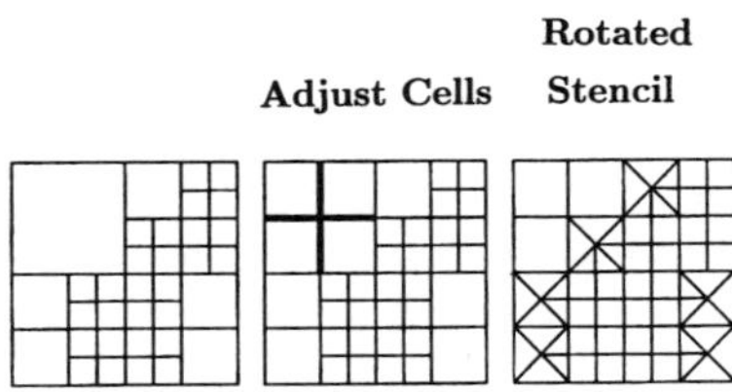

Fig. 8. Various types of grid points that one encounters in a rectangular model of a reservoir.

A similar technique may be used in 3D, although the extension is not straight-forward. Then, the governing equations at such nodes are transformed to this new coordinate system. The neighboring nodes for these nodes will be the corner points of the coarser blocks.

5.2. *Wavelet scale-up of fractured porous media*

As discussed earlier, many FSPM contain fractures at several length scales, and therefore it is of crucial importance that the scale-up method be applicable to models of such media. Here, we briefly discuss how the wavelet method can be extended to fractured FSPM. Up-scaling of fractured FSPM should be done in two steps.

Step 1: Scale-up of a single fracture. A considerable amount of research has established that a rock fracture is not just a pair of parallel flat plates with smooth surfaces, as has been assumed in many of the past modeling efforts, but, because of its rough internal surface, it is a complex network of voids and contact areas.[4,5,59] Therefore, proper modeling of flow through a single fracture entails its representation by a network of channels through which a fluid flows. This network contains high- and low-flow rate zones, and experiments and simulations[4,5,59] indicate that only a small portion of the whole network is actually used for fluid flow. Thus, we first apply the wavelet scale up to the individual fractures to identify the high-flow rate areas. The method for this is exactly the same as the one described above. This results in a considerable saving of the computational time for simulation of flow and transport in the entire fractured reservoir, as it identifies the key areas in the fractures, so that one will not need to use a large number of unneeded grid points within the fractures.

Step 2: Scale-up of the fractured porous medium. Once the key high flow rate zones within the fractures are identified, we proceed to the next stage in which the entire porous medium must be scaled up. The wavelet scale-up method is ideally suited for this purpose because, (1) the fractures represent the high-permeability, high-flow rate paths in the porous medium (identified in Step 1), and hence are resolved automatically during the scale-up process, as their corresponding wavelet *scale coefficients* are much larger than those of the porous matrix; (2) the *interface* between the fractures and the matrix is also resolved automatically, since this is where there is a large permeability contrast between the matrix and the fractures, and thus the corresponding wavelet *detail coefficients* are large. Therefore, the scale-up process automatically preserves and/or carefully resolves this region.

6. Wavelet Analysis of Pressure-Transient Data

After the scaled up model of FSPM is prepared, one employs it to carry out simulation of flow and transport and predict the behavior of the system. For oil reservoirs even after this is accomplished, there would remain an additional task. Typically, the rate of oil production decreases after a few years, and thus one must inject water or gas into the reservoir in order to increase the production, or at least keep it at some reasonable level. This requires *reservoir management* and planning, that is, one must first update the oil reservoir model (since oil production changes the distribution of the original water, oil and gas in the reservoir, and may also change its structure). In order to update the reservoir's model, additional new wells may be drilled, new seismic data may be collected, and then they are all combined with the transient data for the reservoir that have been collected over the years. Typical of such data are the *production history* of the reservoir, which provides information about the rate of oil production from the reservoir over a period of several years. This type of information, which is collected for *each* of the producing oil wells, may provide insight into the structure of the reservoir, particularly around the producing wells for which the data have been collected. Thus, one may use[60] WT in order to analyze such data by a method that was discussed above for denoising and interpreting the direct data, in order to uncover special features that the data may contain.

7. Computational Efficiency of Wavelet Method

One of the most important aspects of any numerical analysis method, in addition to its accuracy, is its computational efficiency. This is particularly important when one has to deal with a large amount of data and, in the case of FSPM, a large geological model that must be efficiently treated and scaled up. Here we point out two crucial points to demonstrate the efficiency of the wavelet method: (1) Calculation of WTs is done by completely localized operations (since the wavelets have compact supports), ideal conditions for parallel computations, a very useful virtue if huge amounts of data are to be treated. This is particularly important for up-scaling of the geological model. *None* of the previous scale-up methods are easily parallelizable, if at all; (2) With a data array of size n, the computation time of calculating the wavelet coefficients scales as $n \log n$, whereas practically all other methods of data analysis scale at least as n^2.

8. Summary

We have discussed application of wavelet transformations to data processing, and in particular to characterization of field-scale porous media which is a particularly huge problem of data processing, and data construction and reconstruction. We have argued that WTs provide an effiecient, unified, and highly accurate method for this important task. In addition, since the same method is used at every stage of the data processing and model construction, the possibility of error or inconsistencies is minimized.

Acknowledgments

I would like to thank my wife Mahnoush and my son Ali who put up with the long hours that I spent in the Regional Summer School in Zanjan, Iran. I would also like to thank Dr. M. R. H. Khajehpour who insisted that I should write this article, Dietrich Stauffer who made it clear to me that I either deliver this article on time "or else," and the Institute for Advanced Studies on Basic Sciences for warm hospitality during the Regional Summer School. The preparation of this article has been supported in part by the Petroleum Research Fund, administered by the American Chemical Society.

References

1. E. L. Lehmann, *Testing Statistical Hypothesis* (Wiley, New York, 1959).
2. H. Scheffe, *The Analysis of Variance* (Wiley, New York, 1959).
3. R. G. Miller, *Simultaneous Statistical Inference* (McGraw-Hill, New York, 1967).
4. M. Sahimi, *Rev. Mod. Phys.* **65**, 1393 (1993).
5. M. Sahimi, *Flow and Transport in Porous Media and Fractured Rock* (VCH, Weinheim, Germany, 1995).

6. A. B. Kersting, D. W. Efurd, D. L. Finnegant, D. J. Rokop, D. K. Smith, and J. L. Thompson, *Nature* **397**, 56 (1999).

7. R. A. Behrens, M. K. Macleod, T. T. Tran, and A. O. Alimi, *SPE Reservoir Eval. & Eng.*, 122 (April, 1998).

8. D. E. Lumley and R. A. Behrens, *SPE Reservoir Eval. & Eng.*, 528 (December, 1998).

9. C. Rossini, F. Brega, L. Piro, M. Rovellini, and G. Spotti, *J. Pet. Techno.*, 979 (November, 1994).

10. T. A. Hewett, Society of Petroleum Engineers Paper 15386 (New Orleans, LA, 1986); T. A. Hewett, and R. A. Behrens, *SPE Formation Eval.* **5**, 217 (1990).

11. M. Sahimi and S. Arbabi, *Phys. Rev. Lett.* **68**, 608 (1992); *ibid.* **77**, 3689 (1996).

12. C. C. Barton and P. A. Hsieh, *Physical and Hydrological-Flow Properties of Fractures*, Guidebook T385 (American Geophysical Union, Las Vegas, Nevada, 1989).

13. M. Sahimi, M. C. Robertson, and C. G. Sammis, *Phys. Rev. Lett.* **70**, 2186 (1993); M. C. Robertson, C. G. Sammis, M. Sahimi, and A. J. Martin, *J. Geophys. Res.* **B100**, 609 (1995).

14. J. L. Jensen, D. W. Hinkley, and L. W. Lake, *SPE Formation Eval.* **2**, 461 (1987).

15. P. R. King, *Transport in Porous Media* **4**, 37 (1989).

16. C. A. Kossack, J. O. Aasen, and S. T. Opdal, Society of Petroleum Engineers Paper 18436 (Houston, Texas, 1989).

17. G. Le Loćh, in *Geostatistics*, Vol. 2 (Kluwer, London, 1989), p. 557.

18. A. H. Muggeride, in *Reservoir Characterization II*, eds. L. W. Lake *et al.* (Academic Press, San Diego, 1991).

19. A. Jones, P. R. King, C. McGill, and J. Williams, *Proc. 3rd International Reservoir Characterization Tech. Conference* (Tulsa, Oklahoma, 1991).

20. H. N. J. Poulisse, in *The Mathematics of Oil Recovery*, ed. P. R. King (Clarendon, Oxford, 1992), p. 699.

21. P. R. King, A. H. Muggeride, and W. G. Price, *Transport in Porous Media* **12**, 237 (1993).

22. E. Kasap and L. W. Lake, *SPE Adv. Tech. Ser.* **1**, 27 (1993).

23. G. E. Pickup, P. S. Ringrose, J. L. Jensen, and K. S. Sorbie, *Math. Geol.* **26**, 227 (1994).

24. N. Saad, A. S. Cullik, and M. M. Honarpour, *Proc. 2nd Norwegian Meeting on Quantification and Modeling of Heterogeneities* (1994).

25. N. Saad, C. T. Kalkomey, and A. Quenes, in *Proc. 4th European Conference on the Mathematics of Oil Recovery* (1994).

26. D. Li, A. S. Cullik, and L. W. Lake, *J. Pet. Sci. Eng.* **14**, 1 (1995).

27. L. J. Durlofsky, R. A. Behrens, R. C. Jones, and A. Bernath, Society of Petroleum Engineers Paper 30709 (Dallas, Texas, 1995).

28. A. Kumar, C. L. Farmer, G. R. Jerauld, and D. Li, Society of Petroleum Engineers Paper 38744 (San Antonio, Texas, 1997).

29. J. Morlet, G. Arens, E. Fourgeau, and D. Girad, *Geophysics* **47**, 203, 222 (1982).

30. A. Grossmann and J. Morlet, *SIAM J. Math. Anal.* **15**, 723 (1984).

31. A. Goupillaud, A. Grossmann, and J. Morlet, *Geoexploration* **23**, 85 (1984).

32. I. Daubechies, *Commun. Pure Appl. Math.* **XLI**, 901 (1988).

33. S. Mallat, *IEEE Trans. Pattern Anal. Mach. Intel.* **11**, 674 (1989).

34. S. Mallat, *IEEE Trans. Accoust. Speech Signal Anal.* **37**, 2901 (1989).

35. I. Daubechies, *IEEE Trans. Inf. Theory* **36**, 961 (1990).

36. I. Daubechies, *Ten Lectures on Wavelets* (SIAM, Philadelphia, 1992).

37. C. K. Chui, *Wavelet Analysis and its Applications*, Vols. 1 and 2 (Academic, San Diego, 1992).

38. Y. Meyer, *Wavelets and Compact Operators* (Cambridge University Press, New York, 1992).

39. M. Holschneider, *Wavelets: An Analysis Tool* (Oxford University Press, New York, 1995).

40. G. W. Wornell, *Signal Processing with Fractals: A Wavelet-Based Approach* (Prentice Hall, Englewood Cliffs, New Jersey, 1995).

41. C. Meneveau, *Phys. Rev. Lett.* **66**, 1450 (1991); *J. Fluid Mech.* **232**, 469 (1991).

42. M. Yamada and K. Ohkitani, *Prog. Theor. Phys.* **86**, 799 (1991).

43. L. H. Hudgins, C. A. Friehe, and M. E. Mayer, *Phys. Rev. Lett.* **71**, 3279 (1993).

44. D. L. Donoho and I. M. Johnstone, *Biometrika* **81**, 525 (1994).

45. A. R. Mehrabi, H. Rassamdana, and M. Sahimi, *Phys. Rev.* **E56**, 712 (1997).

46. P. Flandrin, *IEEE Trans. Inf. Theory* **38**, 910 (1992).

47. G. W. Wornell, *IEEE Trans. Inf. Theory* **36**, 859 (1990).

48. M. Sahimi and M. Hashemi, *Geophys. Res. Lett.* (to be published).

49. L. J. Pyrak-Nolte and D. D. Nolte, *Geophys. Res. Lett.* **22**, 1329 (1995).

50. S. Roy and L. J. Pyrak-Nolte, *Geophys. Res. Lett.* **22**, 2773 (1995).

51. H. J. Grubb and A. T. Walden, *Geophys. Propect.* **45**, 183 (1997).

52. N. Bethoux, G. Ouillon, and M. Nicolas, *Geophys. J. Int.* **135**, 177 (1998).

53. G. Ouillon, D. Sornette, and C. Castaing, *Nonlinear Proc. Geophys.* **2**, 158 (1995).

54. G. Ouillon, C. Castaing, and D. Sornette, *J. Geophys. Res.* **B101**, 5477 (1996).

55. P. Gaillot, J. Darrozes, and M. de Saint Blanquat, *Geophys. Res. Lett.* **14**, 1819 (1997).

56. J. L. Jensen, L. W. Lake, P. W. M. Corbett, and D. J. Goggin, *Statistics for Petroleum Engineers and Geoscientists* (Prentice Hall, New Jersey, 1997).

57. A. R. Mehrabi and M. Sahimi, *Phys. Rev. Lett.* **79**, 4385 (1997).

58. M. Sahimi and A. R. Mehrabi, *Physica* **A266**, 136 (1999).

59. L. Moreno, Y. W. Tsang, C. F. Tsang, F. V. Hale, and I. Neretnieks, *Water Resour. Res.* **24**, 2033 (1988).

60. J. Kikani and M. He, Society of Petroleum Engineers Paper 48966 (New Orleans, Louisiana, 1998).

Annual Reviews of Computational Physics VIII (pp. 113–132)
Edited by Dietrich Stauffer

PORE-SCALE CHARACTERIZATION OF POROUS ROCKS: EVIDENCE OF CORRELATED HETEROGENEITY AND IMPLICATIONS TO FLUID DISPLACEMENT PROCESSES

MARK A. KNACKSTEDT

Department of Applied Mathematics,
Research School of Physical Sciences and Engineering,
Australian National University, Canberra ACT 0200, Australia

School of Petroleum Engineering,
University of New South Wales, Sydney, NSW 2052, Australia

1. Introduction

The study of flow phenomena in sedimentary rocks is of interest in the physics community[1–3] and is relevant to many problems of industrial importance including transport of nonaqueous contaminants in groundwater and the production of oil and gas from sedimentary reservoirs. For example, oil and gas recovery from petroleum reservoirs depends on residual saturation — the fraction of the pore space occupied by oil or gas which cannot be recovered because it is trapped or bypassed in the reservoir rock by the combined effects of capillary forces and heterogeneity. Recovery is usually very inefficient leaving, on the average, two or more barrels trapped for every barrel produced to the surface. Estimating realistic residual saturations for different reservoir systems and production options is a critical problem. Residual saturations cannot be predicted *a priori* and current industry practice is to make measurements on small core plugs (cm length scale) and use these to calibrate measurements made by the analysis of borehole data at the meter length scale. Increasingly accurate statistical descriptions of heterogeneity in sedimentary rocks have become available through the analysis of borehole data.[4–9] These studies have shown that long-range correlations in properties exist from the kilometre scale to the metre scale. The recognition that natural porous media frequently have long-range spatial correlations in their properties has encouraged studies of percolation in correlated property maps.[3,10–13]

Past attempts to relate flow properties to the pore-scale microstructure of the porous materials have been more limited. Most methods use oversimplified representations of the structure for which properties can be more simply evaluated. Capillary pressure curves, measured on rock samples, are routinely used to relate macroscopic transport coefficients to microstructural information. Sample volumes measured are usually on the order of $\simeq 1$ cm^3. At this scale it is assumed that the porous medium is statistically homogeneous, that is, only small fluctuations in properties occur as a function of sample volume. Theoretical studies of capillary pressure measurements therefore either assume that the pore space is spatially disordered[14–16] or include spatial correlation at a local scale; for example, between pore body and pore throat sizes.[17,18] While it is now widely recognised that macroscopic properties in geological fields are strongly correlated it is still unclear whether this behavior persists at smaller scales. To a large extent this is due to an inability to characterize rock structure at the pore scale.

In this review we describe experimental results that indicate the presence of correlated heterogeneity at the pore scale. We show that the correlations can be described by fractional Brownian motion with a cutoff. We use invasion percolation[19] to simulate rate-controlled mercury injection experiments on porous media displaying both uncorrelated and correlated disorder and illustrate that the introduction of correlations has a marked effect on the nature of the capillary pressure curve. We then show that it is only possible to account for the behavior of the experimental data for sedimentary rocks by *including* correlated heterogeneity. This result brings into question the use of ordinary percolation concepts for modeling flow behavior at the pore scale in porous rocks. We then use the invasion percolation model to investigate the effect of correlated heterogeneity on capillary dominated displacements in porous media. The residual saturations are shown to be strongly sensitive to the degree of correlation. For example, correlated heterogeneity leads to substantially lower residual saturations than observed on random lattices.

The plan of the paper is as follows. In the next section we describe the simulation methods and the generation of lattices with correlated heterogeneity. In Sec. 3 we review experimental data which indicates that rocks exhibit correlated heterogeneity. In Sec. 4 we present results of a modified invasion percolation simulation for rate-controlled mercury injection experiments on porous media displaying both uncorrelated and correlated disorder and compare to experiment. In Sec. 5 we measure residual saturations from IP simulations on lattices with varying degrees of correlated heterogeneity. The final section discusses the implications of these results to the interpretation of multiphase flow studies on sedimentary rocks.

2. Numerical Methods

2.1. *Invasion percolation simulation*

We consider the invasion percolation (IP) model in three dimensions. Classical IP simulations begin by assigning an uncorrelated random number to each site on the lattice from an arbitrary distribution. Initially the lattice is filled with the defending phase, and the invading phase occupies one edge or face of the lattice. At each step in the simulation the site with the largest value on the interface between the invading and defending phases is occupied by the defender. Two main variants of IP have been studied. In the first, compressible IP (CIP), the defending fluid is compressible and the invading fluid can potentially invade any region on the interface occupied by defending fluid. In the second, trapping IP (TIP), the defending fluid is incompressible and can be trapped when a portion of it is surrounded by the invading fluid.

We utilise a new algorithm which allows rapid simulation of invasion percolation.[20] In conventional algorithms,[1,10-12] the search for the trapped regions is done after every invasion event using a Hoshen–Kopelman algorithm,[21] which traverses the entire lattice, labels all the connected regions, and then only those sites that are connected to the outlet face are considered as potential invasion sites. A second sweep of the lattice is then done to determine which of the potential sites is to be invaded in the next time step. Thus, each invasion event demands $O(N^2)$ calculations, where N is the number of sites in the lattice. This is highly inefficient for two reasons. First, after each invasion event only a small local change is made to the interface; implementing the global Hoshen–Kopelman search is unnecessary. Secondly, it is wasteful to traverse the entire network at each time step to find the most favorable site (bond) on the interface since the interface is largely static.

We tackle the first problem by searching the neighbors of each newly invaded site (bond) to check for trapping. This is ruled out in almost all instances. If trapping is possible, then several simultaneous breadth first "forest-fire" searches are used to update the cluster labeling as necessary. This restricts the changes to the most local region possible. Since each site (bond) can be invaded or trapped at most once during an invasion, this part of the algorithm scales as $O(N)$. This cluster searching method has some similarities with the "perimeter scouting" algorithm for 2D clusters. In this algorithm one checks whether the most recently invaded sites could have been trapped in the interior of the cluster. If so, oriented walks are started on the just-invaded site, pointing away from it to the neighboring sites which are those that neither belong to the cluster nor are candidates for invasion. The walks continue until all but one of them have again reached the site of origin. The growth sites visited by these walks are then eliminated from the list of active sites. This method is effective in 2D but not as efficient in 3D. Our method differs from it by searching cluster volumes rather than perimeters, and incorporating local checking to minimize cluster searching and is thus equally effective in 3D.

The second problem is solved by storing the sites on the fluid–fluid interface in a list, sorted according to the capillary pressure threshold (or size) needed to invade them. This list is implemented using a balanced binary search tree, so that insertion and deletion operations on the list can be performed in $\log(n)$ time, where n is the list size. The sites that are designated as trapped using the procedures described above are removed from the invasion list. Each site (bond) is added and removed from the interface list at most once, limiting the computational effort of this part of the algorithm to $O[N\log(n)]$. Thus, the execution time for N sites is dominated (for large N) by list manipulation and scales at most as $O[N\log(N)]$.

While the execution time is approximately $O[N\log(n)]$, in practice the time and memory requirements depend on the total number of lattice sites and those forming the cluster boundary. For example, we find empirically that for 3D TIP, the execution time scales as $M^{1.24}$, and the memory used is 20 bytes for each lattice site plus 64 bytes for each cluster site. On a 500 MHz 21164A Alpha microprocessor, a trapped cluster of 2×10^5 sites is grown on a $181 \times 181 \times 181$ lattice in 12.0 s, using 120 Mb of memory, while in 2D a cluster of 5×10^5 sites is grown on a 2000×2000 lattice in 12.0 s, using only 52 Mb. Complete details of the algorithm, which can be used for arbitrary networks, are given elsewhere.[20]

2.2. *Generation of correlated lattices*

Heterogeneity in geological formations exists at all length scales. Hewett[4] has shown that on the large (field) scale heterogeneity is correlated and follows fractional Brownian motion (fBm). Following Hewett, we generate correlated fields based on fBm. A feature of a network of pores with a statistical description based on a fBm is that it contains long-ranged correlations in pore sizes, with the variance of the pore size given by $\langle r(\mathbf{x}) - r(\mathbf{x}_0) \rangle = C_0 |\mathbf{x} - \mathbf{x}_0|^{2H}$, where C_0 is a constant and the type and extent of the correlations can be tuned by varying H. This contrasts with random lattices where adjacent pore sizes are independent. The fBm has not been used or tested for representing the correlations at the pore scale, and we are not aware of any experimental evidence that supports the introduction of correlated heterogeneity extending to the full grid size (which is generated by the fBm). We therefore introduce a cutoff length scale ℓ_c where $\langle r(\mathbf{x}) - r(\mathbf{x}_0) \rangle = C_0 |\mathbf{x} - \mathbf{x}_0|^{2H} : |\mathbf{x} - \mathbf{x}_0| < \ell_c$ and $\langle r(\mathbf{x}) - r(\mathbf{x}_0) \rangle = C_0 |\ell_c|^{2H} : |\mathbf{x} - \mathbf{x}_0| > \ell_c$. The introduction of ℓ_c allows us to choose an appropriate length scale for correlations at the pore scale.

3. Experimental Evidence for Correlated Heterogeneity

Characterizing the pore space of complex porous media requires the ability to examine the microstructure of the material. Modern imaging techniques now allow scientists and engineers to observe extremely complex material morphologies in 3D. X-ray computed tomography (CT) is a nondestructive technique for visualizing features in the interior of opaque solid objects and for resolving information on their 3D geometries. Conventional CT can be used to obtain the porosity map of a piece of sedimentary rock at scales down to a millimetre.[22] High-resolution CT[23] and laser scanning confocal microscopy[24] has enabled the measurement of geometric properties at scales as small as a few microns.

We have obtained millimetre-scale CT images of Berea sandstone obtained in our laboratories.[25] Heterogeneity in the porosity distribution is evident from visual inspection. We give in Fig. 1 a plot of the variance in the porosity distribution at different scales. As one can see, there is a power law in the variance of the porosity: the distribution can be described by fBm. Statistical analysis of the porosity distribution indicate a Hurst exponent $H \simeq 0.5$. We use this value for the simulations described throughout this paper.

Image facilities can now provide 1024^3 voxel images of porous materials at a voxel resolution of under 6 microns.[23] In Fig. 2 we compare two series of six consecutive slices of a crossbedded sandstone sampled at 10 micron resolution. The two series of images are separated by *less* than 1 mm. One can see a large change in the porosity of the material with this small change in depth — pore sizes, throat sizes and other geometric properties of the rock are also

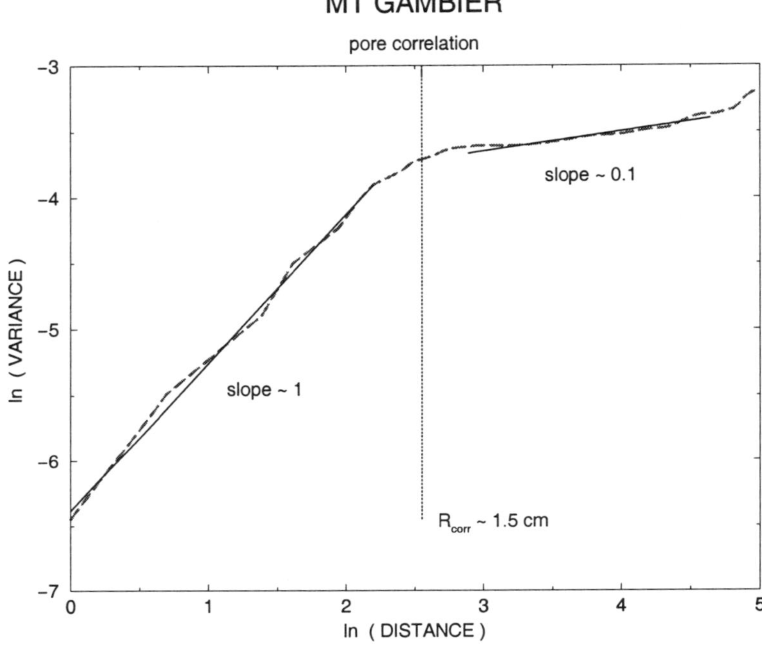

Fig. 1. Variance in the porosity distribution showing power-law (fBm) behavior.

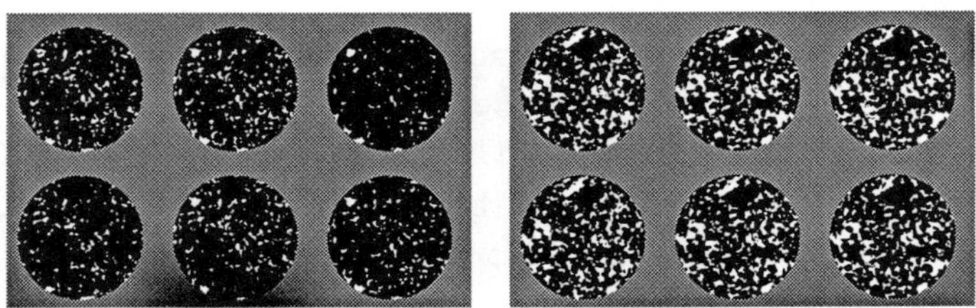

Fig. 2. Comparison of two sets of six consecutive slices of a crossbedded sandstone at 10 μm spacing. In the first set the porosity is < 10%, while the second set which is less than 1 mm away, the porosity is $\simeq$ 20%.

very different despite the images being only two grain diameters apart. We show in Fig. 3 a trace of 660 values of the porosity measured at a separation of 10 μm. A preliminary statistical analysis of the data indicates that the description of correlated heterogeneity used to describe rock properties at the meter scale through borehole analysis[4-9] may also describe the properties at

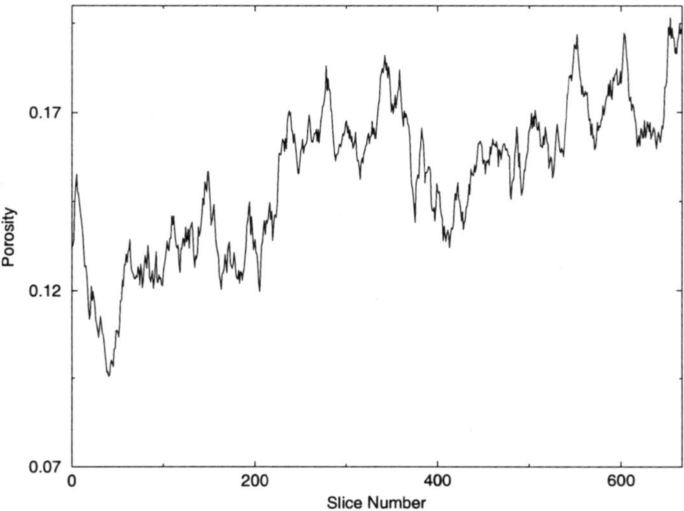

Fig. 3. Porosity trace shows fBm-like behavior at the 10 μm scale.

this pore scale. This result brings into question the common assumption that on the pore scale rock properties are randomly distributed and that ordinary percolation concepts can be used for modeling flow behavior at the pore scale in porous rocks.

4. Simulation of Rate-Controlled Mercury Injection Experiments

To investigate this issue further we use a modified IP model to simulate rate-controlled mercury injection experiments on porous materials displaying both correlated and uncorrelated disorder and compare with experimental data for sedimentary rocks. Rate-controlled mercury injection experiments provide far more information on the statistical nature of pore structure than conventional porosimetry.[26] Fluid intrusion under conditions of constant-rate injection leads to a sequence of jumps in capillary pressure which are associated with regions of low capillarity.[26] While the envelope of the curve is the classic pressure controlled curve, the invasion into regions of low capillarity adds discrete jumps onto this envelope. In Fig. 4(a) we show an example of a capillary pressure curve obtained in our laboratories for Berea Sandstone under rate-controlled conditions.[25] The detailed geometry of the jumps in the capillary pressure curve over different saturation ranges is shown in Figs. 4(b)–4(d).

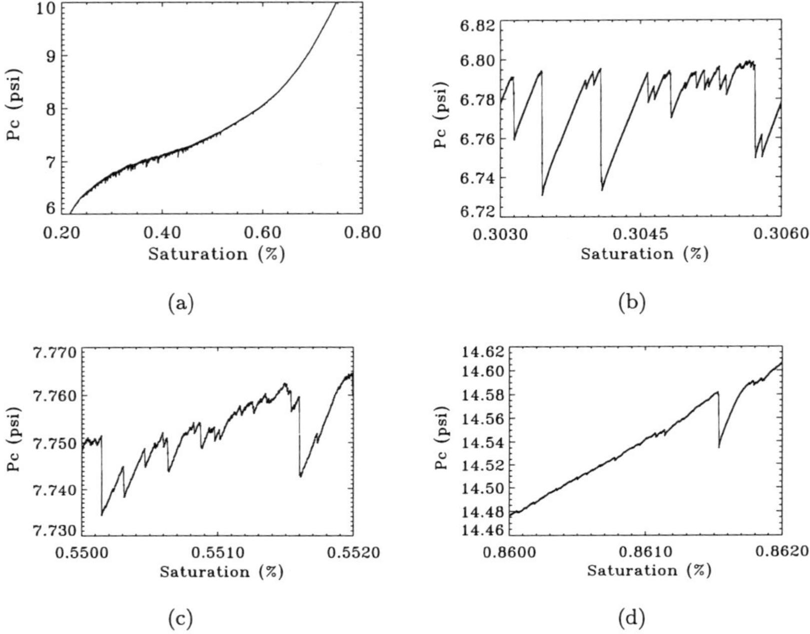

Fig. 4. Experimental constant volume porosimetry curves for Berea sandstone. (a) Over large saturation range; (b)–(d) Detailed curves over different saturation ranges.

This process is naturally mapped onto the IP model. In the conventional IP, the porous medium is represented by a lattice of sites and bonds. Each site or bond i is assigned a random number r_i which represents an effective size for a pore body or a pore throat located at i. Initially all sites are considered empty. The invading fluid is injected into the medium and fills the pores at each time step by identifying all the sites (or bonds) on the interface between the invading fluid and the empty portion of the pore space, and occupying the site (or bond) with the largest r_i. As there is no defending phase in mercury porosimetry, we consider CIP.

IP models of capillary pressure have previously been used to model the constant-*pressure* curve alone.[1,27] However, since IP models a quasistatic process, one can naturally recover the full constant-volume porosimetry curve from a simulation. We model constant-volume porosimetry on both random and correlated grids. The conventional IP algorithm requires minimal modifications to realistically mimic a capillary pressure experiment. The geometry of the conventional IP algorithm considers invasion from one face of the lattice,

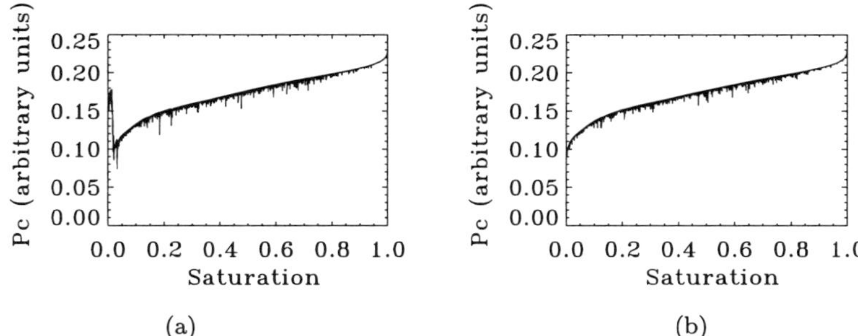

Fig. 5. Effect of the modification of the IP algorithm on a constant-volume and conventional porosimetry curve. In this case we consider an *identical* cubic sample of size 128^3. (a) Invasion from one side; note the large downward jump in the capillary pressure due to the inlet effect. The constant-pressure envelope is therefore very flat. (b) Invasion from all six sides.

with a defending phase exiting from the opposite face. In mercury porosimetry the geometry of the displacement is altered. The core is placed in a cell and the mercury completely surrounds the sample. To mimic this process we allow the invader to enter the pore space from all sides. In Fig. 5 we show the effect of altering the boundary condition on the constant-volume capillary pressure curve for a correlated grid. When injection comes from one side only, the capillary pressure curve is often punctuated with extremely large drops at small to intermediate pressures. The effect on the conventional capillary pressure curve is even more dramatic. These features are not observed in experiments. When we modify the IP algorithm to allow the correct condition of invasion from all sides, these large drops in pressure are no longer noted.

The volume of a sample studied by constant-volume porosimetry is of the order of 1 cm^3 which, assuming a rock with a grain size of $\simeq 100$ μm, gives one a porous medium with up to 1 million individual grains/pores. The simulations were therefore performed on grids of comparable size (128^3). The statistical data were based on a minimum of 1000 runs at this size. When comparing correlated and uncorrelated systems, the pore throat distribution is the *same*. Choosing throat radii from the same distribution insures that any differences in the simulated curves is due solely to the presence of correlations.

Figure 6 shows the simulated rate-controlled capillary pressure curves for correlated and uncorrelated systems. Qualitatively the curves are distinctly different. The uncorrelated curves show a higher frequency of jumps in capillary pressure and the jumps have a consistent baseline over the whole saturation

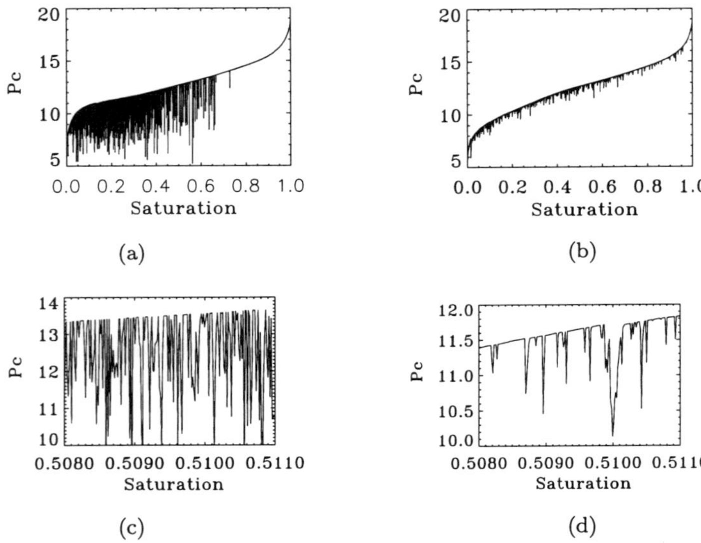

Fig. 6. Volume-controlled capillary pressure curves for (a) and (c) uncorrelated and (b) and (d) fbm grid $\ell_c \to \infty$. (a) and (b) give the curves for the full saturation range and (c) and (d) for a small range of saturation. The signature of the curves is distinct in both cases. (a) and (c) give no resemblance to the data in Fig. 5.

range. In contrast, the porosimetry curve for correlated fields, which exhibits a lower frequency of jumps, is characterized by a more gradual rise in the envelope of the curve, and the baseline of the jumps in the capillary pressure steadily increases with pressure. Comparing Fig. 6 with the experimental curve in Fig. 4 shows that the correlated systems give a better qualitative match, while the uncorrelated case displays no resemblance to the experimental data. This qualitative comparison of the detailed geometry of simulated capillary pressure curves points to the existence of correlated heterogeneity in Berea sandstone.

To evaluate the appropriate length scale ℓ_c of the correlations we consider a quantitative measure used by Yuan and Swanson[26] to characterize the porous rocks; the size distribution of regions of low capillarity over different pressure ranges. The regions of low capillarity measured by constant volume porosimetry can range in size from 1–1000 $n\ell$ — from a single pore volume to hundreds of pore volumes. At low saturations numerous jumps in the capillary pressure curve of various sizes are noted. At higher saturations the number of jumps into regions of low capillarity are less frequent [compare Figs. 4(b) and 4(d)],

Size distribution of Low Capillarity Regions

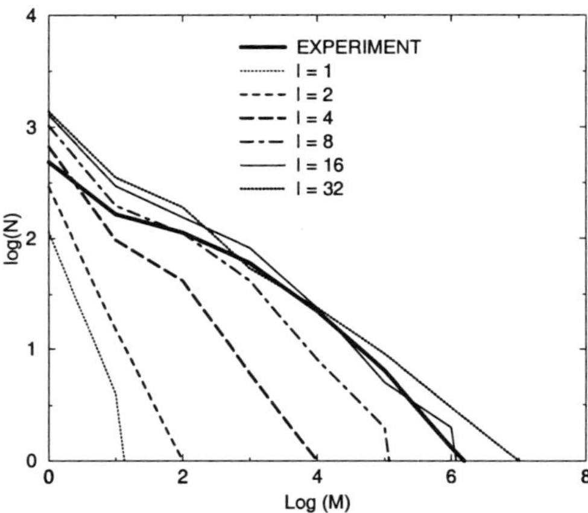

Fig. 7. Size distribution of the low capillarity regions over the saturation range from 60–80%. N is the number of low capillarity jumps measured and M is the size (number of pores) of the jumps.

although large regions of low capillarity are still invaded at high saturations [Fig. 4(d)]. We have measured the size distribution of low capillarity regions on several Berea sandstone samples in our laboratory. We use this measure to obtain a quantitative prediction of the extent of the length scale ℓ_c of the correlated heterogeneity. At lower saturations differences between the predicted size distributions for varying ℓ_c are difficult to discern. At higher saturations differences between the models becomes more evident. In the uncorrelated case, $\ell_c = 1$, for saturations above 60% *no* regions of low capillarity are evident [see Fig. 6(a)]. This disagrees with the experimental data shown in Fig. 4. We plot the size distribution of low-capillarity regions in Fig. 7 for models with varying ℓ_c and compare to experiment. It is clear from this plot that the best fit to the experimental data is consistent with an ℓ_c of 10 or more pores.

More direct evidence for the presence of correlation at the pore scale comes from the experimental work of Swanson.[28] Swanson presented micrographs of the spatial distribution of a nonwetting phase in a range of reservoir rocks including Berea sandstone. He showed that appreciable portions of the rock are still not invaded by the nonwetting phase at low to moderate nonwetting phase saturations. A micrograph of Berea sandstone at 22% saturation showed large unswept regions of more than 2 mm in extent. Assuming a grain size of

100 μm, uninvaded regions of this extent would contain *thousands* of pores. The experiments of Swanson showed, however, that at higher saturations > 50% the extent of the uninvaded regions is significantly smaller than observed at lower saturations.

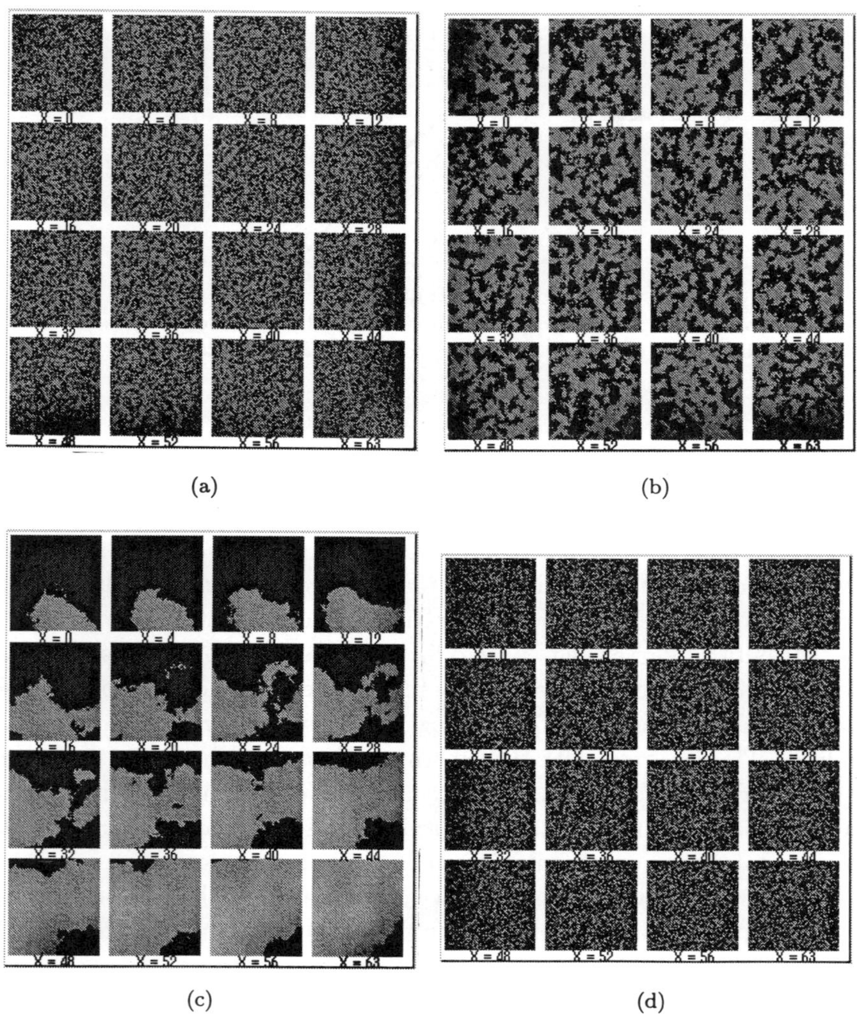

(a)   (b)

(c)   (d)

Fig. 8. A number of slices through a 3D 64^3 simulation illustrating the distribution of the nonwetting phase in network simulations after (a) and (c) 25% and (d) and (f) 75% saturation. (a) and (d), $\ell_c = 1$: (b) and (e), $\ell_c = 8$: (c) and (f) $\ell_c \to \infty$.

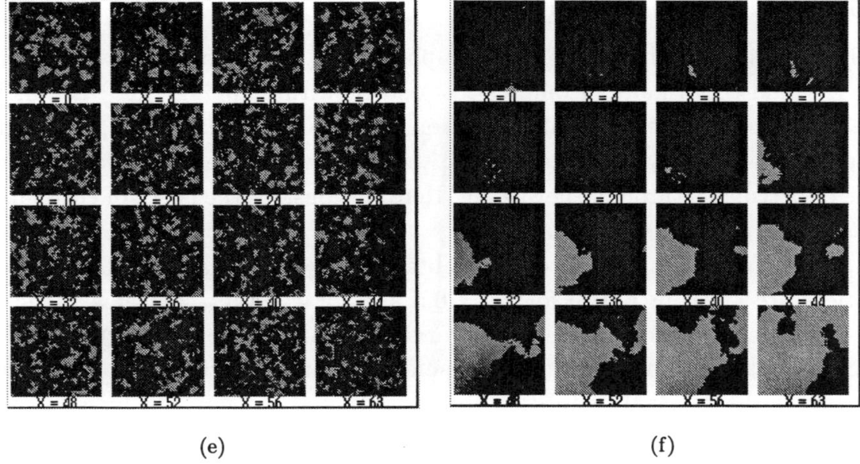

(e)   (f)

Fig. 8. (*Continued*)

We visualize the distribution of nonwetting phase during drainage and find that the experimental observation of Swanson can be accounted for if the pore space is correlated with a cutoff length ℓ_c of approximately 10 pores. We show in Figs. 8(a)–8(c) a simulation of a displacement for uncorrelated and correlated grids at 25% saturation. The morphology of the displacement on the uncorrelated grid spans much of the lattice and has invaded the majority of the pore space. No large unswept regions are evident. In the two correlated cases, large regions of the pore space remain untouched by the invading fluid in agreement with the observations of Swanson. In Figs. 8(d)–8(f) a simulation of a displacement for uncorrelated and correlated grids at 75% saturation is shown. In our simulation for the fBm grid $\ell_c \to \infty$ [Fig. 8(f)] the regions of the grid uninvaded by the nonwetting fluid remain large. The observations of Swanson are consistent with the simulation in both cases for the cutoff grid [Figs. 8(b) and 8(e)].

5. Implications of Correlated Heterogeneity to Two-Phase Flow

Having verified experimentally that correlated heterogeneity exists at the pore scale in the most homogeneous sandstones, we now consider the effect of correlated heterogeneity on conventional two-phase capillary dominated displacement processes. We consider the effect on both the measurement of the residual saturation and to the distribution of the trapped regions of the displaced fluid.

5.1. *Residual saturations*

When considering the problem of two phase flow in porous materials the displaced fluid is normally incompressible and IP with trapping (TIP) is a more realistic model. The residual saturation p_r in TIP is related to p_c of ordinary percolation[19] as it corresponds to the point where the displaced phase no longer percolates through the system. Here we consider the effect of correlated heterogeneity on p_r.

Values of the residual saturation obtained for uncorrelated and correlated lattices are given in Fig. 9 for $H = 0.5$ and for different ℓ_c over a range of L. The spread in the data is due to variation in the pore size distribution for correlated lattices and is not due to insufficient numerical sampling. Finite size scaling of the residual saturation for uncorrelated lattices was fitted to the relationship

$$p_r(L) = p_r(\infty) + cL^{-a}. \tag{1}$$

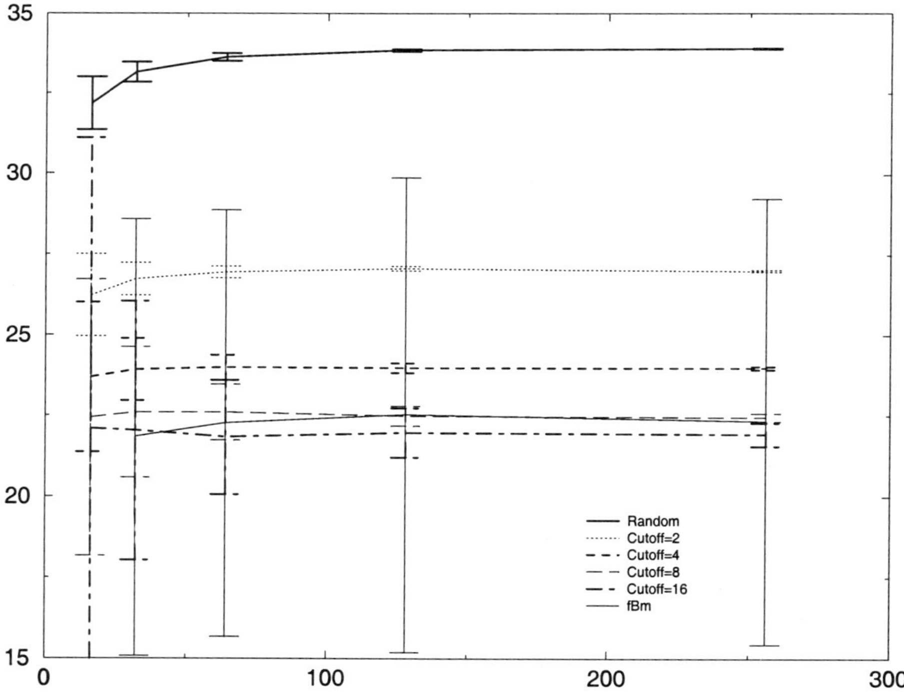

Fig. 9. Residual saturations for correlated lattices for a range of ℓ_c.

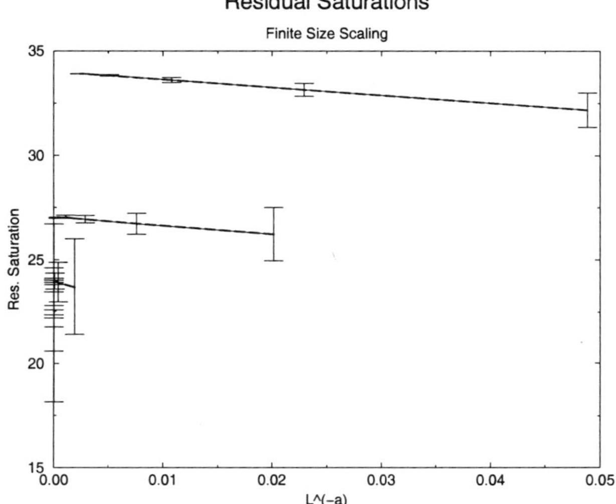

Fig. 10. Finite size scaling of the residual saturation for correlated grids. The upper curve is for an uncorrelated lattice. Curves for $\ell = 2$, $\ell = 4$, $\ell = 8$, $\ell = 16$, and $\ell = \infty$ follow from upper left to bottom right. The fits to the values are good. The values of the asymptotic saturations and the scaling exponent a are given in Table 1.

For the uncorrelated system we find $a = 1/\nu = 1.14 \pm 0.02$ in good agreement with the exponent for ordinary percolation without trapping ($\nu = 0.88$). The finite size scaling relationship allows one to predict the residual saturation of an infinite system; we obtain a value of $p_r(L \to \infty) = 0.3400 \pm 0.0003$. The finite-size scaling behavior for the correlated lattices is also evaluated at scales up to $L = 128$ and an example is shown in Fig. 10 for $H = 0.5$. The asymptotic predictions of $p_r(L \to \infty)$ are given in Table 1 along with the value of the scaling exponent a. It should be noted that the exponent associated with the finite size scaling is significantly different to $1/\nu$ observed for RP.

From the results for the residual saturations we make the following observations. First, the introduction of correlations leads to a large drop in the observed residual saturation. The value of the residual phase saturation generally decreases with increasing ℓ_c. However the residual can show a minimal value for finite ℓ_c. This small increase of the residual at larger cutoff scales *may* be due to the possibility of trapping very large regions of the defending phase at larger ℓ_c. Remarkably, the change in p_r is significant for correlations even at a *small* scale. For example, for a system with only a nearest-neighbor

Table 1. Residual phase saturations for correlated lattices with various ℓ_c. Numerical predictions are given along with the value of exponent a in Eq. (1). For comparison, the value of p_r for a random lattice is 0.3400 with $a = 1.14$.

ℓ_c	$p_r(\infty)$	a
2	0.271 ± 0.0003	0.93
4	0.240 ± 0.0006	1.41
8	0.225 ± 0.0012	2.25
16	0.222 ± 0.0035	3.85

correlation $\ell_c = 2$, the residual drops from 0.34 to 0.27, a variation of approximately 20%. Small-scale correlations clearly have a profound effect on resultant saturations even at large scales.

5.2. *Cluster size distribution*

The cluster size distribution of the trapped phase is also of great interest in the study of immiscible displacement processes. We have studied the size distribution of trapped clusters at the residual saturation point and find that the distribution is strongly affected by the type and extent of correlation. We consider the number of residual clusters of size s for a correlated field with $H = 0.5$ and different values of the cutoff length ℓ_c. From percolation theory[29,30] one expects $n(s)$, the number of clusters of size s, to follow

$$n_s(s) \propto s^{-\tau}, \tag{2}$$

where $\tau = 2.18$. A more accurate way of measuring the cluster size statistics is[31] by investigating $N_s(s) = \sum_{s'>s} s' n_{s'}$, the average total number of clusters with a size greater than a given size s. In general one expects to have

$$N_s(s) \propto s^{2-\tau}. \tag{3}$$

If there are no long-range correlations in the system, percolation theory predicts that the exponent τ is universal. Since the 3D spanning cluster for TIP with no long-range correlations has the same fractal dimension as RP we expect that the uncorrelated case and the cases with a finite ℓ_c will show this scaling behavior. This is observed in Fig. 11. For the system with long-range

Cluster Size Distribution

H=0.5 : L=512

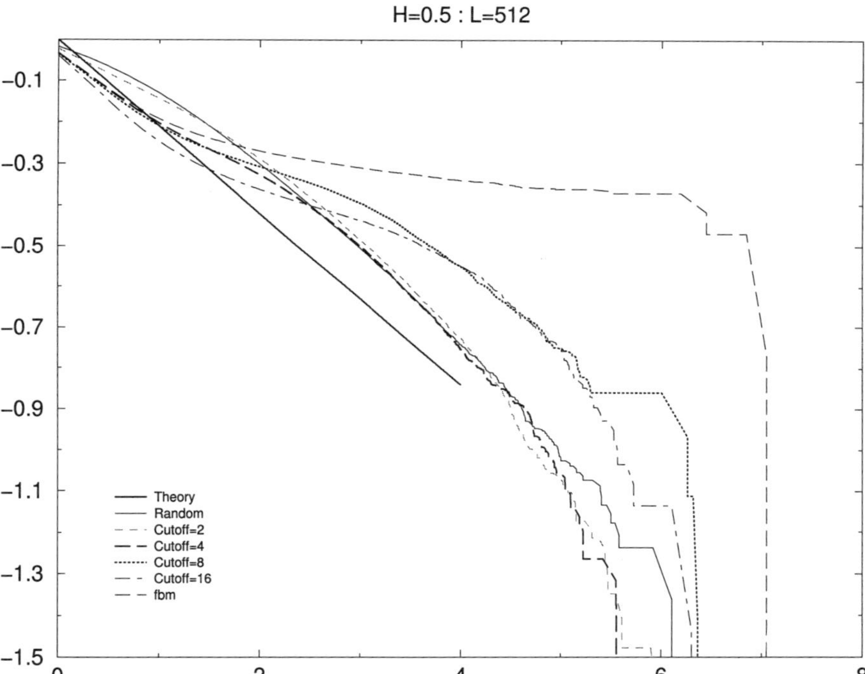

Fig. 11. Size distribution of the trapped clusters as a function for $H = 0.5$ as a function of ℓ_c. Plotted on the y-axis is the log of the cumulative cluster size distribution $\log(N(s))$ versus $\log(s)$, where $N_s(s) = \sum_{s>} sn_s$, the average total number of clusters with a size greater than a given size s. In the uncorrelated case and for finite ℓ_c we see that the scaling predicted from RP holds. For the case of infinite correlations, the scaling behavior differs strongly. Note that for the fBm field one trapped cluster has a huge proportion of the trapped phase and more than half the trapped phase lies within a few trapped clusters. However, the proportion of small trapped clusters is also large.

correlations, d_f is nonuniversal and depends on H.[32] We therefore observe nonuniversal values of τ that depend on H for $\ell_c \to \infty$.

We can however, make some observations on the size of the trapped clusters as a function of ℓ_c. For small ℓ_c there is little effect on the distribution of the trapped phase. At intermediate and larger values of ℓ_c we observe a higher proportion of trapped sites lie in larger trapped clusters. For the case $\ell_c \to \infty$ one single trapped cluster dominates over 30% of the residual phase.

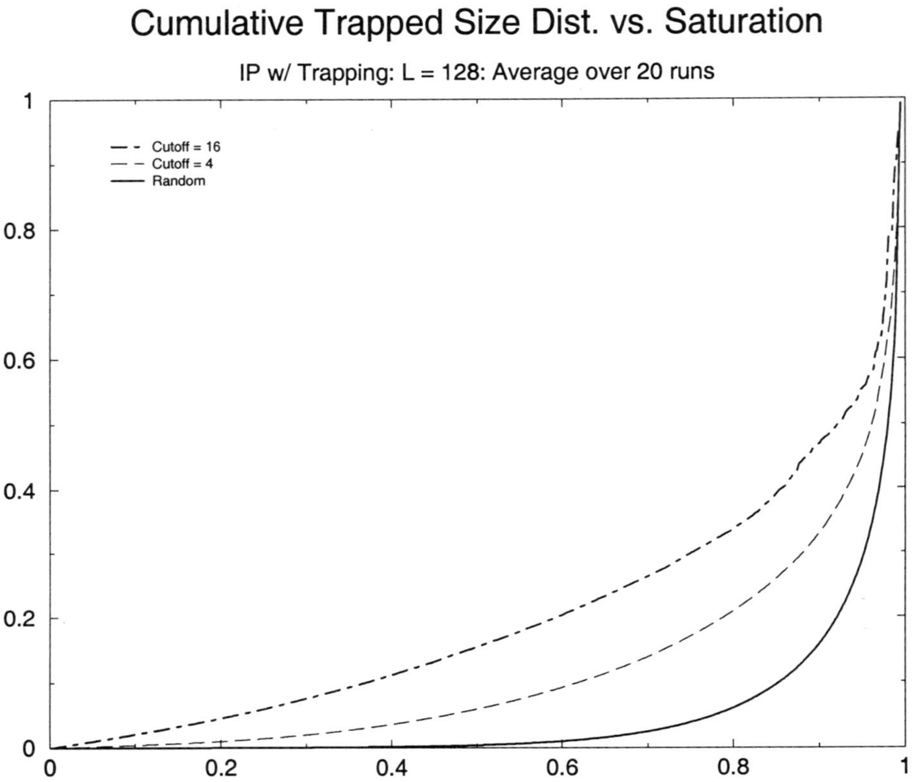

Fig. 12. Proportion of trapped phase as a function of total invading phase saturation. In the random system all trapping occurs at the end of the invasion. In contrast for finite ℓ_c much of the defending phase is trapped at earlier stages of the invasion.

However, for finite ℓ_c there is also a higher number of small trapped regions evident.

The dynamics of the trapping is also strongly dependent on the presence of correlated heterogeneity. We plot in Fig. 12 the total proportion of trapped clusters as a function of saturation. We see very different dynamics when comparing random and correlated systems. In the random system, trapping occurs *only* near the end of the flood — when over 80% of the total invading phase is present, less than 5% of the defender is trapped. In contrast, when we introduce a cutoff of $\ell_c = 16$, over 30% of the defending phase is trapped at 80% invader saturation.

6. Discusssion

Our experimental results suggest that correlated heterogeneity exists down to the pore scale even in a rock like Berea sandstone, which is generally considered to be homogeneous and to exhibit no correlations in its pore size distribution. Moreover direct comparison of experiments on Berea sandstone and simulation of porosimetry on correlated grids provides compelling evidence that correlations *do* persist beyond one or two pore lengths, and is quite extended. Most of the previous studies of fluid displacement in porous media[1,27] have used random and invasion percolation models for modeling the displacement processes. Our results suggest that the use of random percolation concepts to derive the pore size distribution from mercury porosimetry without considering such extended spatial correlations may neglect an essential aspect of the physics of sedimentary rocks and hence yield misleading results. We illustrate this by considering the effect of correlated heterogeneity on the two phase flow properties. The introduction of correlated heterogeneity leads to lower residual phase saturation than observed on random lattices and has a strong effect on the resultant distribution of clusters of the trapped phase.

Our results highlight the need to incorporate realistic descriptions of pore-scale heterogeneity in the scale up of residual saturation measurements. The very high computational effort involved with TIP simulations has limited previous studies to small networks and random fields — networks too small to study the effects of long range correlations. Having relaxed the limitation of small networks by the development of more efficient numerical algorithms for TIP will allow us to examine scaling behavior from the pore to meter scales. Future work will consider the scale-up behavior of rocks. The work has the potential to influence the manner in which industry carries out laboratory measurements and the procedures used to relate these measurements to the log and reservoir scales. Modest improvements in our understanding of these areas would significantly reduce the economic risk associated with new oil and gas developments.

Acknowledgments

MAK is supported by the Australian Research Council. He thanks his many collaborators on this project: S.-J. Marrink, Val Pinczewski, Muhammad Sahimi, Tim Senden, Adrian Sheppard, and Rob Sok.

References

1. M. Sahimi, *Rev. Mod. Phys.* **65**, 1393 (1993).
2. H. A. Makse *et al.*, *Physica* **A233**, 587 (1996).
3. L. Paterson, S. Painter, M. Knackstedt, and W. V. Pinczewski, *Physica* **A233**, 619 (1996).
4. T. Hewett, Technical report, SPE Paper 15836 (unpublished).
5. S. Painter, G. Beresford, and L. Paterson, *Geophysics* **60**, 1187 (1995).
6. S. Painter and L. Paterson, *Geophys. Res. Lett.* **21**, 2857 (1994).
7. S. Painter, *Water Resources Res.* **32**, 1183 (1996).
8. K. Liu, P. Boult, S. Painter, and L. Paterson, *AAPG Bulletin* **80**, 1850 (1996).
9. H. A. Makse *et al.*, *Phys. Rev.* **E54**, 3129 (1996).
10. S. Prakash, S. Havlin, M. Schwartz, and H. Stanley, *Phys. Rev.* **A46**, 1724 (1992).
11. C. Du, C. Satik, and Y. Yortsos, *AIChE J.* **42**, 2392 (1996).
12. M. Sahimi and S. Mukhopadhyay, *Phys. Rev.* **E54**, 3870 (1996).
13. S. J. Marrink, L. Paterson, and M. Knackstedt, *Physica* **A280**, 207 (2000).
14. A. Katz and A. Thompson, *Phys. Rev.* **B34**, 8179 (1986).
15. A. Katz and A. Thompson, *J. Geophys. Res.* **92**, 599 (1987).
16. A. Heiba, M. Sahimi, L. Scriven, and H. Davis, *SPE Reserv. Engin.* **7**, 123 (1992).
17. C. D. Tsakiroglou and A. C. Payatakes, *J. Colloid Int. Sci.* **146**, 479 (1991).
18. M. A. Ioannidis and I. Chatzis, *J. Colloid Int. Sci.* **161**, 278 (1993).
19. D. Wilkinson and J. Willimsen, *J. Phys. A: Math. Gen.* **16**, 3365 (1983).
20. A. P. Sheppard and R. M. Sok, *Transport in Porus Media*, submitted.
21. J. Hoshen and R. Kopelman, *Phys. Rev.* **B14**, 3438 (1976).
22. P. J. Hicks, H. A. Deans, and K. R. Narayanan, SPE formation evaluation **293**, 235 (1992).
23. P. Spanne *et al.*, *Phys. Rev. Lett.* **73**, 2001 (1994).
24. J. Fredrich, B. Menendez, and T. F. Wong, *Science* **268**, 276 (1995).
25. W. Mahmud, unpublished.
26. H. H. Yuan and B. F. Swanson, SPE formation evaluation, 17 (1989).
27. M. Sahimi, *Flow and Transport in Porous Media and Fractured Rock*, 1st edition (VCH, Weinheim, Germany, 1995).
28. B. F. Swanson, *J. Petroleum Tech.* 10 (1979).
29. D. Stauffer and A. Aharony, *Introduction to Percolation Theory*, 2nd edition (Taylor and Francis, London, 1994).
30. M. Sahimi, *Applications of Percolation Theory*, 1st edition (Taylor and Francis, London, 1994).
31. M. Dias and D. Wilkinson, *J. Phys.* **A19**, 3131 (1986).
32. M. A. Knackstedt, M. Sahimi, and A. P. Sheppard, *Phys. Rev.* **E61**, 4920 (2000).

Annual Reviews of Computational Physics VIII (pp. 133–156)
Edited by Dietrich Stauffer

DIRECTED PERCOLATION, THE FIXED SCALE TRANSFORMATION AND THE RENORMALIZATION GROUP

AYŞE ERZAN

Department of Physics, Faculty of Sciences and Letters,
Istanbul Technical University, Maslak, Istanbul 80626,
Turkey Feza Gürsey Institute, P.K. 6 Çengelköy, Istanbul 81220, Turkey

The fixed scale transformation (FST) is a rather general method for computing a steady state, self-similar probability measure on an infinite set of configurations generated by evolution rules which typically incoporate both deterministic and stochastic aspects. A basic feature of our approach consists of projecting the state space onto a set of scale invariant product states. We can thus obtain our measure in terms of a finite number of distinct amplitudes which can be determined self-consistently from the evolution rules by means of a steady state condition. For non-Markovian processes lattice path sums enable us to cast the problem in Markovian form. We incorporate these features of the FST into a renormalization group calculation for the correlation length exponents of directed percolation.

1. Introduction

The fixed scale transformation $(\text{FST})^{1-3}$ is a conceptually new framework which enables us to characterize complex patterns arising in nonequilibrium phenomena. Nonequilibrium phenomena are distinguished, among other things, by their ability to display spontaneous pattern formation. Open systems in which a flux of energy (or matter) is maintained are known to exhibit complex self organization.[4] Moreover, given an appropriate set of conservation laws,[5] or "rigidity"[6,7] they have been shown to exhibit self-organized criticality.[8]

Pattern formation involves the spontaneous breaking of translational invariance in space or time, or both. The challenge presented by such diverse phenomena as diffusion-limited aggregation to chemical instabilities and reaction-diffusion systems to fully developed turbulence, is the definition of a measure on highly inhomogeneous, coherent structures arising from coexisting instabilities. Statistical descriptions such as afforded by path integral approaches are plagued by the fact that the resulting field theories can be nonlocal in space and/or time.[9–12]

This battery of problems has certain aspects in common with dynamical critical phenomena. Key ingredients in any theory of dynamical critical phenomena[13] are the identification of a natural set of basis states on which to project the fluctuations and the separation of static and dynamical properties. (Note that one can determine the scaling behavior of the spatial fluctuations in dynamical critical phenomena from the equilibrium critical properties.) Our approach will also involve these two basic ingredients. We will narrow down the scope of our investigation to seeking stationary probability distributions. The problem of nonlocality in space and time arises even for Markovian processes, when the phase space is projected on fewer-dimensional states.[13] This problem is solved in the present case by performing lattice path sums which enable us to obtain Markovian processes connecting asymptotic states.

I would like to illustrate the above notions by referring to so-called fractal growth problems, such as irreversible aggregation.[14–17] Strictly speaking, criticality is achieved here only in the limit of infinite growth, when the size of the system, and with it, the correlation length ξ go to infinity. If time is measured in the number of aggregated particles, the usual mass to radius scaling relation becomes

$$t \sim \xi^{d_F} , \tag{1}$$

where now d_F, the fractal dimension, plays the role of the dynamical critical exponent. It is only in the limit of $t \to \infty$, moreover, that the aggregate attains criticality, which we would like to subject to a statistical description. We would like to note here that as in many other nonequilibrium problems, the adoption of such boundary conditions as will lead to the establishment of a steady state will enable us to make a separation between the time-like and the space-like directions in the problem. (It turns out that this separation can always be made at least locally, in practice.) Then, a finite size scaling analysis in the space-like direction leads to $M \sim L^D$, where L is the size of the channel and $d_F = D + 1$.

The other key ingredient in our approach is the choice of a natural set of "basis states" for the description of broken translational invariance at all scales. Note that here one has to make a radical departure from mean field like approaches, where the lowest order description of the system starts from a homogenous state on which a uniform probability density is defined. Our state space consists of an ensemble of random Cantor sets, generated by the iterative use of a probabilistic fine graining rule. The problem then becomes one of defining a probability measure on this ensemble by determining, via the dynamics, the probability distribution of internal states conditional on a given cell state. In the present approach, we determine the asymptotic probability distribution of configurations in the space-like dimension, from the steady state condition which implies that the scaling behavior is invariant under translation in the time-like direction.

For nonlinear dissipative systems or fractal growth, no Gibbs measure parameterized by a set of interaction constants is available with which to weight the internal states. The obvious solution which suggests itself is to weight internal configurations by the probabilities of the paths leading to each configuration, and to simultaneously seek the scale invariant dynamics. The idea would be then to identify the scale invariant distribution with that yielded by the scale invariant probabilities along each step of the relevant paths. However, the fragile analogy between Boltzmann weights and, say, growth probabilities cannot be pushed very far. One of the problems is due to nonlocality. In the case of DLA and related problems, the paths leading to the asymptotically stable configurations are typically not confined to the cell whose internal states are under consideration, and taking only these into account leads to a greatly distorted description of the probability weights of the final states[18] *even if the scale invariant dynamics are known exactly.* An alternative approach[19] that focusses on the asymptotic shape of the resulting structure avoids this problem and also leads to the "correct" fixed point for the growth rule (at least for one particular way in which the growth rule can be perturbed, namely by the introduction of the so-called noise reduction parameter) but fails to provide a good estimate of the fractal dimension, apparently because the picture of an asymptotically stable envelope simply does not describe the true situation in a radial geometry.[20] The only direct connection between the mass scaling exponent and the dynamics is via the singularity spectrum of the growing front, i.e. $d_F = 1 + \alpha_{\min}$.[21–24] Recently, a very exciting new renormalization group idea has been introduced by Levitov and Hastings,[26] and taken up by Procaccia and co-workers,[27] where an iterated conformal mapping is used to "grow" the Laplacian cluster. This formalism is able to yield very good values for the generalized dimensions of the harmonic measure, as well as relationships between them.

It should also be remarked that the generally held view that fractal growth corresponds to an attractive fixed point of the RG flows is somewhat misleading. In seeking the scale invariant dynamics, one may perturb away from the given microscopic rules in many different ways and it has been found for many fractal growth problems that the fixed point to which the growth rules flow under, say, large numerical simulations depends, sometimes continuously, on the starting point[28] or some parameter.[26] In other words, the phase space for "relevant" directions is much richer than that encountered in second order phase transitions, and so is the structure of the critical manifold.

In Sec. 2, I will try to present a coherent if somewhat refractory account of what the fixed scale transformation (FST) approach is and what it is good for, in a language which is intended to facilitate comparison with field theoretic approaches to similar problems. What I would like to convey are what I think are the novel ideas which could perhaps be developed further in the future. I introduce the product states and their Fock space representation and outline the self-consistent determination of the relevant amplitudes. In Sec. 3, I will present an FST approach to the determination of the fractal dimension of the directed percolation cluster. In Sec. 4, I review a recent renormalization group approach, incorporating the FST, for the same problem.

2. The Fixed Scale Transformation Formalism

I will consider the problem of fractal growth in a channel with periodic boundary conditions in the transverse direction. To this class of objects one may include such "equilibrium" processes as percolation, Ising or Potts clusters and droplets (where each bond adjacent to the growing cluster is sampled only once and filled with a probability p; adjacent sites get added on if they are eventually touched by an unbroken chain of occupied bonds) as well as models of irreversible aggregation such as DLA,[14,15] (where each bond is sampled, in principle, infinitely many times, with the probability of occupation decreasing exponentially with the distance from the tip of the nearest branch and that from the nearest, etc.). In channel growth, this exponential decay of the growth probability (screening) leads, in practice, to the freezing of the growth in a given region after a finite time. Within the frozen region, then, we may examine the spatial correlations independently of time. To do this, we separate the forward and the transverse directions. (It is well known that many models lead to self-affine structures where the scaling behavior in these two directions are different from each other.) The forward direction we will call time-like, and the transverse direction, space-like.

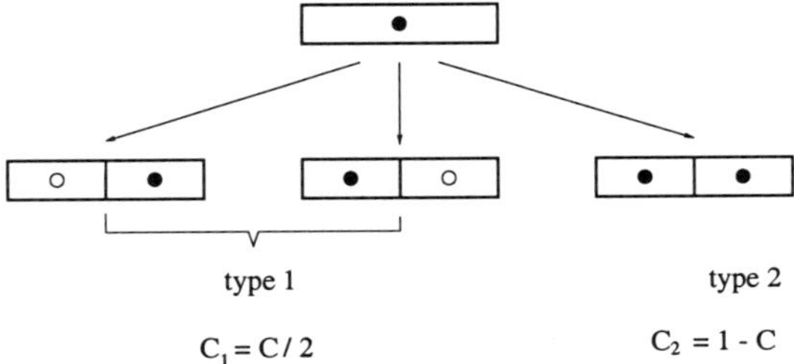

Fig. 1. The generator for the random Cantor set corresponding to a transverse subset of the incipient infinite cluster. Cells of type-1 and type-2 participate in the fine-graining process with corresponding probabilities C_1 and C_2. We represent the activated sites with black circle and inactive sites with white circle.

To model the long-range, self-similar correlations in the transverse direction, we project the system on an ensemble of random Cantor sets generated by the iteration of a stochastic rule. It is self-evident that any realization of the intersection of a fractal with a $d - 1$ dimensional manifold can be generated as a particular realization of such a set at any given length scale. In two dimensions, this scheme is particularly easy to envisage. The simplest iterative scheme leading to a random Cantor set can be constructed by fine graining each nonempty interval by a factor of 2. The internal states can be labeled according to the number of nonempty subintervals (1,2) with the probabilities C_1, C_2, at any given level of the hierarchy (see Fig. 1). Taking the C_i to be independent of the level in the hierarchy gives rise to a scale invariant set. A probability measure can be associated with this ensemble by weighting each random Cantor set with the product over the probabilities of the internal states of each cell at each level. The resulting probability trees are complex, with nonuniform branching ratios.

Clearly, there is an approximation made here in assuming that the internal states are distributed independently in each subinterval. Nevertheless, the product states constructed in this way capture an essential feature of our problem, namely the highly inhomogenous, clustering nature of the asymptotic configurations. This latter fact is crucial in that, together with the dynamics, we will use it to obtain self-consistent equations for the distribution of the internal states. This is the basic strategy of the fixed scale transformation.

I would like to introduce some formal considerations here, which I hope will facilitate a comparison between the present formalism and more standard field theoretic approaches, for those who may be motivated to do so. Let us adopt a Fock space representation for the transverse states. At any given length scale

138   A. Erzan

$\ell = 2^{-k}$, the microstates in the transverse direction can be represented by a set of occupation numbers $\underline{n}^{(k)} = (n_1, n_2, \ldots, n_r, \ldots)$, which can take on the values of 0 or 1. The "macroscopic state" of the system can be represented as

$$|\Phi(y)\rangle = \sum_{\underline{n}^{(k)}} \phi^{(k)}(\underline{n}^{(k)}; y)|\underline{n}^{(k)}\rangle \ , \tag{2}$$

where y indicates the time-like direction.

The scaling ansatz then becomes

$$\left\langle \sum_r n_r \right\rangle \sim \ell^{-D} \ . \tag{3}$$

A coarse-graining operation on the occupation numbers can be defined via

$$n_{\tilde{r}} = n_r + n_{r+1} - n_r n_{r+1} \tag{4}$$

for $\tilde{r} = (r+1)/2$ for odd r. The mass-scaling factor Z is then given by

$$Z \equiv \frac{\langle \sum_r n_r \rangle}{\langle \sum_{\tilde{r}} n_{\tilde{r}} \rangle} = 2^D \ . \tag{5}$$

The assumption that we may expand in the basis states obtained from the random Cantor set implies that the weight associated with any given state factorizes and one has

$$\phi^{(k+1)}(n_1^{(k+1)}, n_2^{(k+1)}, \ldots) = \phi^{(k)}(n_1^{(k)}, n_2^{(k)}, \ldots) \prod_{\tilde{r}} \psi(n_r^{(k+1)}, n_{r+1}^{(k+1)}|n_{\tilde{r}}^{(k)}), \tag{6}$$

where ψ is the weight associated with the internal states of the cell at $\tilde{r}$, conditional on the cell occupation number $n_{\tilde{r}}^{(k)}$. (We will drop the superscripts indicating the level of the hierarchy where the meaning is clear). Note that by construction,

$$\begin{aligned}
\psi(0, 0, |0) &= 1 \, , \\
\psi(1, 1|0) &= \psi(1, 0|0) = \psi(0, 1|0) = 0 \, , \\
\psi(1, 1|1) &= C_2 \, , \\
\psi(0, 1|1) &= \psi(1, 0|1) = C_1/2 \, .
\end{aligned} \tag{7}$$

The microstates $|\underline{n}^{(k)}\rangle$ may correspondingly be written as product states

$$|n_1^{(k)}, n_2^{(k)}, \ldots\rangle = \prod_{\otimes \tilde{r}} |n_r^{(k)}, n_{r+1}^{(k)}\rangle \, , \tag{8}$$

where we have introduced the two-site states $|n_r^{(k)}, n_{r+1}^{(k)}\rangle$.

Now, from Eq. (4) we have

$$n_r + n_{r+1} = n_{\tilde{r}}(n_r + n_{r+1}) \,. \tag{9}$$

Using the relation $p(A|B)p(B) = p(A, B)$ for conditional probabilities, we get $\langle AB \rangle = \sum_B B p(B) \langle A \rangle_B$, where the $\langle \rangle_B$ denotes an expectation value conditional upon the value of B. Therefore,

$$\left\langle \sum_r n_r \right\rangle = \sum_{\tilde{r}} \langle n_{\tilde{r}} \rangle \langle n_r + n_{r+1} \rangle_1 \,. \tag{10}$$

Finally, if we assume that the $\psi(n_r, n_{r+1}|n_{\tilde{r}})$ do not depend on $\tilde{r}$ except through $n_{\tilde{r}}$, we get

$$\left\langle \sum_r n_r \right\rangle = \langle n_1 + n_2 \rangle_1 \sum_{\tilde{r}} \langle n_{\tilde{r}} \rangle \,, \tag{11}$$

where $\langle n_1 + n_2 \rangle_1$ just denotes the average occupation number of a nonempty cell. From Eqs. (5) and (11) one finds that the scale factor Z is given by

$$Z = \langle n_1 + n_2 \rangle_1 = \sum_{n_1, n_2} \psi(n_1, n_2|1)(n_1 + n_2) \,. \tag{12}$$

I would like to underline the fact that Eq. (12) *is not a renormaliztion group equation*. It simply follows from the definition of the mass scaling exponent D under a coarse-graining by a factor of 2, and the expansion in terms of the basis states which obey a hierarchical structure. It is nothing but an identity until we find a way to determine the scale invariant probability distribution given by the C_i. As we have emphasized above, we do not have a renormalization group equation for the C_i in the general case. We therefore employ an alternative strategy to determine them, namely, translational invariance in the time-like direction in the steady state regime.

The requirement of invariance under translation in the time-like direction may be written $dZ/dy = 0$. To determine the evolution of $\psi(n_1, n_2|1)$ in the y direction, we would like to find $|\Phi\rangle$ such that

$$\langle \underline{m}|(U - 1)|\Phi(\underline{n}; y)\rangle = 0 \,, \tag{13}$$

with $\langle \underline{m}|\underline{n} \rangle = \prod_r \delta_{m_r, n_r}$ and U is the one-step evolution operator in the y direction. This equation in itself does not involve a scale change or a coarse-graining procedure, therefore we have come to call it the *fixed scale transformation*.

It should be noted that growth probabilities depend, in general, on correlations in both the space-like and time-like directions. Therefore the time evolution operator acts on the whole object embedded in d dimensions. The tranverse states are then to be projected out, once growth in that region has ceased. If T is the one-*time-step* operator acting on the whole cluster, we may write symbolically,

$$U_N \equiv P(y+1)[T^N - 1] , \tag{14}$$

where the $P(y)$ is the projection operator on to transverse state at y and the scale in the time-like direction has been appropriately chosen. Strictly speaking, $U = \lim_{N\to\infty} U_N$. However, screening ensures that within a finite region of size ℓ only a finite number of steps of the order of ℓ^d will suffice for the matrix elements to converge. In all the models in which we will be interested, the empty state is the absorbing state, and growth propagates via nearest neighbor links only. Thus, T^N may be constructed as a sum over lattice paths,

$$T^N = \prod_{\text{paths} \in S} \prod_{r,r' \in \text{path}} T_{r,r'} , \tag{15}$$

with

$$T_{r,r'} = p_r(\pi_r - 1)\pi_{r'}a_{r'}(s - \pi_r a_r) + 1 , \tag{16}$$

where π_r and a_r are the usual creation and annihilation operators at r,[10,11] and p_r, the growth probability, is in general itself an operator. Here S is the complete lattice neighboring the cluster. In practice, we have considered states at y to be frozen and restricted r to sites in S with $y' \geq y+1$. The last factor, with $s = 1$ would just inhibit the multiple occupation of a site; $s > 1$ acts as a noise reduction parameter. It is understood that $T^2 = \sum_{\underline{n}} T|\underline{n}\rangle\langle\underline{n}|T$.

Due to screening, the matrix elements in Eq. (13) depend on the state $|\underline{n}^{(k)}\rangle$ most strongly through the distribution of the gap sizes. Note that neighboring growth effectively prevents information about the far side of the chain from reaching a growth site, so that if there is growth in the immediate neighborhood, we can make the approximation of taking periodic boundary conditions in the computation of the growth probabilities. On the other hand, although the $\phi(\underline{n})$ are constructed purely from products over ψ, through the conditionality requirements they involve nontrivial correlations and yield detailed predictions as to the distribution of the gap sizes.[1,2,29] In particular, the probability $Pr(\lambda)$ that a nonempty cell at any level k of the hierarchy is neighbored, say on the right, by a gap of size $\lambda \sim 2^{-k'}$, $k' < k$ may be computed[29] in terms of the ψ in a straightforward manner.

These facts allow us to approximately factorize Eq. (13) and get a Markovian transformation over only a finite number of states, involving the $\psi(n_1, n_2)$ alone. We may expand[30,31] T^N in the following way:

$$T^N \simeq \prod_{\text{paths} \in S_{\tilde{r}}} \prod_{r,r' \in \text{path}} T_{r,r'} \equiv 1 + \sum_{\tilde{r}} u_{\tilde{r}} . \tag{17}$$

Here $S_{\tilde{r}}$ denotes the part of the lattice lying above the box situated at $\tilde{r}$. The order of the expansion, corresponds to the order of the computation. Consider now, at a generic scale k,

$$. \quad (T^N - 1)|\Phi(\underline{n}; y)\rangle \simeq \sum_{\underline{n}^{k-1},\underline{n}^k} \phi^{(k-1)}(n_1^{(k-1)}, n_2^{(k-1)}, \ldots) \sum_{\tilde{r}'} u_{\tilde{r}'}$$

$$\times \prod_{\otimes \tilde{r}} \psi(n_r^{(k)}, n_{r+1}^{(k)} | n_{\tilde{r}}^{(k-1)}) | n_r^{(k)}, n_{r+1}^{(k)} \rangle . \tag{18}$$

The sum over $\underline{n}^{(k)}$ is of course conditional on the $\underline{n}^{(k-1)}$. Let us introduce more compact notation be labeling the state of the 2-site boxes at $\tilde{r}$ by the total occupation number $j_{\tilde{r}} = n_r + n_{r+1}$ so that instead of $|n_r^{(k)}, n_{r+1}^{(k)}\rangle$ we may write $|j_{\tilde{r}}\rangle$, $j_{\tilde{r}} = 0, 1, 2$, and $\psi(n_r^{(k)}, n_{r+1}^{(k)} | n_{\tilde{r}}^{(k-1)}) \equiv \psi(j_{\tilde{r}})$. Then, with

$$u_{\tilde{r}}|j_{\tilde{r}}\rangle = \sum_{\mu_{\tilde{r}}} M_{\mu_{\tilde{r}}, j_{\tilde{r}}} |\mu_{\tilde{r}}\rangle , \tag{19}$$

we get

$$(T^N - 1)|\Phi(\underline{n}; y)\rangle \simeq \sum_{\underline{n}^{k-1}} \phi^{(k-1)}(n_1^{(k-1)}, n_2^{(k-1)}, \ldots)$$

$$\times \sum_{\{j_{\tilde{r}}\}} \sum_{\tilde{r}'}{}' \left(\prod_{\tilde{r} \neq \tilde{r}'} \psi(j_{\tilde{r}})|j_{\tilde{r}}\rangle \right) \psi(j_{\tilde{r}'}) \sum_{\mu_{\tilde{r}'}} M_{\mu_{\tilde{r}'}, j_{\tilde{r}'}} |\mu_{\tilde{r}'}\rangle \tag{20}$$

Taking a scalar product of this from the left with any one of the state vectors at $y' = y + 1$, namely one step removed *in the time-like direction*, then gives

$$\langle i_{\tilde{r}''}| \prod_{\otimes \tilde{r}''} P(y+1)[T^N - 1]|\Phi\rangle$$

$$\simeq \sum_{\tilde{r}'}{}' \phi(\underline{m}^{(k-1)}) \left(\prod_{\tilde{r} \neq \tilde{r}'} \psi(i_{\tilde{r}}) \right) \sum_{j_{\tilde{r}'}} \psi(j_{\tilde{r}'}) M_{i_{\tilde{r}'}, j_{\tilde{r}'}} . \tag{21}$$

Notice $u|00\rangle = 0$ by definition so that the sum over $\tilde{r}'$ now runs over the nonempty boxes only. In Eq. (7) we have already made the assumption of homogeneity for the ψ. Therefore in the final sum, we may drop the spatial dependence except for the fact that the matrix elements $M_{i_{\tilde{r}'},j}$ depend on $\tilde{r}'$ through the boundary conditions. From Eq. (13), together with

$$\langle \underline{m}^{(k)}|\Phi\rangle = \phi(\underline{m}^{(k-1)}) \prod_{\tilde{r}} \psi(i_{\tilde{r}}),\qquad (22)$$

we get, dropping the prime over $\tilde{r}$,

$$\sum_{\tilde{r}} \psi^{-1}(i_{\tilde{r}}) \sum_{j} \psi(j) M_{i_{\tilde{r}},j} = 1.\qquad (23)$$

Grouping the terms of the sum over $\tilde{r}$ into those for which $i_{\tilde{r}} = 1$ and $i_{\tilde{r}} = 2$, we may rewrite it as a sum over gap sizes λ. Thus,

$$\sum_{\lambda} \psi(1)^{-1} Pr(\lambda) \sum_{j} \psi(j) M_{1,j}(\lambda) + \sum_{\lambda} \psi(2)^{-1} Pr(\lambda) \sum_{j} \psi(j) M_{2,j}(\lambda)$$

$$= \psi(1) + \psi(2).\qquad (24)$$

Now, in each term, $Pr(\lambda)/\psi(i)$ is the probability of encountering a gap of size λ neighboring the box at $\tilde{r}$, irrespective of the internal state of the box. With the definitions in Eq. (7), a solution of this equation is

$$C_i = \sum_{j=1,2} \langle M_{ij}\rangle C_j,\qquad (25)$$

where the brackets indicate an average over the gap size distribution. With the normalization condition $C_1 + C_2 = 1$, this is a fully nonlinear equation that determines the C_i. In practice, we have come to approximate this distribution with a bimodal one, namely, $Pr(\lambda) \to P^{cl}\delta(\lambda) + P^{op}(1 - \delta(\lambda))$. One finds[2,29]

$$P^{cl} = \frac{2C_2}{3 - C_2}.\qquad (26)$$

Notice that Eq. (25) is now in Markovian form.

3. Directed Percolation

Directed percolation[32] may be treated as a dynamical problem in discrete time. It is then the simplest kind of "fractal growth" problem imaginable since the time-like direction corresponds in this case directly to the time direction.

Directed percolation has been studied extensively[32-38], since it plays very important role in a large variety of nonequilibrium systems with a single absorbing state.[39] A wide array of dynamic processes as fluid flow through a porous medium in an external field,[40] forest fires[41,42] or epidemic growth models,[43] reaction-diffusion systems,[44-46] damage spreading,[47] self-organized criticality,[48] models of growing surfaces with roughening transition,[49-51] etc. fall into the same universality class as directed percolation. There is no exact result, say from conformal invariance, for the critical exponents characterizing the dynamical phase transition separating the absorbing steady state from the active phase. The correlation length exponents in the longitudinal and transverse directions, defined via $\xi_\| \sim |p - p_c|^{-\nu_\|}$ and $\xi_\perp \sim |p - p_c|^{-\nu_\perp}$, and the percolation threshold p_c have been found to great accuracy by series expansion methods[33,34] to be $\nu_\| = 1.733$, $\nu_\perp = 1.097$ and $p_c = 0.6447$, respectively. The correlation length exponents in the transverse and forward directions and the susceptibility exponent are conjectured[33] to be given by $\nu_\perp = 79/72$, $\nu_\| = 26/15$, and $\gamma = 41/8$ and the order parameter exponent[33,34] by $\beta = 199/720$.

Given an initial state $|\underline{n}\rangle$ in $d-1$ dimensional space, where each site is either singly occupied or empty, it evolves according to the rules: (i) each occupied site remains occupied with probability p and annihilates with probability $1-p$, (ii) each occupied site may independently give rise to an occupied site at each of its $d-1$ neighbors in the "increasing" space directions, with probability p and fail to do so with probability $1-p$. If this site is already occupied at time t and remains occupied at the next stage (time $t+1$), contributions from neighbors will have no effect, (iii) vacant sites remain vacant with probability 1 unless they are filled by their neighbors. The empty sites are the absorbing states. Above the threshold there is a finite probability that the growth process is maintained indefinitely. Of this incipient infinite cluster, the subset I_t of vertices which fall on a constant t will have the fractal dimension $d_\perp$. In terms of critical exponents, $d_\perp = 1 - \beta/\nu_\perp$, with the numerical value from series results being $d_\perp = 0.748$.

The FST approach consists of requiring that as $t \to \infty$, the infinite transverse subsets I_t, $I_{t+\delta t}, \ldots$ are statistically similiar. In particular, they are modeled by generalized Cantor sets generated by a random sequence of fragmentations obeying a one-parameter, scale invariant distribution, as illustrated in Fig. (1). The fractal dimension depends on the scale-invariant probability C

for the creation of voids, via

$$d_\perp = \frac{\ln(2 - C)}{\ln 2}.$$

Note that this literally calls for a simultaneous sampling of all the directed bonds issuing from each point in the $d - 1$ dimensional lattice. Let us confine ourselves to $d = 2$. We thus have[30] for the one-time step evolution operator T,

$$T = \prod_r T_r T_{r+1} , \tag{27}$$

where T_r and T_{r+1} are the evolution operators acting along the bonds $(r, t) \to (r, t+1)$ and $(r, t) \to (r+1, t+1)$, respectively. From an inspection of the rules (i)–(iii),

$$T_r = [p + (1 - p)a_r]\pi_r a_r + (1 - \pi_r a_r) = (1 - p)(a_r - 1)\pi_r a_r + 1 , \tag{28}$$

and similarly,

$$T_{r+1} = p[\pi_{r+1} + (1 - \pi_{r+1})\pi_{r+1}a_{r+1} - 1]\pi_r a_r + 1 . \tag{29}$$

Notice we have taken care that the sites remain singly occupied, at the expense of introducing higher order terms. Conventional wisdom would tell us to expand the product in Eq. (27) up to the lowest order terms and then to obtain the evolution operator by applying the Trotter formula,[10] taking $\lim_{N \to \infty}(1 + tL_0/N)^N$, where now

$$L_0 = \sum_r (1 - p)(a_r - 1) + p[\pi_{r+1} - 1)(1 - \pi r + 1a_{r+1})]\pi_r a_r . \tag{30}$$

Instead, we will restrict ourselves to the two-site growth cells as in Eq. (17).

The operators corresponding to the "open" boundary conditions (gap to the right of the two-site box) are then

$$T_{\tilde{r}} = \prod_{r=1,2} T_r T_{r+1} . \tag{31}$$

For directed percolation, the only other contribution in a one-step process can come from an occupied nearest neighbor on the right. This can be taken care of by extending the product in Eq. (31) over $r = 1, 2, 3$, to consider

$$\sum_{n_3'} \langle n_1' n_2' n_3' | \left(\prod_{r=1,2,3} T_r T_{r+1} - 1 \right) \psi(n_0 n_1 n_2 | 1) | n_1 n_2 n_3 \rangle = 0 , \tag{32}$$

where reduced conditional probabilities for three-site states, $\psi(n_1 n_2 n_3|1)$ have been introduced. The primes on the bra indicate that it is to be taken at the $t + 1$st time step.

The hierarchical ansatz in Eq. (6) can be used to express the three-site reduced probabilities once more in terms of the two-site reduced probabilities. In particular, notice that the probabilitiy that the third site (nearest neigbor on the right) is occupied, conditional to the sites (1,2) being nonempty, is given by Eq. (26).

Given a hierarchical partitioning of the sets I_t into cells of size 2^{-k}, a moment's reflection will show that the probability of encountering a cell of *type 1* (or *type 2*) (see Fig. 1) among nonempty cells at a fixed arbitrary scale at time t must also be given by the probability C (or $1 - C$). The fixed scale transformation is given by

$$\begin{pmatrix} C' \\ 1 - C' \end{pmatrix} = \begin{pmatrix} M_{11} & M_{12} \\ M_{21} & M_{22} \end{pmatrix} \begin{pmatrix} C \\ 1 - C \end{pmatrix},$$

where C' is the relative frequency of cells of *type 1* at some time $t + n$. The matrix elements M_{ij}, with $\sum_j M_{ij} = 1$, are the transition probabilities that a given configuration of type i at time t "grows" into a configuration of type j at some later stage. Here n may be regarded as the order of the computation as will become clear below. The steady-state distribution is determined from the detailed balance condition,

$$C = \left[1 + \frac{M_{12}}{M_{21}} \right]^{-1}. \tag{33}$$

We now proceed to compute the M_{i2}. Since we are interested in the statistics of the incipient infinite cluster, the initial and final cells will be connected with probability one. Thus we first enumerate all (nonbranching) paths originating in the initial cell and terminating in the corresponding cell at $t + n$. Conditional to the occupied sites in the initial cell, one first chooses, with uniform probability, how the first bond of the backbone is to be placed. Each subsequent step is weighted by $1/v_k$ where v_k is the valency (in the increasing time direction) of the vertex encountered at the kth step. It should be noted that the resulting distribution is not uniform, due to the anisotropy of the embedding graph. The backbone leads to one occupied site in the cell at stage $t + n$. We now ask for the probability that the second site in this cell

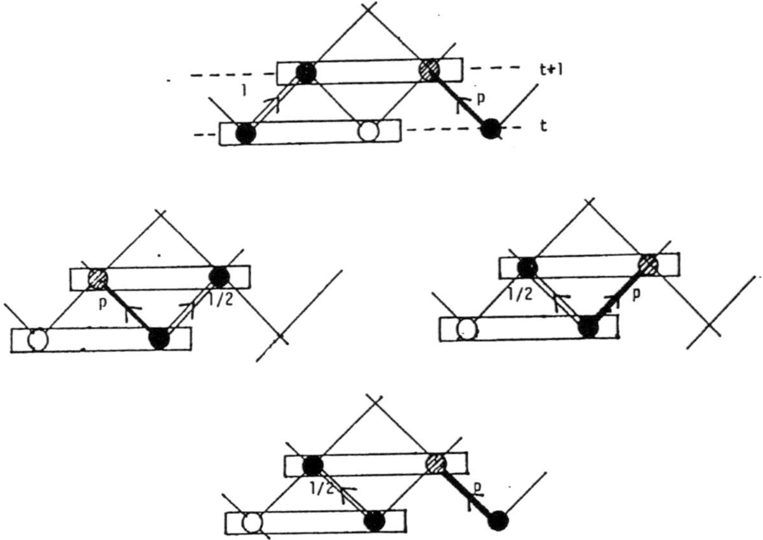

Fig. 2. Graphs, with their respective weights, for the matrix element M_{12} of the FST in the lowest order (see text). The backbone is shown as a double bond; the target site is shaded. The graph is further decorated by placing arrows on lines emanating from occupied vertices, each independently, with probability p (and blocking them with probability $(1-p)$). Vertices at which no bonds terminate are empty, and lines emanating from empty vertices are empty with unit probability.

also become occupied according to the growth rule stated above, by placing arrows on lines emanating from occupied vertices, each independently, with probability p, and leaving them empty with probability $(1 - p)$. We then sum over all backbone configurations, and in the case of an initial cell of *type 1*, average over the R, L orientations with equal weights.

First order computation. For $n = 1$, in the case of "open" boundary conditions (no neighbors to the right), the result is

$$M_{12}^{\mathrm{op}} = \frac{1}{2}p \,, \tag{34}$$

$$M_{22}^{\mathrm{op}} = \frac{4}{3}p - \frac{1}{3}p^2 \,. \tag{35}$$

The right neighbor to the initial cell is the only site, which, if occupied, can contribute to the state of occupation of the target cell (see Fig. (2)). The matrix elements conditional to an occupied right neighbor ("closed" b.c.) are

$$M_{12}^{\text{cl}} = \frac{5}{4}p - \frac{1}{4}p^2 , \tag{36}$$

$$M_{21}^{\text{cl}} = (1 - p)^2 . \tag{37}$$

We now use Eq. (26) for the relative weight of the "closed" configurations. Averaging the above set of expressions for the matrix elements with respect to boundary configurations and substituting in Eq. (33) yields a self-consistent equation for C, namely,

$$A_1 C^2 + B_2 C + D_1 = 0 , \tag{38}$$

where

$$A_1 = M_{12}^{\text{cl}} + M_{21}^{\text{cl}} - \frac{3(M_{12}^{\text{op}} + M_{21}^{\text{op}})}{2} , \tag{39}$$

$$B_1 = - \left(M_{12}^{\text{cl}} + 2M_{21}^{\text{cl}} - \frac{3}{2}M_{21}^{\text{op}} \right) , \tag{40}$$

$$D_1 = M_{21}^{\text{cl}} . \tag{41}$$

In terms of p this finally gives

$$C = \frac{2 - 3p + 5p^2 - \sqrt{36 - 108p + 109p^2 - 30p^3 + 9p^4}}{2p^2 + 4p - 4} , \tag{42}$$

up to first order in the FST approach. Substituting $p_c = 0.644071$ for p, we find to this order

$$C^{(I)} = 0.2283 , \tag{43}$$

$$d_\perp^{(I)} = 0.8251 . \tag{44}$$

Considering C as a function of p, however, we may use it to weight the different configurations in an RG cell. Thus we obtain a fixed point equation for p, which also allows us to determine the percolation probability p_c, as we shall see below.

Since in this case all graphs are directed in the time direction, the order of the computation can only be increased by taking further number of time steps. This means that to compute the matrix elements, we have to consider contributions from further neighbors to the right, i.e. basis states with a greater number of sites, and it seems as if one faces a closure problem. However, the hierarchical ansatz can again be used to compute the probabilities for all the respective configurations that may occur within the right neighborhood inside the "causality" cone, in terms of the two-site probabilities alone. The probablity to have the boundary configurations $\{11\}$, $\{10\}$, and $\{01\}$, where a "1" indicates occupied and a "0" empty, as right neighbors to a nonempty box can readily be derived to be $(1 - C_1)P^{\mathrm{cl}}$, $1/2C_1P^{\mathrm{cl}}$ and $1/2C_1P^{\mathrm{cl}}$ respectively. Details of the computation are given in Ref. 52. The fractal dimension found to third order in the computation is 1.747, to be compared with the best series results 1.748.[34]

4. Renormalization Group Approach to DP

Renormalization group ideas have been used in conjunction with the FST approach, in order to identify the scale invariant growth rules in terms of which the finite cell fixed scale transformation matrix should be computed, with good results. Forest fires[53] (which fall into the same universality class as directed percolation) and the sandpile model[54,55] have been studied by means of the "dynamically driven renormalization group",[56] which combines real space renormalization group ideas with a dynamical steady state condition reminiscent of the fixed scale transformation approach.

Here I would like to review a novel position space renormalization group (PSRG) treatment[61] of the directed percolation problem. This approach modifies conventional PSRG methods in two ways: (i) the weights of different initial states in the RG cell are computed from the steady-state distribution found from the fixed point of the FST; (ii) a "dynamical" coarse-graining procedure is defined which allows for the appearance of two different scale factors in the longitudinal (time-like) and transverse (space-like) directions, thus taking into account the self-affine nature of the problem. These scale factors are determined independently, without having to make any additional assumptions.

4.1. *A dynamical RG transformation*

We now outline a renormalization group procedure which takes into account fluctuations in regions larger than the renormalization group cell by using a steady-state distribution of initial conditions, obtained from the FST. It also introduces self-consistently determined rescaling lengths, rather than preset scale factors between the original and coarse-grained lattices, to allow for the self-affine nature of the problem.

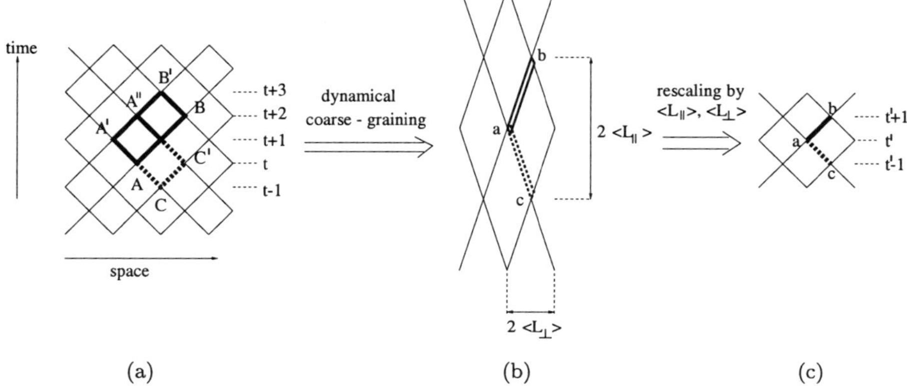

Fig. 3. The dynamical coarse-graining procedure. (a) The RG cells $AA'BB'$ and $CC'A'A''$ (the bold lines) in the original lattice are coarse-grained, (b) to the bonds ab and ac respectively, which are rescaled in the next step, (c) to preserve the lattice angles.

Our renormalization group cell is shown in Fig. 3(a). The boxes $AA'BB'$ and $CC'A'A''$ will coarse grain, under a dynamic coarse-graining procedure we describe below, to the bonds ab and ac, as shown in Fig. 3(b). This process conserves the transverse and longitudinal directions, but does not conserve the lattice angles, since these two directions are scaled differently. Rescaling by the appropriate longitudinal and transverse scale factors will restore the original lattice [Fig. 3(c)]. The lattice spacing is taken to be $\sqrt{2}$ for convenience, yielding $\ell_\perp = 1$ and $\ell_\parallel = 1$ for the transverse and perpendicular distances between the nearest neighbors, such as AA' and BB'. Clearly, we only have to consider one of the boxes, e.g., $AA'BB'$ for our renormalization procedure, by symmetry.

To obtain the renormalization transformation for the bond occupation probability p, we compute the total probability $P(p)$ that a spanning path exists across the box $AA'BB'$, by considering all the different initial configurations (states of A and A') and the spanning configurations that can be obtained from them. A spanning configuration is defined as one where a path starts from either A, or A' or both, and ends on B or B' or both.

4.2. *Steady-state distribution of initial conditions*

Since there are two possible origins which can be active both together or by themselves, four different initial configurations have to be considered as illustrated in Fig. 4. Any spanning configuration may have more than one path

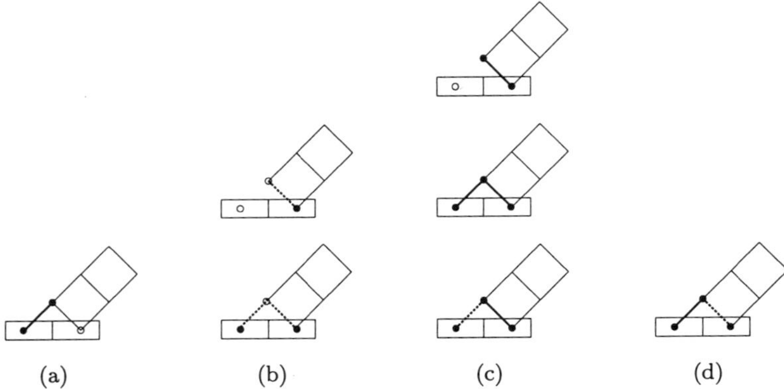

Fig. 4. Four different initial configurations are represented. In computing their respective weights (see text), the bold lines in the each configuration are replaced with the probability p and the dashed lines with $1 - p$.

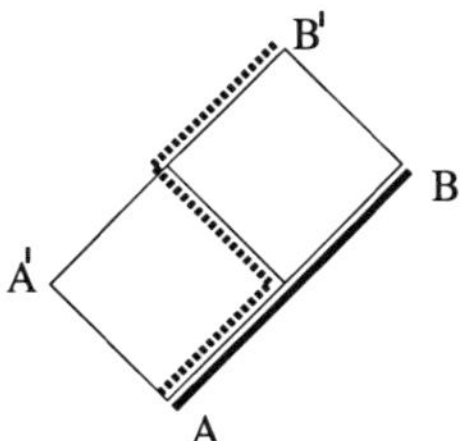

Fig. 5. Two different paths belonging to the same spanning configuration.

which connects the origins to the end points. For instance, two different paths are possible for connecting the sites A and B' in the spanning configuration shown in Fig. 5.

In order to compute the probabilities of finding each initial configuration, say W_i, $i = 1, \ldots, 4$, we make use of the FST to first order and directly use Eq. (42).

In terms of C and p, we have

$$W_1 = p\,C/2\,, \tag{45}$$

$$W_2 = (1 - p)\,C/2 + (1 - p)^2\,(1 - C)\,, \tag{46}$$

$$W_3 = p\,(1 - C/2)\,, \tag{47}$$

$$W_4 = p\,(1 - p)\,(1 - C)\,. \tag{48}$$

In the following subsection we proceed to obtain the renormalization transformation in terms of which the critical value of p can be determined.

4.3. *The fixed point equation for* p

Depending on the initial configurations $i = 1, \ldots, 4$, the total number of possible spanning configurations will be different. One sees that in Fig. 4(a) only one; 4(b) seven; and 4(c)–4(d) eighteen spanning configurations each are possible. It should also be noticed that some spanning configurations can be observed in more than one different initial configuration. For example, the spanning configuration which contains only one path which starts from A' and ends at B' can be observed in all the initial configurations except the second one (see Fig. 4). The total probability f_i of the spanning cluster for the i'th initial configuration is given by

$$f_1(p) = p^2 \,, \tag{49}$$

$$f_2(p) = p^2 + p^3 - p^4 \,, \tag{50}$$

$$f_3(p) = 2p^2 + p^3 - 3p^4 + p^5 \,, \tag{51}$$

$$f_4(p) = f_3(p) \,. \tag{52}$$

The renormalization group transformation for p is then found to be

$$P(p) = W_1 f_1(p) + W_2 f_2(p) + W_3 f_3(p) + W_4 f_4(p) = p' \,. \tag{53}$$

The fixed point of this transformation gives the threshold value, $p_c = 0.6443$ which is in agreement up to the third digit with the series expansion[33,34] result, namely 0.6447.

4.4. *The Affine transformation in the longitudinal and transverse directions*

The system has two independent correlation lengths $\xi_\parallel$ and $\xi_\perp$, parallel and perpendicular to the time direction, respectively, which diverge with different exponents as $\xi_\parallel \sim |p - p_c|^{-\nu_\parallel}$ and $\xi_\perp \sim |p - p_c|^{-\nu_\perp}$. To compute these exponents, we use the eigenvalue equations,

$$\left. \frac{dp'}{dp} \right|_{p=p_c} = b_\parallel^{1/\nu_\parallel} \,, \qquad \left. \frac{dp'}{dp} \right|_{p=p_c} = b_\perp^{1/\nu_\perp} \,. \tag{54}$$

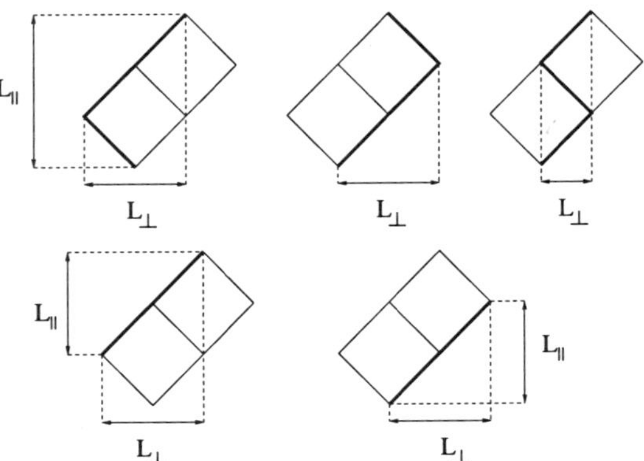

Fig. 6. All possible paths and their extremal projections contributing to the dynamical rescaling factors $\langle L_\parallel \rangle$ and $\langle L_\perp \rangle$. See Fig. 3.

We determine the appropriate rescaling lengths $b_\parallel$ and $b_\perp$ from

$$b_\parallel = \frac{\langle L_\parallel \rangle}{l_\parallel}, \qquad b_\perp = \frac{\langle L_\perp \rangle}{l_\perp}, \tag{55}$$

in terms of the average projected lengths, $\langle L_\parallel \rangle$ and $\langle L_\perp \rangle$, of the spanning paths onto the time and transverse directions as shown in Fig. 6. These quantities are the amounts by which the coarse-grained lattice [see Fig. 3(b)] has been dilated in the "dynamical coarse-graining" step. Since we have taken the lattice constant to be $\sqrt{2}$, the projections of a single bond on the original lattice onto the time and space directions are $\ell_\perp = \ell_\parallel = 1$.

In the time direction, one has to take into account the fact that under coarse-graining, different time steps collapse onto each other; that is why we consider each path originating from A or A' and terminating on B or B' as contributing seperately to $L_\parallel$. While enumerating the possible paths over which the average is taken, it makes a difference if there is a bond between A and A', since the paths change in case A' is activated by A. The same thing holds for the end sites B and B'. Note that the result of taking the extremal projections first and then averaging is different from finding the "average path" and then taking its projections. This point will be further discussed below.

For $X \equiv \{\perp, \parallel\}$, we have

$$\langle L_X \rangle = \sum_i W_i \sum_\beta q_{i,\beta} \sum_\alpha L_X^{i,\beta,\alpha}, \tag{56}$$

where $q_{i,\beta}$ is the relative probability of any spanning path β belonging to an initial state i, and is to be found from f_i by giving equal weights to the distinct paths in any spanning configuration equally. Substituting the value of p_c found from Eq. (10) into Eqs. (2) and (3), we find

$$\langle L_\perp \rangle = 1.7996 \tag{57}$$

and

$$\langle L_\| \rangle = 2.5561 \,. \tag{58}$$

These values yield, together with Eq. (11), the correlation length exponents to be $\nu_\| = 1.719$ and $\nu_\perp = 1.076$, which are comparable with the best known results[33,34] $\nu_\| = 1.733$ and $\nu_\perp = 1.097$.

4.5. *Discussion*

There have been some earlier studies of directed (therefore self-affine) systems via position space renormalization group (PSRG) techniques. Introduction of two different scaling factors $b_\|$ and $b_\perp$ due to existence of two independent scaling directions was firstly suggested by Dhar and Phani,[57] where they employed a decimation transformation. However, their results are far from the accepted values of the critical exponents and dependent upon their choice of the RG cell, which determines $b_\|/b_\perp$. Very large-cell PSRG calculations by Zhang and Yang[58] are able to accurately reproduce the self-affine behavior of directed self-avoiding walks, but are more appropriately in a class with Monte Carlo RG. A bond-moving and decimation transformation for anisotropic directed bond percolation in arbitrary dimension[59] and its generalization for other directed systems[60] by da Silva and Droz give the critical fugacity very accurately in all dimensions for directed self-avoiding-walks. For two dimensional directed percolation, it yields good results for the threshold value p_c and $\nu_\|$, but is not as good for $\nu_\perp$. The approach in these papers is in fact very close to ours; however, the way the projections of paths are taken to compute the scaling factors in the two different directions are different.

One may consider two different ways of projecting a path. The approach of da Silva and Droz[59,60] is to draw a vector between the origin and the end point, and take its two components to be the projections of the path. Our is to take the projection to be the transverse or longitudinal distance between the *extremal* points of the path, as shown in Fig. 6.

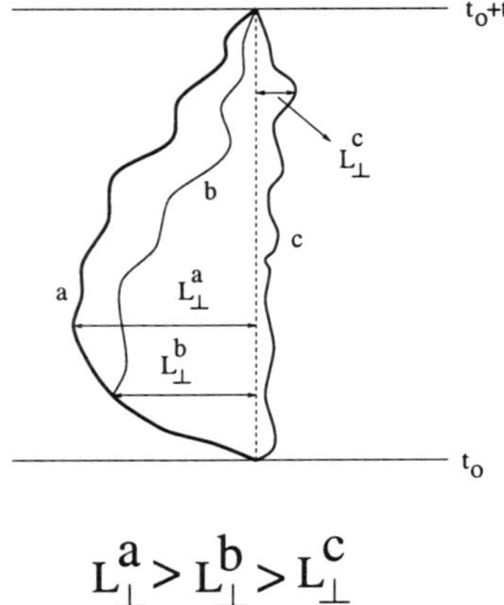

$$L_\perp^a > L_\perp^b > L_\perp^c$$

Fig. 7. Macroscopic paths with identical end-to-end vectors, but strongly differing extremal projections.

We believe that our method yields a result which corresponds more closely to what is meant by the size of the transverse and longitudinal fluctuations, in that it measures more accurately the actual size of the region over which a coherent flow takes place. This becomes more evident if we consider the incipient infinite cluster of our system for $p \sim p_c$. In this cluster, there can be many paths which connect some point at time t_0 to an end point at $t_0 + t$. Even though the total number of bonds in these different paths are the same, the spatial size of the spanned regions can be radically different (see Fig. 7). Since the transverse (longitudinal) correlation length corresponds to the spatial (time-like) size of the fluctuations, it is more appropriate to take the average over the extremal transverse (longitudinal) extent of each path.

In conclusion, by incorporating the FST fixed point condition in the determination of our distribution of initial configurations, and by a new, dynamical coarse-graining procedure, which makes use of averages over extremal projections of spanning paths, we have succeeded in computing the percolation threshold and the correlation function exponents much more accurately than before.

Acknowledgments

I would like to thank L. Pietronero, C. J. G. Evertsz, A. Vespignani and H. Kaya, with whom I have had the pleasure of collaborating on these matters. Support by the Turkish Academy of Sciences is gratefully acknowledged.

References

1. A. Erzan, L. Pietronero, and A. Vespignani, *Rev. Mod. Phys.* **67**, 545 (1995).
2. L. Pietronero, A. Erzan, and C. Evertsz, *Phys. Rev. Lett.* **61**, 861 (1988); *Physica* **A151**, 207 (1988).
3. A. Erzan, *Physica* **A185**, 66 (1992).
4. G. Nicholis and I. Prigogine, *Self-Organization in Nonequilibrium Systems* (Wiley-Interscience, New York, 1977).
5. T. Hwa and M. Kardar, *Phys. Rev. Lett.* **62**, 1813 (1989).
6. R. Cafiero, V. Loreto, L. Pietronero, A. Vespignani, and S. Zapperi, *Europhys. Lett.* **29**, 111 (1995).
7. A. Vespignani, S. Zapperi, and L. Pietronero, *Phys. Rev.* **E51**, 1711 (1995).
8. P. Bak, C. Tang, and K. Wiesenfeld, *Phys. Rev. Lett.* **59**, 381 (1988); *Phys. Rev.* **A38**, 364 (1988).
9. G. Parisi and Y.-C. Zhang, *J. Stat. Phys.* **41**, 1 (1985).
10. L. Peliti, *J. Physique* **46**, 1469 (1985).
11. C. DeDominicis and L. Peliti, *Phys. Rev. Lett.* **38**, 505 (1977).
12. C. De Dominicis and L. Peliti, *Phys. Rev.* **B18**, 353 (1978).
13. J. D. Gunton, "Mode coupling theory in relation to the dynamical renormalization group theory method," in *Dynamical Critical Phenomena and Related Topics*, ed. C. P. Enz, Lecture Notes in Physics Vol. 104 (Springer Verlag, Heidelberg, 1979).
14. T. A. Witten and L. M. Sander, *Phys. Rev. Lett.* **47**, 1400 (1981).
15. L. Niemeyer, L. Pietronero, and H. J. Wiesmann, *Phys. Rev. Lett.* **52**, 1038 (1984).
16. P. Meakin, *Phys. Rev. Lett.* **51**, 1119 (1983).
17. T. Vicsek, *Fractal Growth Phenomena* (World Scientific, Singapore, 1992).
18. H. Gould, F. Family, and H. E. Stanley, *Phys. Rev. Lett.* **50**, 686 (1983).
19. J. P. Eckmann, P. Meakin, I. Procaccia, and R. Zeitak, *Phys. Rev.* **A29**, 3185 (1989); *Phys. Rev. Lett.* **65**, 52 (1990).
20. B. Mandelbrot, A. Vespignani, and H. Kaufmann, *Europhysics Lett.* **32**, 199 (1995).
21. L. Turkevich and H. Scher, *Phys. Rev. Lett.* **55**, 1026 (1985); *Phys. Rev.* **A33**, 786 (1986).
22. T. C. Halsey, M. H. Jensen, L. P. Kadanoff, I. Procaccia, and B. Shraiman, *Phys. Rev.* **A33**, 1141 (1986).
23. T. Nagatani, *J. Phys.* **A20**, L381 (1987); *Phys. Rev.* **A36**, 5812 (1987).
24. X. R. Wang, Y. Shapir, and M. Rubinstein, *J. Phys.* **A22**, L207 (1989); *Phys. Rev.* **A39**, 5974 (1989).
25. T. C. Halsey and M. Leibig, *Phys. Rev.* **A46**, 7793 (1992).
26. M. B. Hastings, *Phys. Rev.* **E55**, 135 (1997); M. B. Hastings and L. S. Levitov, *Physica* **D1845**, 1 (1998).
27. B. Davidovitch, H. G. E. Hentschel, Z. Olami, I. Procaccia, L. M. Sander, and E. Somfai, *Phys. Rev.* **E59**, 1368 (1999).

28. R. De Angelis, L. Pietronero, A. Vespignani, and H. J. Wiesmann, *Europhys. Lett.* **16**, 417 (1991).
29. R. R. Tremblay and A. P. Siebesma, *Phys. Rev.* **A40**, 5377 (1989).
30. Y. Shapir and Y.-C. Zhang, *J. Physique Lett.* **46**, L-529 (1986).
31. J. L. Cardy and R. L. Sugar, *J. Phys.* **A13**, L423 (1980).
32. W. Kinzel, in *Percolation Structures and Processes* (Ann. Israel Phys. Soc. **5**), eds. G. Deutcher, R. Zallen, and J. Adler (Bristol, Adam Hilger, 1983), 425ff.
33. R. Baxter and A. J. Guttman, *J. Phys.* **A21**, 3193 (1988).
34. J. Essam, A. J. Guttmann, and K. De 'Bell, *J. Phys.* **A31**, 3815 (1988) and refs. therein.
35. D. Ben-Avraham, R. Bidaux, and L. S. Schulman, *Phys. Rev.* **A43** 7093 (1991).
36. H. Chaté and P. Manneville, *Phys. Rev.* **A38**, 4351 (1988).
37. F. Schloegel, *Z. Phys.* **253**, 147 (1972).
38. T. M. Liggett, *Interacting Particle Systems* (Springer Verlag, New York, 1985).
39. P. Grassberger, *Z. Phys.* **B47**, 365 (1982).
40. S. R. Broadbent and J. M. Hammersley, *Proc. Camb. Phil. Soc.* **53**, 629 (1957).
41. P. Bak, K. Chen, and C. Tang, *Phys. Rev. Lett.* **A147**, 297 (1990).
42. B. Drossel and F. Schwabl, *Phys. Rev. Lett.* **A69**, 1629 (1992).
43. J. L. Cardy, *J. Phys.* **A 16**, L709 (1983).
44. R. M. Ziff, E. Gulari, and Y. Barshad, *Phys. Rev. Lett.* **56**, 2553 (1986).
45. F. Schlogl, *Z. Phys.* **252**, 147 (1972).
46. G. Grinstein, Z. W. Lai, and D. A. Browne, *Phys. Rev.* **A40**, 4820 (1989).
47. P. Grassbeger, *J. Stat. Phys.* **79**, 13 (1995).
48. M. Paczuski, S. Maslov, and P. Bak, *Europhys. Lett.* **27**, 97 (1994).
49. L.-H. Tang and H. Leschhorn, *Phys. Rev.* **A45**, R8309 (1992).
50. H. Leschhorn and L.-H. Tang, *Phys. Rev.* **E49**, 1238 (1994).
51. Z. Olami, I. Procaccia, and R. Zeitak, *Phys. Rev.* **E52**, 3402 (1995).
52. A. Erzan and L. Pietronero, *Europhys. Lett.* **20**, 595 (1992).
53. V. Loreto, L. Pietronero, A. Vespignani, and S. Zapperi, *Phys. Rev. Lett.* **75**, 465 (1995).
54. L. Pietronero, A. Vespignani, and S. Zapperi, *Phys. Rev. Lett.* **72**, 1690 (1994).
55. A. Vespignani, S. Zapperi, and L. Pietronero, *Phys. Rev.* **E51**, 1711 (1995).
56. A. Vespignani, S. Zapperi, and V. Loreto, *J. Stat. Phys.* **88**, 47 (1997).
57. M. K. Phani and D. Dhar, *J. Phys.* **C15**, 1561 (1982).
58. Z. Q. Zhang and Y. S. Yang, *J. Phys.* **A17**, 1267 (1984).
59. J. Kamphorst Leal da Silva and M. Droz, *J. Phys.* **C18**, 745 (1985).
60. J. Kamphorst Leal da Silva and M. Droz, *J. Phys.* **A20**, 1865 (1987).
61. H. Kaya and A. Erzan, *Physica* **A265**, 53 (1999).

Annual Reviews of Computational Physics VIII (pp. 157–203)
Edited by Dietrich Stauffer

STATISTICAL MECHANICS AND SCALING THEORIES OF MACROMOLECULES

TANNIEMOLA B. LIVERPOOL*

Laboratoire de Physico-Chimie Théorique,
Ecole Superieure de Physique et de Chimie Industrielles,
10 rue Vauquelin, 75231 Cedex 05, Paris, France

We review the statistical mechanics and scaling properties of macromolecular (polymer) solutions. A pedagogic introduction to the modern theory of polymer solutions is given focusing on the link to experimental results. A short review of the analytic and numerical techniques used to study these systems is presented. Whilst the classical approach to polymer physics has been to focus on universal properties, recent work has tended to look at specific properties and how they change the relatively well understood universal properties. We describe in detail some new work on two facets of this nonuniversal behavior, namely the effect of electrostatics and rigidity on polymer conformations.

1. Introduction

Polymers are large complex *chain* molecules made up of "small", "simple" sub-units (*monomers*).[1-4] They come in a variety of shapes and sizes. *Linear polymers* which are the most commonly studied and produced are essentially a single open strand of monomers, but there has also been a lot of research on the properties of polymers with different back-bone structures such as branched polymers, ring polymers. We shall be interested here only in linear polymers. Polymers for which all the monomers are the same we refer to as *homopolymers* and those with chemically different monomers as *hetero-polymers*. In our discussion of polymers they will always be in solution, that is, dissolved in a liquid (solvent). When in solution, the polymers can also be charged and are called *polyelectrolytes* (PEs). Applications of polymers in solution are widespread — they are essential in the chemical industry and *biopolymers* are crucial for

---

*Permanent address: Condensed Matter Theory, Blackett Laboratory, Imperial College of Science, Technology and Medicine, Prince Consort Road, London SW7 2BZ, U.K.

numerous biological processes at the cellular scale.[5] Examples of polymers are polymethylmethacrylate $[CH_2 - C(CH_3)COOCH_3]_N$, which is used to make plexiglass or polyethylene $[CH_2 - CH_2]_N$, which is what the plastic bags that one gets at the supermarket are made of. Polyelectrolyte gels have an important application as super-absorbers. DNA, RNA are *charged* polymers of nucleotides present in the cells of all living organisms as well as numerous polymeric cytoskeletal proteins such as actin or microtubules.

Naturally, to understand the properties of polymeric materials or the functioning of a complex intercellular process, we must understand how many polymers interacting with each other behave. Nonetheless we must begin first to understand how one single polymer behaves before we can start studying more complex scenarios.

There has been a lot of work on the properties of neutral (uncharged) polymers on which there is a general consensus in the scientific community and some rather good agreement between experiment and theory.[2,1] Scaling theories in polymer physics have been particularly successful.[2] A very useful concept for the understanding of the behavior of polymer solutions is *universality*: one finds that polymers with very different microscopic (chemical) structure all have the same quantitative *macroscopic* behavior, the only difference between chemically different polymers being different pre-factors in front of scaling functions. This is due to a separation of length-scales (the microscopic or chemical length-scale and the size of the random coil). For neutral polymers scaling theories and renormalization group calculations have worked hand in hand to produce some beautiful results.

Recently there has been renewed interest in the so-called *nonuniversal* aspects of polymer behavior in particular with respect to biological applications and the effect of structure on physical properties and hence function.[5] The effect of rigidity i.e. resistance to bending and twisting of polymers has been particularly well studied.

Another important aspect of polymer theory is the effect of electrostatics on the behavior of polymers in solutions as many useful industrial and biological polymers are charged. Charged polymers though long under study have not been able to produce the agreement between experiment and theory found for neutral polymers. There are a number of different but comparable lengths in the polyelectrolyte system in typical aqueous solutions. As a result it is difficult to find universal behavior and construct unambiguous scaling theories. Nonetheless prompted by experiment and the expanding capacity of modern computers, the study of polyelectrolytes has become fashionable again. Charge affects both the microscopic and macroscopic structure of polymers so that electrostatics can lead to *new* universality classes of behavior and *new* scaling theories.

Our plan is as follows: in the next section we review the properties of neutral polymers in solution focusing on universal properties and scaling theories. Several very good introductory texts[1-4] have been used as the basis for this introduction. In Sec. 3 we describe the statistical physics of semiflexible polymers. We discuss some aspects of polyelectrolyte behavior in Sec. 4 and to round up we make our conclusions in Sec. 5.

2. Neutral Flexible Polymers

A linear polymer in solution on a large enough coarse-grained scale can be pictured as essentially a random coil. All linear polymers, no matter what their microscopic structure if they are long enough, can be considered to be *flexible* polymers. At very low temperatures, $T \to 0$, the polymer would be a rigid rod but of course would not dissolve in the solvent. The coarse-gaining procedure required to consider the polymer as a flexible coil is shown in Fig. 1.

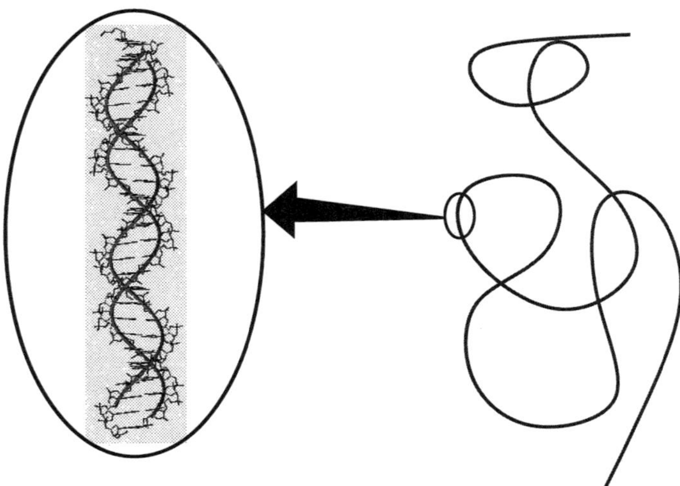

Fig. 1. DNA on microscopic and on a macroscopic scale. An example of a coarse-grained **polymer as** a random coil.

2.1. *Random walk models — ideal chains*

2.1.1. *Freely jointed chain*

As a start, we consider a polymer essentially as a random walk (RW) of bonds between monomers.[1] Each bond's orientation is totally uncorrelated with all the other bonds. The RW consists of N bonds of length b_0 joining "monomers," positions $\{\mathbf{R}_n\}$ with the bond vectors $\{\mathbf{r}_n \equiv \mathbf{R}_n - \mathbf{R}_{n-1}\}$. There are no interactions other than those between nearest-neighbor bonds. In Sec. 3 we will associate b_0 with the *persistence length* of the polymer. We ignore interactions between the polymer segments for now. We generalize to d dimensions not forgetting that we live in $d = 3$.

We can define a normalized bond distribution function

$$\psi(\mathbf{r}) = \frac{1}{\Omega_d b_0^{d-1}} \delta^d(|\mathbf{r}| - b_0)\,, \tag{1}$$

where $\Omega_d = 2\pi^{d/2}/\Gamma(d/2)$ is the surface area of a unit radius d-dimensional hyper-sphere and from that the polymer distribution function

$$\Psi[\{\mathbf{r}_n\}] = \prod_{n=1}^{N} \psi(\mathbf{r}_n)\,, \tag{2}$$

from which we can calculate averages $\langle F[\{\mathbf{r}_n\}] \rangle = \int \prod_{n=1}^{N} d^d r_n F[\{\mathbf{r}_n\}]\, \Psi[\{\mathbf{r}_n\}]$ such as that of the end-to-end distance $R = R_N - R_0$,

$$\langle R^2 \rangle = N b_0^2\,. \tag{3}$$

Similarly, the end-to-end distribution function, $P^{(2)}(\mathbf{R}, N)$, which is defined as $P^{(2)}(\mathbf{R}, N) = \langle \delta^d(\mathbf{R}_N - \mathbf{R}_0 - \mathbf{R}) \rangle$ is

$$P^{(2)}(\mathbf{R}, N) = \int \frac{d^d k}{(2\pi)^d} e^{i\mathbf{k}\cdot\mathbf{R}} \left(\frac{\Omega_{d-1}}{\Omega_d} \int_{-1}^{1} dx (1 - x^2)^{(d-3)/2} \exp\{ikb_0 x\} \right)^N . \tag{4}$$

2.2. *The Gaussian model*

If we consider the limit of a long chain $N \to \infty$, the integral in Eq. (4) will be dominated by the regime $k \to 0$, and after performing an expansion in k and rexponentiating we obtain

$$\lim_{N \to \infty} P^{(2)}(\mathbf{R}, N) \simeq \left(\frac{d}{2\pi N b_0^2} \right)^{d/2} \exp\left[\frac{-dR^2}{2N b_0^2} \right] . \tag{5}$$

This is an example of the *Central Limit Theorem* (CLT) which states that all distributions tend to become more and more Gaussian if samples are taken over larger and larger sets. As a result of the CLT, other random chain models with *only* short range interactions along the backbone all give the same distribution but with a different effective bond length.[1] We refer to all such chains as *ideal*.

This allows us to make a technical simplification; rather than use the freely jointed chain we work with a monomer distribution function which is already Gaussian (and technically easy to handle) which gives the same distribution as the freely jointed chain (and all other random chain models) in the long chain limit. We can consider a monomer distribution function

$$\psi_g(\mathbf{r}) = \left(\frac{d}{2\pi b_0^2}\right)^{d/2} \exp\left(-\frac{d\mathbf{r}^2}{2b_0^2}\right),$$

and as a result a polymer distribution function

$$\Psi_g[\{\mathbf{R}_n\}] = \left(\frac{d}{2\pi b_0^2}\right)^{dN/2} \exp\left[-\frac{d}{2b_0^2}\sum_{n=1}^{N}(\mathbf{R}_n - \mathbf{R}_{n-1})^2\right].$$

We can take the limit of continuous n

$$\sum_n \to \int dn \Rightarrow \Psi_g[\{\mathbf{R}(n)\}] \propto \exp\left[-\frac{3}{2b^2}\int_0^N dn \left(\frac{\partial \mathbf{R}(n)}{\partial n}\right)^2\right].$$

The statistical mechanics of a Gaussian polymer is described by a partition function Z and free energy F,

$$\exp\{-\beta F\} = Z = \int \mathcal{D}[\mathbf{R}(s)] \exp\{-\beta H_p[\mathbf{R}]\}, \tag{6}$$

where

$$\beta H_p = \frac{d}{2b_0^2}\int_0^N ds|\partial_s \mathbf{R}(s)|^2, \tag{7}$$

and where $\partial_s A = \partial A/\partial s$ and $\beta = 1/k_B T$.

Another useful quantity is the average radius of gyration, R_g, defined by

$$R_g^2 \equiv \frac{1}{2N^2}\sum_{i,j=1}^{N} \langle(\mathbf{R}_i - \mathbf{R}_j)^2\rangle. \tag{8}$$

$$k = 4\pi\lambda^{-1}\sin\theta/2$$

Fig. 2. Scattering from a polymer solution.

For the Gaussian chain

$$R_g \sim \frac{1}{\sqrt{6}}bN^{1/2}. \tag{9}$$

2.3. *Experiments*

Scattering experiments (see **Fig. 2**) are a good way to measure the structure of the polymer chain. Using x-rays, neutrons, or visible light one can obtain information about the structure on different length scales.[2,6]

The scattering function is calculated,

$$g(\mathbf{k}) \equiv \frac{1}{N}\sum_{i,j=1}^{N} \langle \exp[i\mathbf{k}\cdot(\mathbf{R}_i - \mathbf{R}_j)]\rangle. \tag{10}$$

One of most measured quantities in dilute solution is the radius of gyration of single polymers. R_g is obtained from the small k region of $g(k)$; $g(\mathbf{k}) \simeq g(0)(1 - (k^2/3)R_g^2 + \cdots)$.

Using measurements of the viscosity of dilute solutions one can determine a "hydrodynamic radius", R_η,

$$\eta = \eta_s \left[1 + 2.5\frac{4\pi c}{3n}R_\eta^3 + \cdots\right] \qquad (c \to 0). \tag{11}$$

The diffusion constant, D of the polymers in solution is obtained from dynamic light scattering experiments and defines another estimate of the coil size R_D,

$$D = \frac{T}{6\pi\eta_s R_D}. \tag{12}$$

All these experiments measure typically $R \sim N^{0.57} \rightarrow N^{0.60}$ so we can conclude after comparison with Eqs. (3) and (9) that most polymers are not ideal chains. A possible reason is discussed in Sec. 2.4 below: repulsion of the monomers on the chain.

The osmotic pressure Π of polymer solutions is another often measured quantity. Naively one expects $\Pi \sim c$ where c is the concentration of polymers for low concentrations of polymer.[4,7] This is known as Van't Hoff's law (VH). Big deviations from VH are obtained for very low concentrations of long polymers.

2.4. *Excluded volume interaction*

If a chain is in solution because the monomers have a finite volume we expect that chain to be self-repelling. Therefore monomers far from each other along the chain will repel each other. Whilst short range interactions (along the chain backbone) do not change the scaling behavior of R_g, long-range ones do! The random coil will become *swollen*. As a result, $R_g^{(0)} \sim b\sqrt{N}$ will be changed to $R_g(N)$.

A simple argument due to Flory gives a very good estimate of this effect.[3] To model this one must include the interaction energy of each monomer with all the other monomers. If each monomer has a volume v_0, the free energy F can be written

$$F = E_I - T\Delta S = v_0\, N \left(\frac{N}{R_g^d} \right) + \frac{3k_B T}{2b^2 N}(R_g^2). \tag{13}$$

The free energy is minimized w.r.t. R_g to obtain

$$R_g = \left(\frac{v_0 d b^2}{3k_B T} \right)^{1/(d+2)} N^{3/(d+2)} \tag{14}$$

with the Flory exponent $\nu_F(d) = 3/(d+2)$, which is exact in $d = 1, 2, 4$ and very good in $d = 3$. Experiments obtain $\nu \simeq 0.57 - 0.60$ very close to the $\nu_F = 0.6$ for $d = 3$.

2.5. *Simulations*

Many problems in polymer physics are analytically intractable. Numerical simulations are very often a very good way to check if one has understood the underlying physics of the problem. There are essentially two types of simulation techniques widely used for studying the properties of flexible polymers.

2.5.1. *Monte-Carlo*

The technique of Monte Carlo simulation[8] is a very useful way of determining the equilibrium properties of the system. One can consider the ubiquitous Metropolis algorithm as an illustration.[9] The energy E_i of the system is calculated using the effective Hamiltonian. Using the accepted library of dynamical moves, an attempt is made to put the system in a new state and the new energy E_f of the system is calculated. If $\Delta E = E_f - E_i \leq 0$, then the system stays in the new state. If on the other hand, $\Delta E > 0$, a random number between 0 and 1 is generated; if the random number is greater than the Boltzmann factor, $p = \exp\left(-\Delta E/k_B T\right)$, where T is the temperature and k_B Boltzmann's constant, then the system stays in the new state. If the random number is less than the Boltzmann factor then the system goes back to the original state. We always respect the principle of detailed balance. By exploring the phase space this way, one eventually arrives at the equilibrium condition. This can be a very efficient way to find the equilibrium configurations of complex systems. There is usually no way to relate the parameters of the model directly to real parameters and so it is useful only in describing the qualitative behavior of the system. The advantage is that generally, one *can* find the equilibrium.[10]

2.5.2. *Molecular dynamics*

In molecular dynamics simulation, one implements the interaction between monomers using potentials.[11] The forces are given by the gradient of the potentials and by numerically integrating Newton's second law — the system *slowly* evolves in a *physical* way towards its equilibrium state. In this picture, the simulation results can be related to "real" measured parameters. The simulations can also be performed using *explicit* solvent molecules. The disadvantage of this approach is that it is computationally very expensive, requiring long execution times. With present day computers, one can normally only simulate nanoseconds making equilibrium configurations extremely difficult to calculate. For studying dynamical problems it is the best way to proceed. One numerically solves the classical solutions of Newton's equations for the "monomers" positions and velocities $\{\mathbf{R}_i, \mathbf{v}_i\}$ with a force on the monomers due to their potentials $U[\mathbf{R}_j]$ given by $\mathbf{f}_{ij} = \nabla_j U(|\mathbf{R}_i - \mathbf{R}_j|) = -\mathbf{f}_{ji}$. To model the chain connectivity a commonly used potential is a bead-spring model such as the finitely extensible nonlinear elastic (FENE) model

$$U_{bs}(|\mathbf{r}_i|) = \frac{a}{2}\left(\frac{R}{\sigma}\right)^2 \ln\left[1 - \left(\frac{r_i}{R}\right)^2\right], \qquad \sigma \equiv \text{monomer size}. \tag{15}$$

Between all the monomers the excluded volume interaction is modeled by a Lennard–Jones potential ($r_c = 2^{1/6}\sigma$).

$$U_{ex}(|\mathbf{R}_i - \mathbf{R}_j|) = 4\epsilon\left[\left(\frac{\sigma}{|\mathbf{R}_i - \mathbf{R}_j|}\right)^{12} - \left(\frac{\sigma}{|\mathbf{R}_i - \mathbf{R}_j|}\right)^6\right]. \tag{16}$$

To model a finite temperature, the system is coupled to a heat bath

$$m_i\frac{d^2\mathbf{R}_i}{dt^2} = \mathbf{f}_{ji} - \Gamma\mathbf{v}_i + \mathbf{W}_i(t), \tag{17}$$

where $\langle\mathbf{W}_i\rangle = 0$ and $\langle\mathbf{W}_i(t)\mathbf{W}_j(t')\rangle = 6\Gamma k_B T\delta(t - t')\delta_{ij}$.

2.6. *Solvent quality*

The size of a polymer in solution depends strongly on the solvent quality. In a *good* solvent the polymer tends be expand (due to the excluded volume interaction) whilst in a *bad* solvent the polymer shrinks into a compact structure. The effect of solvent quality may be easily described by the simple model below.[4]

We consider a lattice of coordination number z whose sites are occupied either with monomers or with solvent molecules of volume v_0. A schematic is drawn in Fig. 3. The interaction energy between monomers is given by $-\epsilon_{pp}$, monomer and solvent $-\epsilon_{ps}$, and solvent molecules $-\epsilon_{ss}$ where since they are due to van der Waals type interactions $\epsilon_{ij} \propto \alpha_i\alpha_j > 0$ where α_i is the electrical polarizability of species j.

The number of neighboring monomer pairs $N_{pp}^{(i)}$, solvent pairs $N_{ss}^{(i)}$, monomer–solvent pairs $N_{ps}^{(i)}$ respectively. From this we calculate the total interaction energy $E_i = -N_{pp}^{(i)}\epsilon_{pp} - N_{ss}^{(i)}\epsilon_{ss} - N_{ps}^{(i)}\epsilon_{ps}$. We can define the volume fraction $\phi = Nv_0/R_g^d$, which is the probability lattice site occupied.

The average number of pairs is given by $\langle N_{pp}^{(i)}\rangle \simeq zN\phi/2$, $\langle N_{ps}^{(i)}\rangle \simeq zN(1 - \phi)\langle N_{ss}^{(i)}\rangle \simeq N_{ss}^{(0)} - zN\phi/2 - zN(1 - \phi)$ and (*at a mean field level* where we have *ignored* the interconnectivity of the chain) hence the average interaction energy

$$\langle E_i\rangle \simeq -\frac{zv_0 N^2}{R_g^d}\Delta\epsilon; \qquad \Delta\epsilon = \frac{1}{2}(\epsilon_{pp} + \epsilon_{ss}) - \epsilon_{ps}. \tag{18}$$

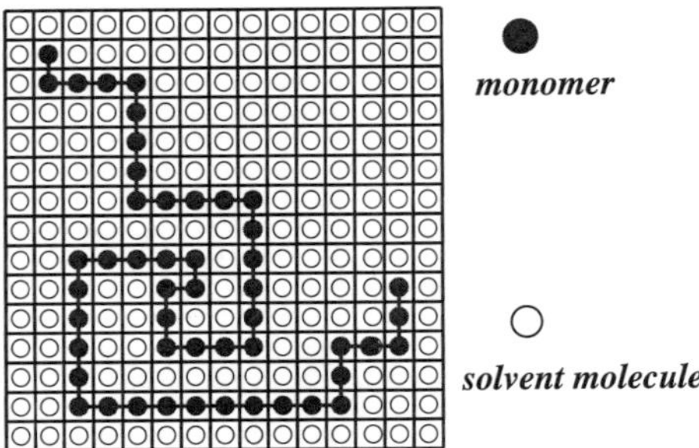

Fig. 3. The solvent–monomer lattice model.

The free energy is given by

$$F = v_0(1 - 2\chi)N \left(\frac{N}{R_g^d} \right) + \frac{3k_B T}{2b^2 N}(R_g^2) \tag{19}$$

with the χ parameter defined by

$$\chi = \frac{z\Delta\epsilon}{k_B T} . \tag{20}$$

By comparison to the free energy in Eq. (13) we can then identify a temperature Θ for which the chains are ideal, i.e. behave like Gaussian chains given by $\Theta = 2z\Delta\epsilon/k_B$. For $T > \Theta$ the chain is swollen or extended (coil) and is referred to as being in a *good solvent* whilst for $T < \Theta$ the chain is compact (globule) and is said to be in a *bad solvent*. At $T = \Theta$ it is evidently in an *ideal solvent*.

2.7. *Semidilute/Concentrated solutions*

We are often interested in systems with many interacting polymers rather than a single isolated chain. A useful concept is that of the *overlap concentration* c_* (see Fig. 4) which is the concentration at which the chains begin to interact with one another. It can be defined as

$$\frac{c^* R_g^d}{N} \sim 1 \Rightarrow c^* \sim b^{-d} N^{1-d\nu} . \tag{21}$$

Using the Flory exponent we obtain $c^* \sim N^{-4/5}$ so we find that long chain polymers $N \to \infty$ are almost always interacting and can hardly ever be considered isolated. One of the quantities most calculated is the osmotic pressure of a polymer solution.

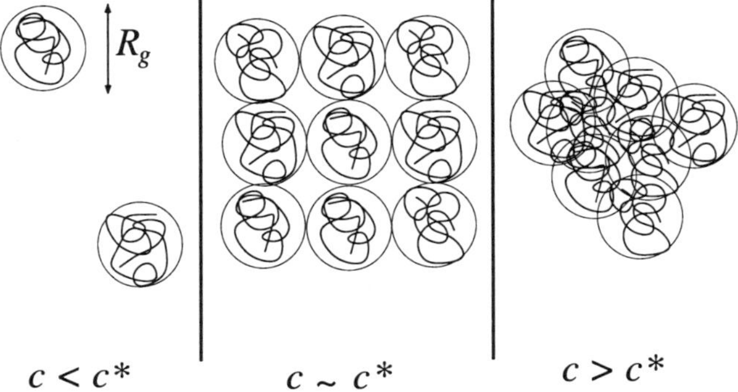

Fig. 4. The overlap concentration c^*.

To study the behavior of concentrated solutions of polymers the mean-field model of Flory and Huggins is used.[2-4] The derivation of the Flory–Huggins free energy is only sketched below but may be found in detail in several standard texts.[2,4]

2.7.1. *Lattice Model: Flory–Huggins*

We have n_p polymers with N "monomers" on a lattice of coordination number z (see Fig. 3). $\Omega \equiv$ total number of lattice sites, and hence the number of solvent sites $n_s = \Omega - n_p N$. We define a volume fraction $\phi = n_p N/\Omega$. The partition function is given by $Z = \sum_{\text{conf}} \exp[-\beta E_{\text{conf}}]$. Assuming small density fluctuations, we can replace E_{conf} by its average $\langle E_{\text{conf}}\rangle/(\Omega z) \simeq -(1/2)\phi^2 \epsilon_{pp} + \phi(1-\phi)\epsilon_{ps} + (1/2)(1-\phi)^2 \epsilon_{ss}$.

The mean field approximation *ignores* the connectivity of the chain and considers a smeared out uniform chain density. The partition function is then approximated by $Z \simeq W \exp[-\beta\langle E_{\text{conf}}\rangle]$ where W is the number of ways of arranging n_p chains. $W = 1/n_p! \prod_{j=1}^{n_p} w_j$, and w_i is the number of ways of arranging the ith chain, The first chain can be placed $w_1 = \Omega z(z-1)^{N-2}$ ways and subsequently the $j+1$th chain $w_{j+1} \simeq (\Omega - Nj)[(z-1)(1-(Nj/\Omega))]^{N-1} \simeq \Omega(z-1)^{N-1}(1-(Nj/\Omega))^N$ ways. The free energy is given by $F(\Omega, \phi) = -k_B T \ln W + \langle E_{\text{conf}}\rangle$ and we use the Fact that $\ln W = \sum_{j=1}^{n_p} \ln(w_j/j) \simeq \int_0^{n_p} dj(\ln w_j - \ln j)$.

The free energy of mixing F_m is calculated from the free energy F by subtracting the part that is due to just polymer or just solvent i.e. $F_m(\Omega, \phi) = F(\Omega, \phi) - F(\Omega\phi, 1) - F(\Omega(1-\phi), 0)$ giving the *Flory–Huggins* free energy[2-4] of mixing per site $f_m = \beta F_m / \Omega$,

$$f_m(\phi) \simeq \frac{\phi}{N} \ln \phi + (1 - \phi) \ln(1 - \phi) + \chi \phi (1 - \phi). \tag{22}$$

2.7.2. *Osmotic pressure*

The Gibbs free energy is given by $G(n_p, n_s, p, T) = F + pV = F + pv_0(n_p N + n_s)$ from which we can calculate the solvent chemical potential which is the change in the Gibbs free energy upon adding one solvent molecule to the system $\mu_s(\phi, p.T) \equiv G(n_p, n_s + 1, p, T) - G(n_p, n_s, p, T)$. Using the expressions for F, n_p, n_s we find

$$\mu_s(\phi, p, T) = \left(\frac{\partial F}{\partial \Omega}\right)_{\phi, T} \left(\frac{\partial \Omega}{\partial n_s}\right)_{n_p} + \left(\frac{\partial F}{\partial \phi}\right)_{\Omega, T} \left(\frac{\partial \phi}{\partial n_s}\right)_{n_p} + pv_0$$

and hence the osmotic pressure Π which is defined as the pressure required to keep the chemical potential constant (i.e. equilibrium) across a semipermeable membrane, $\mu_s(\phi, p + \Pi, T) = \mu_s(0, p, T)$ giving

$$\Pi = \frac{k_B T}{v_0} \left(\phi \frac{\partial f_m}{\partial \phi} - f_m\right). \tag{23}$$

Using Eqs. (22) and (23) and making an expansion for ϕ small, we obtain

$$\Pi = \frac{k_B T}{v_0} \left[\frac{\phi}{N} + \left(\frac{1}{2} - \chi\right) \phi^2 + \frac{\phi^3}{3} + \cdots\right], \tag{24}$$

showing big deviations from "ideal gas" (Van't Hoff's law) behavior even for low concentrations. VH is only true for very small volume fractions $\phi \ll 1/(1/2 - \chi)N$. This deviation had long puzzled polymer scientists but it is now accepted that the mean-field model above captures the essential physics. We see that at the Θ temperature the second virial coefficient of the osmotic pressure vanishes. This is often used in experiments to determine the Θ temperature.

2.8. *Scaling theories*

2.8.1. *Osmotic pressure*

The mean-field approach above explains qualitatively the big deviations from Van't Hoff's law at even very low monomer concentrations but experiments show deviation from the mean-field behavior in semidilute solutions, $\Pi(c) \sim c^\alpha; \alpha > 2$ showing that fluctuations and the chain connectivity which were ignored in the mean field approach are important. A dimensional argument due to des Cloizeaux[12] gives $\Pi(c, N) = ck_BTf(cb^3, N)$. Using c^* as a scaling variable we find $\Pi(c, N) = ck_BTf(\frac{c}{c^*})$ where $\lim_{x \to 0} f(x) = 1 + x + \cdots$. Now Π is independent of N for $c \gg c^*$ as many short chains entangled will have the same behavior as one long chain as long as the monomer density is the same.

$$\lim_{c \to \infty} f \sim \left(\frac{c}{c^*}\right)^{1/(d\nu-1)} , \tag{25}$$

with the Flory exponent $\nu_F(d = 3) = 3/5$, one obtains $\Pi \sim c^{9/4}$.

2.8.2. *Screening length (mesh size)*

In concentrated solutions excluded volume does not swell the chain because there is no free energy gain in being swollen because of all the other chains around. In short the other chains *screen* or reduce the self-repulsion of the chain. We can define a *screening length* above which the excluded volume interaction does not have an effect. The screening length of the concentrated solution (or mesh size) will be of the order of the mean separation of the chains (see Fig. 5) and can be estimated using a scaling argument.[2]

The correlation length ξ is given by $\xi = R_g g(c/c^*)$ where $R_g \sim N^\nu b$. As above it must be independent of chain length for $c > c^*$,

$$\xi \sim R_g \left(\frac{c}{c^*}\right)^{-\nu/(d\nu-1)} . \tag{26}$$

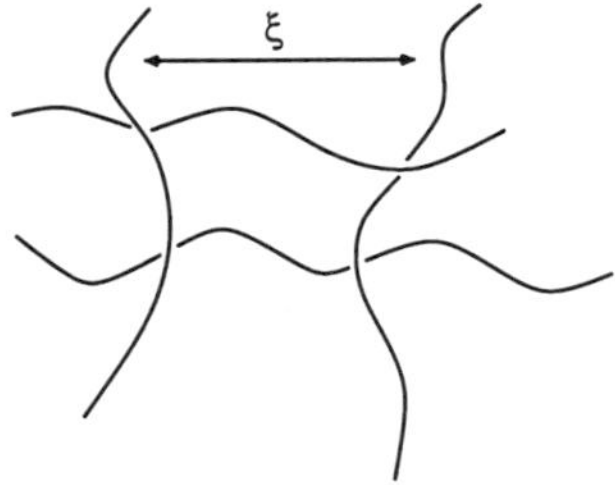

Fig. 5. The mesh size or screening length of polymer solution. We show just the mean path of the polymers whose conformations will be much more convoluted paths around their mean.

Fig. 6. The blobs of a semidilute solution of polymers.

Using the flory exponent $\xi_F \sim c^{-3/4}$. It is interesting to see that the osmotic pressure is then given by

$$\Pi \simeq \frac{k_B T}{\xi^3} . \tag{27}$$

2.8.3. *Blobs*

The idea of the *blob* introduced by Pincus[13] is a very useful concept for understanding the physics of polymer solutions (see Fig. 6). The polymer chain is in an expanded coil (excluded volume or good solvent statistics) until it interacts with other chains after which the excluded volume interaction becomes screened. The blob is the unit of chain which is swollen. We define $g \equiv$ number of monomers per blob and $\xi \equiv$ blob size. The blob size is given by

$$\xi \simeq b g^\nu \Rightarrow g \simeq (\xi/b)^{1/\nu} . \tag{28}$$

The polymer makes a random walk of N/g blobs of size ξ

$$R_g^2 \simeq \frac{N}{g}\xi^2 \sim N b^2 c^{(1-2\nu)/(d\nu-1)} . \tag{29}$$

Using the Flory value for ν in $d = 3$, we obtain $R_g \sim b^{5/4} N^{1/2} c^{-1/8}$.

2.9. *Renormalization group calculations*

In the scaling theories above we used the ν exponent of the excluded volume chain to calculate the scaling behavior of a host of physical quantities. Whilst the Flory exponent is a good approximation to the behavior it is incorrect. A much better approximation to the exponent can be calculated using renormalization group techniques at the expense of *slightly* more involved calculations.

2.9.1. *The Edwards model*

The continuum model of neutral polymers in a good solvent was introduced by Edwards[14] and is the starting point of modern polymer theory with a partition function

$$Z = \mathrm{Tr}_{\mathbf{R}} e^{-\beta H_E[\mathbf{R}]} = \int [\mathcal{D}\mathbf{R}] \exp[-\beta H_E[\mathbf{R}(s)]] \tag{30}$$

where

$$\beta H_E[\mathbf{R}(s)] = \frac{3}{2b^2} \int_0^N ds|\partial_s\mathbf{R}|^2 + v_0 \int_0^N ds' \int_0^N ds\, \delta^d[\mathbf{R}(s) - \mathbf{R}(s')]. \tag{31}$$

The short-range repulsion between all the monomers is described by a δ-function potential. We can rescale the size all the monomers by a factor $1/\lambda$ so that $s \to \lambda s$; $\mathbf{R}(s) \to \lambda^\nu \mathbf{R}(s)$; $\delta^d[\mathbf{R}(s) - \mathbf{R}(s')] \to \lambda^{-\nu d}\delta^d[\mathbf{R}(s) - \mathbf{R}(s')]$. If the chain is Gaussian($\nu = 1/2$) then we find that the short-range interaction becomes "relevant" (i.e. scales with at least the same power of λ as the Gaussian part) for dimensions $d \leq 4$. We can also recover the Flory theory by demanding that both terms grow under rescaling in the same way.

2.9.2. *Mapping to spin model*

De Gennes pointed out that the critical properties of a spin model described also the scaling behavior of long chains.[15] The Laplace transform $G^{(2)}(\mathbf{R}, t) = \int_0^\infty dN \exp[-tN]P^{(2)}(\mathbf{R}, N)$ of the two point distribution function $P^{(2)}(\mathbf{R}, N) \equiv \langle \delta^d[\mathbf{R} - (\mathbf{R}(N) - \mathbf{R}(0))]\rangle$ is equal to the correlation function of a $n = 0$ component spin model

$$\tilde{G}^{(2)}(\mathbf{R}, \tau) = \langle S_1(\mathbf{R})S_1(\mathbf{0})\rangle = \int [\mathcal{D}\mathbf{S}]e^{-h[\mathbf{S}]}S_1(\mathbf{R})S_1(\mathbf{0}) \tag{32}$$

where

$$h[\mathbf{S}] = \frac{1}{2} \int d^d r[|\nabla\mathbf{S}|^2 + \tau\mathbf{S}^2 + u|\mathbf{S}^2|^2], \tag{33}$$

where $\mathbf{S}$ is an $n = 0$ component field. The scaling properties of the correlation function can be calculated (near $\tau = \tau_c$ the correlation length of the spin model diverges like $\xi \sim |\tau - \tau_c|^{\nu})$ and

$$\langle R^2 \rangle = \left. \frac{dP^{(2)}(\mathbf{k}, N)}{dk^2} \right|_{k=0} \sim b^2 N^{2\nu}, \tag{34}$$

where $P^{(2)}(\mathbf{k}, N) = \int d^d r e^{i\mathbf{k}\cdot\mathbf{R}} P^{(2)}(\mathbf{R}, N)$. The exponents are calculated using standard RG techniques[16] (which we will describe later in Sec. 4) and in $d = 3$, $\nu = 0.588$. Using this value of ν we can calculate better estimates of all the scaling quantities above. The most accurate values of the exponents have been calculated using this method.

2.9.3. *Direct RG*

This method was pioneered by des Cloizeaux[17] and here one performs directly a perturbation expansion of Eq. (30) in v_0 which is then reorganized using the RG to obtain the same exponent and scaling behavior.[6,16]

3. Semiflexible Polymers

If we look at a polymer on intermediate length-scales, the chain structure becomes important. One of the most simple structural properties that affects the macroscopic behavior of polymer solutions is the rigidity or stiffness of a polymer.

3.1. *Worm-like chain*

The simplest model is the worm-like chain (WLC) model of Kratky and Porod.[18] The polymer can be considered locally to behave like a rigid rod.[19] We have a potential energy which places a penalty on any local bending of the polymer. A continuum model can be written for a chain of length L.

$$Z_{wlc} = \sum_{\mathrm{conf}} \exp[-\beta H_{wlc}] = \int \mathcal{D}[\mathbf{R}] \exp\{-\beta H_{wlc}[\mathbf{t}(s)]\} \tag{35}$$

where

$$H_{wlc}[\mathbf{t}(s)] = \frac{\kappa}{2} \int_0^L ds \left(\frac{\partial \mathbf{t}}{\partial s}\right)^2 ; \qquad \ell_p = \frac{\kappa}{k_B T}, \tag{36}$$

with $\mathbf{t}(s)$ the unit tangent vector to the chain at point s (see Fig. 7).

$$\mathbf{t}(s) = \left(\frac{\partial \mathbf{R}}{\partial s}\right) ; \qquad |\mathbf{t}(s)|^2 = 1. \tag{37}$$

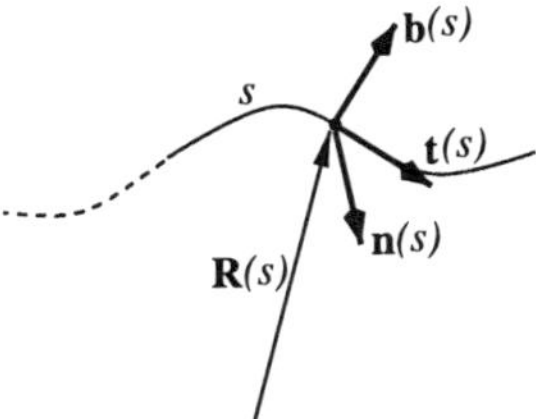

Fig. 7. The orthonormal trihedron of vectors describing the rod-like chain. For the worm-like chain we keep track of only the tangent vector $\mathbf{t}(s)$.

Note that from the Frenet–Seret formulae,[20] $|\partial_s \mathbf{t}|^2 = \mathcal{K}^2$ where $\mathcal{K}(s)$ is the "curvature" of the chain at point s. The potential uses the simplest quadratic form (by symmetry it must be at least quadratic) that places a penalty on bending.

We can write a restricted 2-point probability distribution function of the polymer having a tangent $\mathbf{T}$ at $s = L$ given a tangent vector $\mathbf{T_0}$ at $s = 0$,

$$P[\mathbf{T}, L; \mathbf{T_0}, 0] = \int [\mathcal{D}\mathbf{t}]_{\mathbf{t}^2=1} \delta(\mathbf{T} - \mathbf{t}(L)) \delta(\mathbf{T_0} - \mathbf{t}(0)) \exp[-\beta H_{wlc}], \quad (38)$$

and w.l.g. we can set $\mathbf{T_0} = (0, 0, 1)$. This corresponds to diffusion on a sphere (or quantum mechanics of a spinning sphere),

$$\mathbf{t}(s) = (\sin\theta(s)\cos\phi(s), \sin\theta(s)\sin\phi(s), \cos\phi(s))$$

$$\frac{\partial P}{\partial s} = \frac{1}{2\ell_p} \nabla_{\hat{\mathbf{t}}}^2 P,$$

where $\nabla_{\hat{\mathbf{t}}}$ is the gradient operator on a sphere. We can write the solution in terms of spherical harmonics

$$P(\theta, \phi, s) = \sum_{n,m} Y_{nm}(\theta, \phi) Y_{nm}(0, 0) \exp\left[-\frac{n(n+1)s}{2\ell_p}\right]. \quad (39)$$

This gives a tangent correlation function

$$\langle \mathbf{t}(s).\mathbf{t}(s') \rangle = \exp\{-s/\ell_p\} \quad (40)$$

and consequently the mean square end-to-end distance $\mathbf{R} = \mathbf{R}(L) - \mathbf{R}(0)$,

$$\langle R^2 \rangle = \int_0^L ds \int_0^L ds' \langle \mathbf{t}(s)\mathbf{t}(s') \rangle = \frac{\ell_p^2}{2} (\exp[-2L/\ell_p] - 1 + 2L/\ell_p) \quad (41)$$

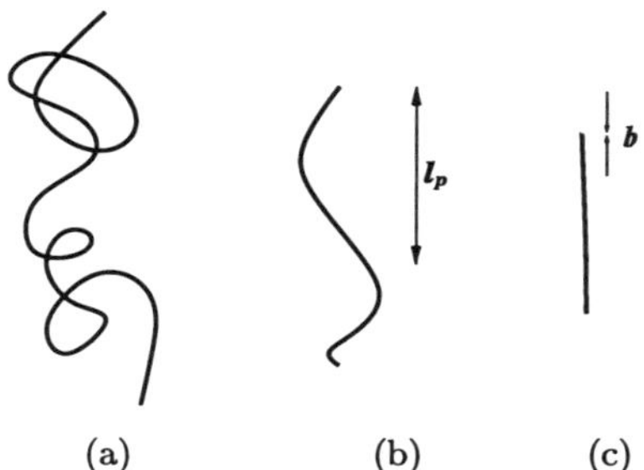

Fig. 8. The ratio of L to ℓ_p. (a) flexible: $L \ll \ell_p \ll b$; (b) semiflexible: $L \sim \ell_p \ll b$ and (c) Rigid $\ell_p \ll L \ll b$.

interpolates (illustrated in Fig. 8) between a rod $\langle R^2 \rangle \sim L^2$ and a "flexible coil" (RW) $\langle R^2 \rangle \sim \ell_p L$ with ℓ_p taking the role of "monomer size" in Sec. 2. We can identify ℓ_p as the persistence length, which is the length scale over which the WLC is rod-like (i.e. retains memory of its orientation).

3.2. *The Rod-like chain*

The rod-like chain (RLC)[21] is a generalization of the worm-like chain which takes into account twisting as well as bending energy. We define a local orthonormal trihedron $\{\hat{\mathbf{e}}_i(s)\} = \{\hat{\mathbf{b}}(s), \hat{\mathbf{n}}(s), \hat{\mathbf{t}}(s)\}$ (see Fig. 7). The conformation of the chain is given by evolution of $\{\hat{\mathbf{e}}_i(s)\}$,

$$\frac{d}{ds}\hat{\mathbf{e}}_i(s) = (\boldsymbol{\Omega}(s) + \omega_0 \hat{\mathbf{t}}(s)) \wedge \hat{\mathbf{e}}_i(s) \,, \tag{42}$$

where ω_0 is the spontaneous twist of the backbone. Using the Frenet–Seret formulae,[20] the antisymmetric tensor $\boldsymbol{\Omega} \overset{\rightarrow}{\wedge} \hat{\mathbf{e}}_i$ is given by

$$\boldsymbol{\Omega} \overset{\rightarrow}{\wedge} \hat{\mathbf{e}}_i \cdot \hat{\mathbf{e}}_j = \begin{pmatrix} 0 & -\mathcal{T}(s) & 0 \\ \mathcal{T}(s) & 0 & -\mathcal{K}(s) \\ 0 & \mathcal{K}(s) & 0 \end{pmatrix} \cdot \hat{\mathbf{e}}_j \tag{43}$$

where $\mathcal{T}, \mathcal{K}$ are the local Frenet–Seret torsion and curvature respectively.[20] We can define the components $\Omega_i = \boldsymbol{\Omega}(s) \cdot \hat{\mathbf{e}}_i(s)$.

The RLC hamiltonian[21] is given by

$$\beta H_{rlc} = \frac{\ell_p}{2} \int ds (\Omega_1^2(s) + \Omega_2^2(s)) + \frac{\ell_{tp}}{2} \int ds \Omega_3^2(s) \,, \tag{44}$$

where $\ell_{tp} = \beta \kappa_t$ is the twist persistence length.

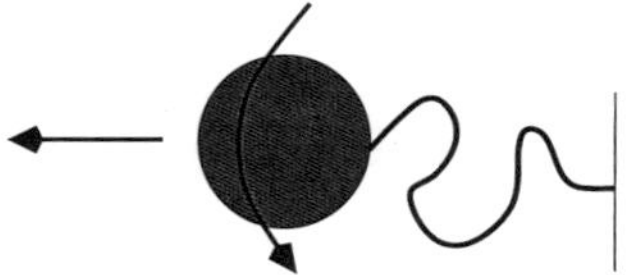

Fig. 9. Pulling and twisting DNA.

For twisted polymer loops or those with fixed ends there is a also a topological linking number constraint[21] $Lk = Tw + Wr$ given by

$$Tw = \frac{1}{2\pi} \int_0^L \Omega_3(s)\mathrm{d}s,$$

$$Wr = \int_0^L \mathrm{d}s \int_0^L \mathrm{d}s' (\hat{\mathbf{t}}(s) \wedge \hat{\mathbf{t}}(s')) \cdot \frac{(\mathbf{R}(s) - \mathbf{R}(s'))}{|\mathbf{R}(s) - \mathbf{R}(s')|^3}.$$

$$(45)$$

The statistical mechanics of the RLC polymer given by

$$Z = \sum_{\mathrm{conf}} \delta(Lk - C) \exp[-\beta H_{rlc}],$$
$$(46)$$

is equivalent to the quantum mechanics in imaginary time L of a symmetric top *with* the added complication that the twisting angle is not 2π periodic.[21] The RLC is used to analyze some data below.

3.3. *Recent experiments on DNA*

3.3.1. *Pulling*

A recent application of the worm-like chain model.

The worm-like chain (WLC) model has been recently very successful in explaining some recent experiments on pulling DNA by Smith *et al.*[22] Using a magnetic bead attached to a strand of double-stranded DNA, acted on by known magnetic and hydrodynamic forces the force-extension relation of the DNA molecules was obtained. The force extension curves which could not be explained by flexible polymer models as was attempted by Smith *et al.*[22] was beautifully explained by Marko and Siggia[23] and the "persistence" length of DNA extracted. The WLC under tension (being pulled) can be described by an effective Hamiltonian $\beta H = \beta H_{wlc} - fz$ where $z \equiv \mathbf{z} \cdot [\mathbf{R}(L) - \mathbf{R}(0)]$ is the end-to-end extension and the force f is Lagrange multiplier to fix z. The probability distribution satisfies an equation $\partial_s P = [1/(2\ell_p)\nabla_{\hat{t}}^2 + f\cos\theta]P$, the eigenstates of the rhs, Ψ_i satisfy $\partial_s \Psi_i = -g_i \Psi \Psi_i(L) = \exp[-g_i L]\Psi_i(0)$. The free energy of WLC is related to lowest eigen value g_0 and the extension computed from $z/L = -\partial g_0/\partial f$. They used 97004 bp DNA in 10 mM Na$^+$ salt solution to obtain $L = 32.8$ μm and $\ell_p = 53$ nm.[23]

3.3.2. *Twisting and pulling*

A magnetic bead attached to DNA was subject to calibrated forces and torques.[24] To measure the twist on a biopolymer, a concept used in the biophysical literature is the *fractional overtwist* $\sigma \equiv 2\pi\Delta Lk/L\omega_0$ where ΔLk is the number of turns applied, ω_0 is the rotation per unit length of the base pair axis $2\pi/\omega_0 = 3.6$ nm, and L is the length of the molecule. The relative extension versus fractional overtwist was measured in the experiments.

A RLC under a force f and Torque τ (torque is Lagrange multiplier to fix Lk) $\beta H = \beta H_{rlc} - fz + 2\pi\tau\Delta Lk$.

Using the analogy to a quantized symmetric top, the experimental data was analyzed by Bouchiat and Mezard[21] where the problem with the unquantized twist angle is taken care of by discretizing the model to obtain $\ell_t/\ell_p = 1.64 \pm 0.04$. The data was also analyzed by Moroz and Nelson[21] who avoided out the problem by looking at chains under strong forces and obtained $\ell_t = 109$ nm and $\ell_p = 49$ nm.

3.4. *Double-stranded semiflexible ribbon polymers*

Modeling of semiflexible polymers, particularly biopolymers, as elastic rods is not necessarily realistic for all scenarios. At large forces and torques new forms of DNA are found[24] obtained Z-DNA; Cluzel *et al.* obtained P-DNA.[24] Micrographs of actin and DNA show inhomogeneous structures on intermediate scales.[5] We now study a more microscopic model specifically looking for qualitative differences between the behavior of such molecules and simple worm-like chains.[18]

3.4.1. *Definitions*

To be precise we study the effect of a double-stranded structure[25,26] on semiflexible polymers. In our approach, we are able to consider polymers embedded in a d-dimensional space for arbitrary d. The system is composed of two semiflexible chains, each with rigidity κ, whose embeddings in d-dimensional space are defined by $\mathbf{r}_1(s)$ and $\mathbf{r}_2(s)$. The Hamiltonian of the system can be written

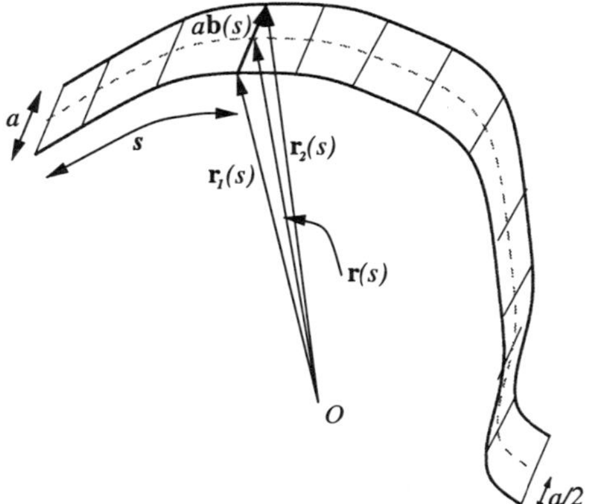

Fig. 10. The schematic of the double-stranded semiflexible polymer of two chains separated by a distance a. Note the bond director field $\mathbf{b}(s)$.

as the sum of the Hamiltonians of two worm-like chains

$$\mathcal{H} = \frac{\kappa}{2} \int \mathrm{d}s \left[\left(\frac{\mathrm{d}^2 \mathbf{r}_1(s)}{\mathrm{d}s^2} \right)^2 + \left(\frac{\mathrm{d}^2 \mathbf{r}_2(s)}{\mathrm{d}s^2} \right)^2 \right]. \tag{47}$$

We assume that the individual strands (that make up the double-stranded polymer) are inextensible: $(\mathrm{d}\mathbf{r}_1/\mathrm{d}s)^2 = (\mathrm{d}\mathbf{r}_2/\mathrm{d}s)^2 = 1$. The ribbon structure is then enforced by having $\mathbf{r}_2(s)$ separated from $\mathbf{r}_1(s)$ by a distance a, i.e. $\mathbf{r}_2(s) = \mathbf{r}_1(s) + a\mathbf{b}(s)$. We have defined a *bond-director field* $\mathbf{b}(s)$, which is a unit vector perpendicular to *both* strands (see Fig. 10). The chains are assumed to have permanent bonds (such as hydrogen bonds) that are strong enough to keep the distance between the two strands constant. (see Fig. 10).

We introduce the "mid-curve" $\mathbf{r}(s)$:

$$\mathbf{r}_1(s) = \mathbf{r}(s) + \frac{a}{2}\,\mathbf{b}\,,$$
$$\mathbf{r}_2(s) = \mathbf{r}(s) - \frac{a}{2}\,\mathbf{b}\,. \tag{48}$$

In terms of the tangent to the mid-curve $\mathbf{t} = \mathrm{d}\mathbf{r}/\mathrm{d}s$, which we call the *tangent-director field*, and the bond-director $\mathbf{b}$, the Hamiltonian of the system can now be written as

$$\mathcal{H} = \frac{\kappa}{2} \int \mathrm{d}s \left[2 \left(\frac{\mathrm{d}\mathbf{t}(s)}{\mathrm{d}s} \right)^2 + \frac{a^2}{2} \left(\frac{\mathrm{d}^2 \mathbf{b}(s)}{\mathrm{d}s^2} \right)^2 \right], \tag{49}$$

subject to the exact (local) constraints

$$\left(\mathbf{t} \pm \frac{a}{2}\frac{d\mathbf{b}}{ds}\right)^2 = 1, \qquad \mathbf{b}^2 = 1,$$
$$\left(\mathbf{t} \pm \frac{a}{2}\frac{d\mathbf{b}}{ds}\right) \cdot \mathbf{b} = 0, \tag{50}$$

This completes the formulation of the model.

3.4.2. *Mean-field approach*

It is well known that the statistical mechanics of semiflexible chains are difficult due to the constraint of inextensibility. Various approximation methods have been devised to tackle the problem. A successful scheme, that somehow manages to capture the crucial features of the problem, is to impose global (average) constraints rather than local (exact) ones.[27] This approximation is known to be good for calculating the average end-to-end length. It can be used for the probability distribution of the end-to-end length *only* if the persistence length is much less than the chain length so that the chain conformation can be considered isotropic. We can get good insight to the approximation scheme, by considering the fact that it corresponds to a saddle-point evaluation of the integrals over the Lagrange multipliers, which are introduced to implement the constraints.[27] In this sense, it is known to be a "mean-field" approximation in spirit. One can then go further by considering the effect of fluctuations on this mean-field result. The power of this approach is that one can easily calculate quantities which turn out to be very difficult if the constraints are required to hold exactly.[28]

With the above discussion as justification, we apply the same approximation scheme to our problem defined in Sec. 3.4.1: the local constraints in Eq. (50) are relaxed to global ones. This can be done by adding the corresponding "mass terms" to the Hamiltonian

$$\frac{\mathcal{H}_m}{k_B T} = \int ds \left[\frac{b}{\ell_p}\left(\mathbf{t} - \frac{a}{2}\frac{d\mathbf{b}}{ds}\right)^2 + \frac{b}{\ell_p}\left(\mathbf{t} + \frac{a}{2}\frac{d\mathbf{b}}{ds}\right)^2 + \frac{ca^2}{4\ell_p^3}\mathbf{b}^2 \right.$$
$$\left. + \frac{e}{\ell_p}\left(\mathbf{t} - \frac{a}{2}\frac{d\mathbf{b}}{ds}\right)\cdot\mathbf{b} + \frac{e}{\ell_p}\left(\mathbf{t} + \frac{a}{2}\frac{d\mathbf{b}}{ds}\right)\cdot\mathbf{b} \right], \tag{51}$$

where b, c, and e are dimensionless constants. The partition function is then given by

$$Z[\mathbf{J}, \mathbf{K}] = \int [\mathcal{D}\mathbf{t}]\,[\mathcal{D}\mathbf{b}]\,\exp\left\{-\frac{\mathcal{H} + \mathcal{H}_m}{k_B T} + \int ds\,[\mathbf{J}(s) \cdot \mathbf{t}(s) + \mathbf{K}(s) \cdot \mathbf{b}(s)]\right\}.$$
$$(52)$$

We next determine the constants self consistently by demanding the constraints of Eq. (50) to hold on average, where the thermal average is calculated by using the total Hamiltonian $\mathcal{H} + \mathcal{H}_m$. Note that in choosing the above form, we have implemented the "label symmetry" of the chains, namely, that there is no difference between two chains. It is convenient to take the limit of an infinitely long chain and perform the functional integrals in momentum space. Details can be found in Refs. 26 and 29.

The next step is to demand self consistently that

$$\langle \mathbf{b}(s)^2 \rangle = 1\,,$$

$$\left\langle \left(\mathbf{t}(s) \pm \frac{a}{2}\frac{d\mathbf{b}(s)}{ds}\right)^2 \right\rangle = 1\,,$$
$$(53)$$

$$\left\langle \left(\mathbf{t}(s) \pm \frac{a}{2}\frac{d\mathbf{b}(s)}{ds}\right) \cdot \mathbf{b}(s) \right\rangle = 0\,,$$

The self-consistency leads to the following set of equations for the constants b, c, and e:

$$\begin{cases} \dfrac{1}{4\sqrt{2b}} + \dfrac{a^2\sqrt{c}}{4d\ell_p^2} = \dfrac{1}{d}\,, \\[2ex] c(b + \sqrt{c}) = \dfrac{d^2\ell_p^4}{2a^4}\,, \\[2ex] e = 0\,. \end{cases}$$
$$(54)$$

The above equations, which are nonlinear and difficult to solve exactly, determine the behavior of b and c as a function of $u = a/\ell_p$. We have solved them numerically in $d = 3$ in Fig. 11. One can solve Eq. (54) analytically in two limiting cases. For $u \ll 1$ we find $b = d^2/32$, and $c = (d/\sqrt{2})^{4/3}u^{-8/3}$, whereas for $u \gg 1$ we find $b = d^2/8$, and $c = 4/u^4$. In Fig. 11, the behavior of b and c is plotted as a function of u. Note that u is proportional to T and can be viewed as a measure of temperature.

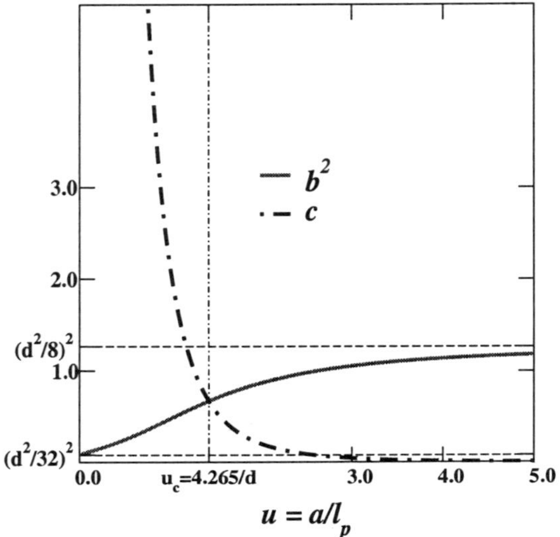

Fig. 11. The solution of the self-consistent equations for the constants b and c as a function of $u = a/\ell_p$ in $d = 3$. The value $u_c \simeq 4.27/d$ corresponds to the transition point.

We can then calculate the correlation functions. For the tangent–tangent correlation one obtains

$$\langle \mathbf{t}(s) \cdot \mathbf{t}(0) \rangle = \frac{d}{4\sqrt{2b}} \exp\left(-\sqrt{2b}\frac{s}{\ell_p} \right), \tag{55}$$

whereas for the bond-director field one obtains

$$\langle \mathbf{b}(s) \cdot \mathbf{b}(0) \rangle = \frac{d\ell_p^2}{2a^2\sqrt{b^2 - c}} \left[\frac{e^{-(b-\sqrt{b^2-c})^{1/2}\frac{s}{\ell_p}}}{(b - \sqrt{b^2 - c})^{1/2}} - \frac{e^{-(b+\sqrt{b^2-c})^{1/2}\frac{s}{\ell_p}}}{(b + \sqrt{b^2 - c})^{1/2}} \right]. \tag{56}$$

The tangent–tangent correlation [Eq. (55)] is exactly what we obtain for a single worm-like chain, and implies uniform behavior for all temperatures. The correlation function of the bond-director field has different behavior below and above the temperature $T_\times \simeq 4.27\kappa/dk_B a$. While it decays purely exponentially for $T > T_\times$, there are additional oscillatory modulations for $T < T_\times$. Equation (56) on the other hand, indicates a change of behavior at $b^2 = c$ for the bond-director correlation. The correlation is *overdamped* for $b^2 > c$ (high temperatures), while it is *underdamped* (oscillatory) for $b^2 < c$ (low temperatures). The interesting point $b^2 = c$ happens for $u_c = 16/(1 + \sqrt{2})^{3/2}d \simeq 4.27/d$, that leads to the value for $T_\times$ quoted above (see Fig. 11). We also find a divergence in the specific heat, $C_V = \partial^2 F/\partial T^2$ where $F = -k_B T \log Z$ at $T_\times$. It should be noted that it is not a thermodynamic phase transition in the sense of long-range ordering and broken symmetry. It is a crossover that appears due to competing effects, and the transition is from a state with some short-range order to a state with a different short-range order. Similar phenomena have been observed in Ising-like spin systems with competing interactions[30] and the crossover (transition) point corresponds to a type of *Lifshitz point* for a 1-d system. The related "bond-persistence length" ℓ_{BP} does not change appreciably at high temperatures, where its scale is again set by ℓ_p alone.

It is interesting to study the bond-director correlation in the limiting case $b^2 \ll c$, that corresponds to relatively low temperatures. In the low temperature phase, however, ℓ_{BP} does show a temperature dependence.

Using the asymptotic forms for b and c, one obtains

$$\langle \mathbf{b}(s) \cdot \mathbf{b}(0) \rangle = \sqrt{2} \exp\left[-\left(\frac{d}{4\ell_p a^2} \right)^{1/3} s \right] \times \sin\left[\left(\frac{d}{4\ell_p a^2} \right)^{1/3} s + \frac{\pi}{4} \right], \quad (57)$$

for very low temperatures. From the above expressions for the correlation functions, one can read off the persistence lengths ℓ_{TP} and ℓ_{BP}, and the pitch H. In particular we find, $\ell_{\mathrm{BP}} \sim \ell_p^{1/3} a^{2/3} \propto T^{-1/3}$ for $T \sim 0$, while $\ell_{\mathrm{BP}} \sim \ell_p$ for $T \sim T_\times$. Similarly, the "pitch" H defined as the period of oscillations in the low temperature regime, changes drastically with temperature, ranging from $H \sim \ell_{\mathrm{BP}} \sim \ell_p^{1/3} a^{2/3}$ near $T = 0$, to $H \sim 0$ near $T = T_\times$. At $T = 0$ we restore a flat ribbon which has true long-range order in both the tangent and bond-director fields. The ribbon is essentially a rigid rod. As we approach $T = 0$, the persistence lengths and the pitch diverge with the scaling $H \sim \ell_{\mathrm{BP}} \sim \ell_{\mathrm{TP}}^{1/3}$.

From the tangent–tangent correlation function, we can calculate the end-to-end distance. It yields

$$\langle (\mathbf{r}(s) - \mathbf{r}(0))^2 \rangle = \int_0^s \int_0^s ds_1 ds_2 \langle \mathbf{t}(s_1) \cdot \mathbf{t}(s_2) \rangle$$

$$= \frac{d\ell_p}{4b} \left[s - \frac{\ell_p}{\sqrt{2b}} (1 - e^{-\sqrt{2b}s/\ell_p}) \right], \quad (58)$$

which is similar to worm-like chains. It interpolates between the limiting behaviors of random walks ($\sim (d\ell_p/4b)s$) for $s \gg \ell_p$ to rods ($\sim (d/4\sqrt{2b})s^2$) for $s \ll \ell_p$. However, it is interesting to note that there is a shrinking in the length of the rod by a factor of $(d/4\sqrt{2b})^{1/2}$, which varies smoothly from 1 at $a \ll \ell_p$ to $1/\sqrt{2}$ at $a \gg \ell_p$. This implies that a polymer made up of two inextensible strands is always "slightly extensible" at any finite temperature, due to the presence of twist fluctuations.

3.4.3. *Simulations*

An intriguing feature of the behavior of this model is that, although the ground state $(T = 0)$ configuration of the system is a flat ribbon, and supports no twists, upon raising the temperature, a twisted structure with short-range *twist* order develops. Extensive molecular dynamics (MD)/Monte Carlo (MC) simulations of double-stranded semiflexible polymers have been performed.[29] A bead-spring model with bending and stretching energies was used. We combined a velocity verlet MD coupled to a heat bath with an off-lattice pivot MC algorithm. The MD was useful for equilibrating the shorter length-scales and MC for the long length-scales.

We used a *triangular* lattice to discretize the ribbon. The position of the ith bead is $\mathbf{r}_i$. The two chains making up the double strands join the odd ($\{1, 3, 5, \ldots, 799\}$) and even ($\{2, 4, 6, \ldots, 800\}$)beads together. The potential energy is given by

$$\frac{U[\{\mathbf{r}_i\}]}{k_B T} = \sum_{i=1}^{N-2} k_s[(\mathbf{r}_{i+1} - \mathbf{r}_i)^2 - \ell^2] + k_s[(\mathbf{r}_{i+2} - \mathbf{r}_i)^2 - \ell^2] - k_b \cos\theta_i , \quad (59)$$

where

$$\cos\theta_i = \frac{(\mathbf{r}_{i+2} - \mathbf{r}_i) \cdot (\mathbf{r}_i - \mathbf{r}_{i-2})}{|\mathbf{r}_{i+2} - \mathbf{r}_i||\mathbf{r}_i - \mathbf{r}_{i-2}|} .$$

We have a bending constant k_b *only* for the springs joining beads on the same chain and a stretching constant k_s for every spring. We also have a short-range repulsion between nearest-neighbor beads. The simulations were performed at $k_B T = 1$. We are in the dissipative regime so we can ignore the inertial term. The friction and noise are chosen so as to satisfy the fluctuation dissipation theorem. The equilibrium bond length was set to $\ell = 1.6$.

We performed in general 10^6 integration time steps followed by 10^4 attempted pivot moves. A pivot move is an attempt to rotate a portion of the chain by a small random angle around a randomly chosen bead. The MC part is done with the usual metropolis algorithm accepting pivot moves with a probability $\exp(-\Delta U/k_B T)$. This mixed MD/MC procedure was repeated 10^3

times until the configurations were equilibrated. Equilibration was checked by starting from crumpled chains and fully extended chains and verifying that the same values for radius of gyration and correlation functions was obtained. We simulated double-stranded ribbon chains of 2×400 monomers. The simulations were performed on a CRAY $T3D$ with 128 processors allowing us to simulate 128 chains in parallel.

3.4.4. *Kink-Rod structure*

We show typical equilibrated conformations above, near, and below $T_\times$ in Figs. 12, 13 and 14. The snapshots of the polymer configurations suggest that at low temperatures the polymer can be viewed as a collection of hard (straight) twisted rods that are connected by some kinks. This structure melts at higher temperatures. This picture can be accounted for using a simple argument. We can model our system of two semiflexible polymers subject to the constraint of constant separation, as a semiflexible ribbon, i.e. a semiflexible linear object

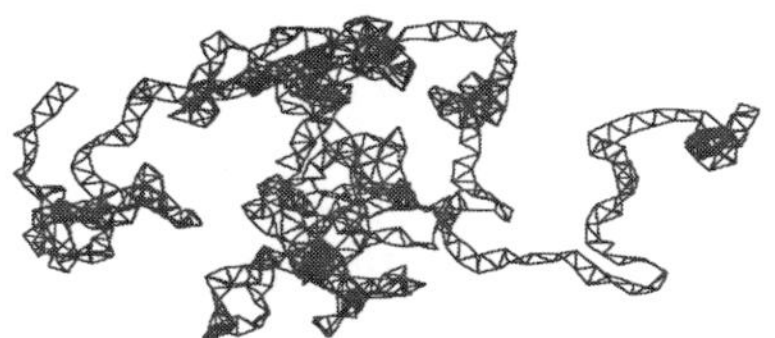

Fig. 12. Typical conformations from MD/MC simulations of a ribbon made up of two chains of 400 monomers above $T_\times$, $k_b = 1$.

Fig. 13. Typical conformations from MD/MC simulations of a ribbon made up of two chains of 400 monomers near $T_\times$, $k_b = 10$.

Fig. 14. Typical conformations from MD/MC simulations of a ribbon made up of two chains of 400 monomers below $T_\times$, $k_b = 100$.

with anisotropic rigidities, whose, Hamiltonian reads

$$\mathcal{H}_{\text{ani}} = \frac{1}{2} \int ds \sum_{i,j} \kappa_{ij} \left(\frac{dt}{ds}\right)_i \left(\frac{dt}{ds}\right)_j, \qquad (60)$$

where $\kappa_{ij} = \kappa_\parallel b_i b_j + \kappa_\perp (\delta_{ij} - t_i t_j - b_i b_j)$ determines the rigidity anisotropy of the ribbon, corresponding to bending parallel or perpendicular to the bond-director field. The ribbon structure would require $\kappa_\parallel \gg \kappa_\perp$. [To be consistent with the hard constraint (see above) of constant separation of the polymers, we should take the limit of infinite $\kappa_\parallel$.] The partition function of a semiflexible ribbon in the $\kappa_\parallel \to \infty$ limit can be written as

$$\mathcal{Z}_{\text{Rib}} = \int [\mathcal{D}t][\mathcal{D}b]\delta \left\{\frac{dt}{ds} \cdot b\right\} \exp \left\{-\frac{\kappa_\perp}{2k_B T} \int ds \left(\frac{dt}{ds}\right)^2 + \mathcal{H}'[t, b]\right\}, \quad (61)$$

in which $\mathcal{H}'[\mathbf{t}, \mathbf{b}]$ controls the dynamics of $\mathbf{b}$, and the functional delta-function enforces the constraint

$$\frac{dt(s)}{ds} \cdot \mathbf{b}(s) = 0, \qquad (62)$$

to hold exactly at *every* point of the ribbon. Recalling $dt/ds = \mathcal{K}(s)\mathbf{n}$ from the Frenet–Seret equations,[20] where $\mathcal{K}(s)$ is the curvature at each point and $\mathbf{n}$ is the unit normal vector to the curve, we can write the constraint as

$$\mathcal{K}(s)\,\mathbf{n}(s) \cdot \mathbf{b}(s) = 0. \qquad (63)$$

This constraint requires that at each point either $\mathcal{K}(s) = 0$, which corresponds to a straight (rod-like) segment that can be twisted, or $\mathbf{n}(s) \cdot \mathbf{b}(s) = 0$, which corresponds to a curved (kink-like) region where the bond-director is locked in to the perpendicular direction to the curve normal, i.e. the *binormal*.

We can argue that the (core) size of the kink regions is very small (of the order of the ribbon thickness) at low temperatures, as observed in Fig. 14. We note that the conformational entropy of the chain is due to the degrees of freedom in the kink regions, whereas the twist entropy comes from the degrees of freedom in the rod segments. The average separation between neighboring kinks is of the order of the persistence length. The ribbon thus tends to keep the rod segments as long as possible to maximally explore the twist degrees of freedom, while it can gain a same conformational entropy (compared to a worm-like chain) from the pivotal moves in the kink regions. This explains the kink-rod structure in low temperatures ($a \ll \ell_p$). As the temperature increases, the kinks get closer to each other, until at some temperature their average separation becomes comparable to their size ($\ell_p \sim a$), and the kink-rod pattern disappears.

This analysis can be understood in the context of the mean-field $(\mathbf{b}, \mathbf{t})$ model above. By observing that at low temperatures $\ell_{\mathrm{BP}} \ll \ell_{\mathrm{TP}}$, one can imagine that there are roughly speaking rod-like (straight) segments of length ℓ_{TP}, each supporting a number of shorter segments of length ℓ_{BP} that are twisted, but de-correlated with one other. One can see that ℓ_{BP} is equal to the length scale λ at which the strands undergo conformational fluctuations of the order of their separation: $a^2 = \langle r^2 \rangle \equiv \int_{1/\lambda} \mathrm{d}q/\ell_p q^4 \simeq \lambda^3/\ell_p$. Hence, segments of length ℓ_{BP} are straight (for $a \ll \ell_p$). However, fluctuations of order a are sufficient to wash out the memory of twist. The anti-correlation in fact comes from the frustration as explained above.

As the temperature is raised, the number of twisted rods in each segment $N = \ell_{\mathrm{TP}}/\ell_{\mathrm{BP}}$ decreases very quickly, until it saturates to unity at $T = T_\times$. For higher temperatures the mechanism changes, and the bond correlations are cut off by the tangent fluctuations. Hence, the short-range twist order does not survive anymore. All the main features of the above picture have also been observed in the simulation.

There are a number of advantages evident in our approach. First, we introduce a microscopic model which remains true to the chemical structure of many biomolecules. Second, our *approximate* method of solving this model also lends itself to the analysis of the fluctuations in the system and to study intermediate-scale behavior as well as the ground state (long length-scale) properties. Finally, this method could be easily be extended to describe multistranded objects.

4. Charged Polymers

The physical properties of charged polymers or polyelectrolytes (PEs) are rather poorly understood and the experimental and theoretical results that have been obtained are often rather controversial.[31,32] This is in contrast to neutral polymers where there is a broad consensus of basic results and some rather good agreement between experiment and theory. There are a number of reasons for this. First, and most fundamentally, there is the long-range Coulomb interaction which causes complications (for theory) which are absent when one deals with local interactions as in neutral polymers. For example, one has to be careful when taking the thermodynamic limit. Second, there are several competing length-scales. This leads to a certain ambiguity when one tries to study polyelectrolyte systems using scaling theories. This in contrast to the case of neutral polymers where a separation of length-scales leads to rather successful scaling theories for which one has good physical intuition. Third, because of the long-range interaction, boundary effects and impurities strongly affect the behavior of the system meaning that different experiments often give quite different results.

Upon dissolving a PE in a polar solvent, it dissociates into a charged macroion and small mobile counterions of opposite sign. If we have a single PE in an infinite system, for entropic reasons we expect the local density of counterions to be zero and so we may consider the PE to be just an isolated charged polymer. In reality there is always some dissociation of the solvent and there is always a finite density of ions in the vicinity of the chain leading to *electrostatic screening*. An unscreened PE is therefore an idealized system. Nonetheless, to better understand the effects of electrostatics, we now consider the statistical mechanics of an unscreened PE before addressing the more physical screened PE. In this and the following we focus on *weakly charged* PEs so that the electrostatic interaction does not affect the local structure of the chain. Strongly charged PEs have new and interesting phenomena such as charge condensation (see the article in this volume by Golestanian and Kardar[33]) which we do not discuss.

4.1. *Unscreened PEs and electrostatic blobs*

We consider an unscreened flexible PE in d dimensions dissolved in an ideal (Θ) solvent with N monomers of size a, a fraction f of which are charged, parametrized by $\mathbf{r}(s)$ where s measures the monomer number. We define a partition function, $Z[\{\mathbf{r}\}]$ is given by

$$Z = \int \mathcal{D}\mathbf{r}(s)\, \exp\{-\beta\mathcal{H}[\mathbf{r}(s)]\} \tag{64}$$

with the Hamiltonian, $\mathcal{H} = (\mathcal{H}_P + \mathcal{H}_C)$, made up of an elastic contribution $\mathcal{H}_P$ (Gaussian chain) and an electrostatic contribution $\mathcal{H}_C$,

$$\beta\mathcal{H}_P = \frac{d}{2a^2} \int_0^N ds \left|\frac{\partial\mathbf{r}}{\partial s}\right|^2 , \tag{65}$$

$$\beta\mathcal{H}_C = \frac{1}{2} f^2 \ell_B^{d-2} \int_0^N \int_0^N ds\, ds' \frac{1}{|\mathbf{r}(s) - \mathbf{r}(s')|^{d-2}} , \tag{66}$$

where $\beta = (k_B T)^{-1}$ and $\ell_B = (\beta e^2/\varepsilon)^{1/(d-2)}$ is the Bjerrum length which measures the strength of the bare Coulomb interaction. In water, $\ell_B = 7.14$ Å. As usual ε is the dielectric constant of the solvent, k_B is the Boltzmann constant and T the temperature. The Coulomb interaction $V_c(\mathbf{r})$ has been generalized to d-dimensions as the solution of $-\varepsilon\nabla^2 V_c = e\delta^d(\mathbf{r})$.

4.1.1. *Flory theory*

A simple Flory theory may be constructed to estimate the radius of gyration of a long chain ($N \to \infty$). Whilst the Flory theory may give wrong values for the exponents in high dimensions, it does show the right qualitative trends and allows one to elucidate the physics. Balancing electrostatic and elastic energies one obtains[32]

$$R_g \sim N^{3/d} f^{2/d} (\ell_B^{d-2} a^2)^{1/d} . \tag{67}$$

The PE is in an extended (rod-like) conformation in $d = 3$.

To get a better physical idea of what is going on it is useful to partition the chain into *electrostatic* "Pincus blobs". A blob may be defined as the chain sub-unit (of g monomers) for which the electrostatic interactions can be considered a weak perturbation. The size of the blob, ξ_e, is determined by a considering a chain section for which the electrostatic energy is of the order of $k_B T$, i.e. $(fg)^2 \ell_B/\xi_e \simeq 1$. Assuming the chain has Gaussian statistics on length-scales less than ξ_e, we also have $\xi_e^2 = ga^2$. We obtain

$$\xi_e \sim a \left[f^2 \left(\frac{\ell_B}{a}\right)^{d-2} \right]^{-1/(6-d)} , \qquad g \sim \left[f^2 \left(\frac{\ell_B}{a}\right)^{d-2} \right]^{-2/(6-d)} . \tag{68}$$

The result given by Eq. (67) may be re-expressed as

$$R_g \sim \left(\frac{N}{g}\right)^{3/d} \xi_e \,, \tag{69}$$

and the PE conformation can be considered to be an extended (rod-like) string of blobs in $d = 3$.

We can therefore already in the absence of screening observe three regimes, (i) if the number of monomers N is much larger than the blob monomer size g, then the PE can be considered to be in an extended (rod-like) regime, (ii) if $N \ll g$ then the PE would have a Gaussian conformation, and (iii) if $N \sim g$ the PE would be swollen in a nontrivial way.

We shall be concerned with the regime (i) and in particular how screening changes the behavior in that case.

4.1.2. *Renormalization group/Variational method*

The Flory theory is quantitatively incorrect as the scaling behavior of the radius of gyration of a single unscreened PE has been calculated exactly for $4 < d < 6$ using renormalization group methods[36,37] and variational techniques.[38] The radius of gyration is given by

$$R_g \sim \ell_B \, f^\nu N^\nu \tag{70}$$

with the exact exponent $\nu = 2/(d-2)$.[36-38] A scaling argument for calculating this exponent is to compare the electrostatic energy with $k_B T$.[39] In $d = 4$, $\nu = 1$ and as ν cannot be greater than 1 (rod), the unscreened PE is therefore considered to be rod-like for all dimensions, $d < 4$. We may consider the PE to be a chain of electrostatic blobs in a similar way to the Flory theory above.

$$R_g \sim \left(\frac{N}{g}\right)^{\nu} \xi_e \,. \tag{71}$$

4.1.3. *Electrostatic stretching*

Since the unscreened PE is rod-like for $d \leq 4$, it is interesting to ask if the electrostatic repulsions would not tear apart. As pointed out by Kantor and Kardar,[39] the stability of a Gaussian PE chain to electrostatic stretching for $d \leq 4$ may be estimated by considering a trial conformation $\mathbf{r}(s) = ms\hat{\mathbf{z}}$ in Eqs. (65) and (66) where $\hat{\mathbf{z}}$ is an arbitrary unit vector and m is a "stretching" factor. The free energy $F = -k_B T \ln Z$, may be minimized with respect to m, and one obtains

$$
\frac{\bar{m}}{a} = \begin{cases}
\left[\dfrac{(d-2)}{2d(d-3)} f^2 \left(\dfrac{\ell_B}{a} \right)^{d-2} \right]^{1/d}, & d > 3, \\[2ex]
\left[\dfrac{\ell_B}{6a} f^2 \ln N \right]^{1/3}, & d = 3, \\[2ex]
\left[\dfrac{(d-2)}{2d(3-d)(4-d)} f^2 (\ell_B/a)^{d-2} N^{3-d} \right]^{1/d}, & d < 3.
\end{cases}
\tag{72}
$$

$\bar{m}$ grows with N for $d \leq 3$, that is, in the limit of long chains $N \to \infty$, a Gaussian chain would be unstable to electrostatic repulsions[39] and a model of a chain with fixed length is required. Nonetheless, if the PE is weakly charged enough and not too long, the Gaussian model would be stable. For example in $d = 3$, $\bar{m}$ would be finite for chains for which $\ln N \leq a/\ell_B f^2$.

4.2. *The Debye–Hückel model of screened PEs*

Because of the large number of different types of interaction occurring in the system there is also some question about the most appropriate model to describe polyelectrolytes. This obviously depends on the problem being studied. As in any physical system there is a need for "simple" mean-field models which can be studied in detail so as to understand (capture) the main features of the system. The simple models can often have very good qualitative agreement with experiment. Once these simple models have been fully understood, one can begin to add details with which it becomes possible to compare the model quantitatively with realistic experimental systems. In our opinion, the Debye–Hückel polyelectrolyte is such a model. It is a mean-field model because it ignores important charge-fluctuation effects. Nevertheless its behavior is highly nontrivial. We study the Debye–Hückel polyelectrolyte in attempt to capture the scaling properties of PE solutions with *two* important and independent length-scales.[40]

4.2.1. *Flexible chains*

We focus here on the single chain behavior. Details of the many chain behavior can be found in Ref. 41. In solution the counterions in the vicinity of the macroion (polymer) "screen" or reduce the range of the Coulombic interaction.

The simplest model of screened flexible polyelectrolytes which we will use as the starting point for our analysis is the Gaussian model of flexible polymers with a Debye–Hückel (DH)/Yukawa potential interaction between the charged monomers of the polyion. We consider a charged polymer in a d-dimensional polar solution. The chain is described by a vector $\mathbf{r}$, parametrized by an arclength (monomer label) s and we define a partition function $Z[\{\mathbf{r}\}]$:

$$Z_{DH} = \int \mathcal{D}[\mathbf{r}] \exp\{-\beta\mathcal{H}_{DH}[\mathbf{r}_j(s)]\} \tag{73}$$

with the Hamiltonian given in d dimensions by

$$\beta\mathcal{H}_{DH} = \frac{d}{2a^2} \int_0^N ds \left(\frac{\partial \mathbf{r}}{\partial s}\right)^2 + \frac{\bar{u}}{2} \int_0^N \int_0^N dsds' V_{DH}(\mathbf{r}(s) - \mathbf{r}(s')), \tag{74}$$

where f is the fraction of monomers charged, $\bar{u} = \Omega_d f^2 \ell_B^{d-2}$, N is the number of Kuhn segments on the chain, and Ω_d is the surface area of a unit d-dimensional hypersphere. The polymer hamiltonian is that of a flexible chain signifying that we are looking at the chain on length-scales much larger than the persistence length. The potential between monomers, $V_{DH}(\mathbf{r})$ is the solution of the linearized Poisson–Boltzmann equation $(-\Delta + \kappa^2)V_{DH}(\mathbf{r}) = \delta^d(\mathbf{r})$ in d dimensions:

$$V_{DH}(\mathbf{r}) = \int \frac{d^dq}{(2\pi)^d} e^{i\mathbf{q}\cdot\mathbf{r}} \frac{1}{q^2 + \kappa^2}. \tag{75}$$

The Debye screening length (range of the interaction) is given by $1/\kappa$ and $\kappa^2 = \Omega_d \ell_B^{d-2} n_q$ is a function of the charge density of screening ions, n_q and the dielectric properties of the solution. Increasing the density of ions thus leads to a reduction of the screening length. The DH model is valid only when the polyelectrolyte is weakly charged and in the presence of salt (see next section). There are no large fluctuations in the counterion/salt density so at least for the Coulomb interaction the DH model may be considered to be *mean-field*. The Debye–Hückel model is derived in Refs. 41 and 34.

4.2.2. *Electrostatic rigidity*

On length-scales of the order of the persistence length we must include the semiflexible nature of the chain. Semiflexible PEs have a much increased persistence length due to electrostatic stiffening[34] as was first shown independently by Odijk *et al.*[35] We consider a WLC with charge separation a,

$$\beta H = \frac{\ell_p}{2} \int ds \left[\frac{d^2 \mathbf{R}}{ds^2}\right]^2 + \frac{\ell_B}{a^2} \int_s \int_{s'} \frac{e^{-\kappa|\mathbf{R}(s)-\mathbf{R}(s')|}}{|\mathbf{R}(s) - \mathbf{R}(s')|},$$

Next we expand conformations about a rod, $\mathbf{R}(x) = [x - r_\parallel(x), \mathbf{r}(x)]$,

$$\mathbf{r}(x') - \mathbf{r}(x) \simeq (x' - x)\partial_x \mathbf{r}(x) + \frac{(x' - x)^2}{2}\partial_x^2 \mathbf{r}(x) + \cdots$$

and

$$|\mathbf{R}(x) - \mathbf{R}(x')| \simeq |x - x'|[1 + \frac{(\mathbf{r}(x') - \mathbf{r}(x))^2}{|x - x'|^2}]^{1/2}.$$

Taking the limit of long chain length $L \to \infty$, we obtain a correction to the persistence length

$$\beta H^{(\text{eff})} \simeq \int dx \underbrace{\left(\ell_p + \frac{\ell_B}{4\kappa^2 a^2}\right)}_{\ell_p^{OSF}} |\partial_x^2 \mathbf{r}|^2, \tag{76}$$

which agrees very well with the experiments.

For flexible polymers we also expect a "stiffening" of the chain due to electrostatic repulsions. It is tempting to model a flexible chain with DH interaction as a semiflexible polymer (WLC) with persistence length $\ell_p^e(\kappa)$. Several variational approaches found an induced electrostatic persistence length $\ell_p^e \sim 1/\kappa$ whilst a "rescaled" version of the Odijk–Skolnick–Fixman theory obtained essentially the same result as for stiff chains $\ell_p^e \sim 1/\kappa^2$.[32]

On the other hand, recent simulations by Micka and Kremer[44] did not find any simple scaling of the effective persistence length of flexible PEs but found a complicated scale-dependent behavior which we will try to explain.

The physics of a flexible DH polyelectrolyte is very rich as there are two *independent* lengths in the system, the size of the chain and the screening length or range of the interaction. In fact, the scaling behavior of the system corresponds to a bicritical field theory with an associated cross-over exponent. As a result, the scaling functions via which the exponents enter the physical properties can be extremely nontrivial and must also be calculated to elucidate the physical behavior.

In order to understand the physics it is instructive to consider the relative values of the screening length, $1/\bar\kappa$ (in units of Kuhn length $\bar\kappa = \kappa a$), and

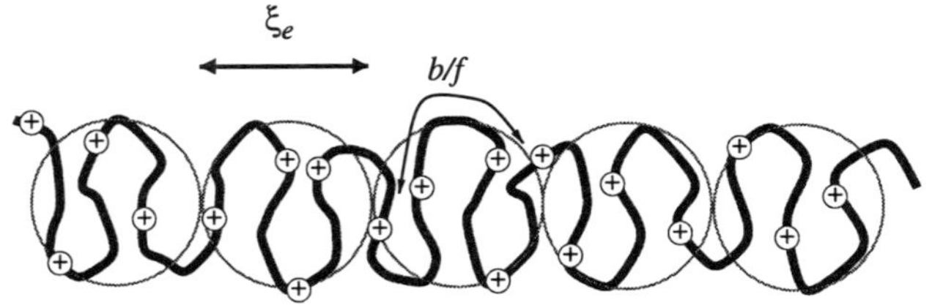

Fig. 15.   Electrostatic blobs in extended chain of monomer size b and charge fraction f.

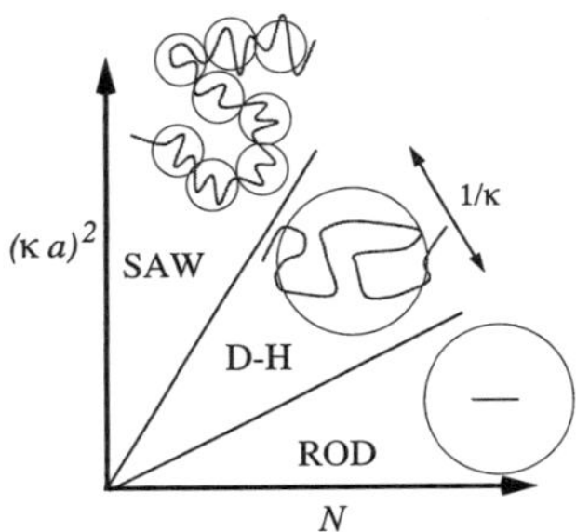

Fig. 16.   The $\kappa - N$ phase diagram. There is the rod-like regime, the DH regime and the SAW regime.

the molecular weight N which is a measure of R_g/a. One may construct a phase diagram in the $1/\bar{\kappa} - N$ plane to describe the conformational behavior of an isolated DH polyelectrolyte (see Fig. 16). The unscreened polyelectrolyte studied by Pfeuty *et al.*[36,37] corresponds to a one-dimensional cross-section of the *two*-dimensional phase diagram, the line $\kappa = 0$. Our task is to explore the rest of the phase diagram.

4.2.3. *Mapping to $n \to 0$ component $O(n)$ field theory*

Following De Gennes and Pfeuty *et al.*,[36,15] the DH Polyelectrolyte can be mapped on to the continuum field theory of an $n = 0$ component $O(n)$ spin-model interacting with a massive scalar field. It is convenient to define the "area" of the chain $L = 2a^2 N/d$.

One may define a 2-point distribution function

$$P_2(\mathbf{R}, L, \kappa^2) \equiv \langle \delta^d(\mathbf{r}(N) - \mathbf{r}(0) - \mathbf{R}) \rangle_{\mathbf{r}} \tag{77}$$

and $\langle \cdots \rangle_{\mathbf{r}}$ corresponds to an average over all chain conformations. We perform a Laplace transform of $P_2(\mathbf{R}, L, \kappa^2)$ to obtain

$$G_2(\mathbf{R}, t_0, \kappa^2) = \int_0^\infty \mathrm{d}L\, e^{-t_0 L} P_2(\mathbf{R}, L, \kappa^2) \tag{78}$$

where $L = 2a^2 N/d$. The main result is

$$G_2(\mathbf{R}, t_0, \kappa^2) \equiv \frac{1}{Z[0,0]} \int \mathcal{D}[\phi]\mathcal{D}[\psi]\phi_1(\mathbf{R})\phi_1(\mathbf{0}) \exp\{-S[\phi,\psi]\} \tag{79}$$

where

$$Z[\mathbf{J}_\phi, J_\psi] = \int \mathcal{D}[\phi]\mathcal{D}[\psi] \exp\left\{-S[\phi,\psi] + \int \mathrm{d}^d r [\mathbf{J}_\phi \cdot \phi + J_\psi \psi]\right\}, \tag{80}$$

and

$$S[\phi,\psi] = \int \mathrm{d}^d r \frac{1}{2} \left[\underbrace{|\nabla\phi|^2 + t_0|\phi|^2 + |\nabla\psi|^2 + \tau_0|\psi|^2}_{S_0} + \underbrace{u_0|\phi|^2\psi}_{S_I} \right] \tag{81}$$

with $\phi(\mathbf{r})$ an $n \to 0$ component field and ϕ_1 its first component, $\psi(\mathbf{r})$ is a scalar (Ising-like) field. The "mass" (reduced temperature) of the ϕ field, $t_0 \equiv$ *the monomer fugacity* is the conjugate variable to the molecular weight (number of monomers), $\tau_0 = \kappa^2$ is the square of the inverse Debye-screening length and $u_0 = i\sqrt{u}$. Note that we have split S into a Gaussian part S_0 and an *interaction* part S_I. A brief proof can be found in Appendix A.

The mean-square end-to-end distance is given by

$$\langle R^2 \rangle = \langle (\mathbf{r}(N) - \mathbf{r}(0))^2 \rangle = -\frac{\partial_{q^2} P_2(q, L, \tau)|_{q=0}}{P_2(q, L, \tau)|_{q=0}}, \tag{82}$$

where $P_2(q, L, \kappa^2)$ is the Fourier transform of $P_2(\mathbf{R}, L, \kappa^2)$.

The scaling behavior of long chains is given by the critical behavior of the field theory. The critical behavior of the field theory is calculated using renormalization group methods. Here we describe the more intuitive momentum shell procedure[42] up to order ϵ. For higher order perturbative calculations, it is more effective to use the field-theoretic perturbative renormalization scheme[43] of which details are available in Ref. 41.

Using dimensional analysis ($S[\phi, \psi]$ must be dimensionless), we obtain the engineering dimensions of the parameters and fields. We define inverse length-scale $[q] = \Lambda$ and obtain the following naive dimensions:

$$[\phi] = [\psi] = \Lambda^{-(d+2)/2},$$

$$[t_0] = [\tau_0] = \Lambda^2, \tag{83}$$

$$[u_0] = \Lambda^{(6-d)/2}.$$

and the dimension below which the effective coupling constant has a positive dimension is $d_c = 6$ identifying the upper critical dimension. Above $d = 6$ the Gaussian fixed point, $u^* = 0$ is stable whilst below $d = 6$ we expect a nontrivial ($u^* \neq 0$) fixed point. Near $d = 6$ we expect u^* to be small i.e. $u^* = O(\epsilon)$ where $\epsilon = 6 - d$.

We are interested in the scaling behavior (if there is any) of the system with κ and N. This corresponds to studying the critical properties of the field theory given by Eqs. (81) and (80) near the bicritical point $\tau_0, t_0 \to 0$. Since there are two independent fields, we have two sets of exponents $\nu_\psi, \nu_\phi; \eta_\psi, \eta_\phi$. To distinguish between the two we use a superscript ψ for the Coulomb terms and a superscript ϕ for the polyion terms.

4.2.4. *Momentum shell procedure*

The momentum shell procedure consists of two steps. We begin with a Lagrangian S_Λ with a high momentum (ultra-violet) cut-off Λ and we choose a rescaling parameter $b > 1$.

(i) The modes of wavelength $\Lambda/b < q < \Lambda$ are integrated out resulting in a Lagrangian with a new cut-off Λ/b.
(ii) All wavelengths are rescaled $\mathbf{q} \to \mathbf{q}' = b\mathbf{q}$, *at the same time* all fields are rescaled

$$\psi(\mathbf{q}) \to \psi'(\mathbf{q}); \psi(\mathbf{q}'/b) = \zeta_\psi \psi'(\mathbf{q}'), \tag{84}$$

$$\phi(\mathbf{q}) \to \phi'(\mathbf{q}); \phi(\mathbf{q}'/b) = \zeta_\phi \phi'(\mathbf{q}'). \tag{85}$$

The exponents η_ϕ, η_ψ can be defined by

$$\zeta_\phi = b^{(d+2-\eta_\phi)/2},$$
$$\zeta_\psi = b^{(d+2-\eta_\psi)/2}. \tag{86}$$

Momentum Shell RG. To perform (i) explicitly, the fields are decomposed into high ">", $(\Lambda/b < q < \Lambda)$ and low "<", $(0 < q < \Lambda/b)$ wavenumber parts

$$\phi(q) = \phi^<(q) + \phi^>(q), \tag{87}$$

$$\psi(q) = \psi^<(q) + \psi^>(q), \tag{88}$$

and because S_0 is harmonic we can split S_0 into high and low wavenumber contributions containing $\phi^>(q), \psi^>(q)$ and $\phi^<(q), \psi^<(q)$ respectively. A perturbative expansion in u_0 is performed and we perform averages over the high wavenumber parts.

We trace over the short wavelength modes

$$e^{-S_{\Lambda/b}} = e^{-S_0^<} \int \mathcal{D}[\phi^>] \int \mathcal{D}[\psi^>] e^{-S_0^> - S_I} \tag{89}$$

and calculate $S_{\Lambda/b}$ perturbatively in S_I to obtain Eq. (90). We now have an effective Lagrangian, $S_{\Lambda/b}$, with a new cut-off Λ/b,

$$S_{\Lambda/b} = S_0^< + \langle S_I \rangle^> - \frac{1}{2}(\langle S_I^2 \rangle^> - \langle S_I \rangle^{>2}) + \frac{1}{3!}(\langle S_I^3 \rangle^> - \langle S_I \rangle^{>3}) + O(u_0^4). \tag{90}$$

From Eq. (90) and the combinatoric (symmetry) factors, we get after rescaling

$$S_\Lambda' - S_\Lambda = -\frac{u_0^2}{2} b^{-d} \zeta_\phi^2 \int_q I_1(q) |\phi(q)|^2 + \frac{u_0^3}{2} b^{-2d} \zeta_\phi^2 \zeta_\psi I_2(0,0)$$

$$\times \int_{q_1,q_2} \phi(\mathbf{q_1}) \cdot \phi(\mathbf{q_2}) \psi(-\mathbf{q_1} - \mathbf{q_2}) + O(u_0^4) \tag{91}$$

where S_Λ' is the Lagrangian after the RG transformation and where

$$I_1(\mathbf{q}) = \int_{\Lambda/b}^{\Lambda} \frac{d^d k}{(2\pi)^d} \frac{1}{(k^2 + \tau)(|\mathbf{k} + \mathbf{q}|^2 + t)}, \tag{92}$$

$$I_2(\mathbf{q_1}, \mathbf{q_2}) = \int_{\Lambda/b}^{\Lambda} \frac{d^d k}{(2\pi)^d} \frac{1}{(k^2 + \tau)(|\mathbf{k} + \mathbf{q_1}|^2 + t)(|\mathbf{k} + \mathbf{q_1} + \mathbf{q_2}|^2 + t)}. \tag{93}$$

Equating the LHS and RHS of Eq. (91) we obtain the effective parameters u', t', τ' after the RG transformation in terms of the original parameters.

To perform step (ii) we now rescale q so that the new rescaled Lagrangian has the same ultra-violet cut-off Λ as the original Lagrangian. We must rescale the fields, ψ, ϕ so that the rescaled Lagrangian has the same form as the original Lagrangian (but in general the value of the parameters will change). Without loss of generality we set the cut-off to unity, $\Lambda = 1$. The rescaling parameter is made infinitesimal, i.e. $b = 1 + \delta\ell$ and the process repeated iteratively. By demanding that the parameters u', t', τ' remain invariant under this infinitesimal rescaling procedure, one arrives at the flow Eqs. (94)–(96).

$$\frac{du}{d\ell} = \frac{\epsilon}{2}u(\ell) + \frac{S_6 u(\ell)^3(\tau(\ell) - t(\ell))}{(1 + t(\ell))^2(1 + \tau(\ell))^2} - \frac{2S_6 u(\ell)^3}{3(1 + t(\ell))(1 + \tau(\ell))^3}, \tag{94}$$

$$\frac{d\tau}{d\ell} = 2\tau(\ell), \tag{95}$$

$$\frac{dt}{d\ell} = 2t(\ell) + \frac{S_6 u(\ell)^2(1 + \tau(\ell) - t(\ell))}{(1 + t(\ell))(1 + \tau(\ell))^2} + \frac{2S_6 u(\ell)^2 t(\ell)}{3(1 + t(\ell))(1 + \tau(\ell))^3}, \tag{96}$$

where $S_d = \Omega_d/(2\pi)^d$ and $\Omega_d = 2\pi^{d/2}/\Gamma(d/2)$ is the area of a d-dimensional unit hypersphere. A $(\phi^2)^2$ term (corresponding to a short range self-avoiding interaction) is also generated but this will be irrelevant compared to the long-range Coulomb interaction.

The fixed points of the renormalization group procedure are obtained by setting the LHS of the flow equations to zero. There is a nontrivial fixed point

$$t^* = -\frac{3\epsilon}{8}, u^{*2} = -\frac{3\epsilon}{4S_6}, \tau^* = 0, \tag{97}$$

in addition to the Gaussian fixed point $u^* = t^* = \tau^* = 0$. The RG flow of τ versus $-u^2$ at $t = t^*$ is shown in Fig. 17(a). The flow of t versus $-u^2$ at $\tau = \tau^*$ is shown in Fig. 17(b). The fixed point is stable only along the plane $\tau = 0$.

From Eq. (86) we obtain the η exponents

$$\eta_\phi = -\frac{\epsilon}{4}, \tag{98}$$

$$\eta_\psi = 0. \tag{99}$$

The anomalous dimensions (and exponents) are obtained by linearizing the recursion relations around the fixed point. The eigenvalues of the flow matrix determine the anomalous dimensions of the parameters determining how they behave under rescaling. The eigenvectors of the RG transformation determines the scaling variables i.e. the combination of the parameters that *scale*. The *subtlety* of the DH system in comparison with neutral polymers

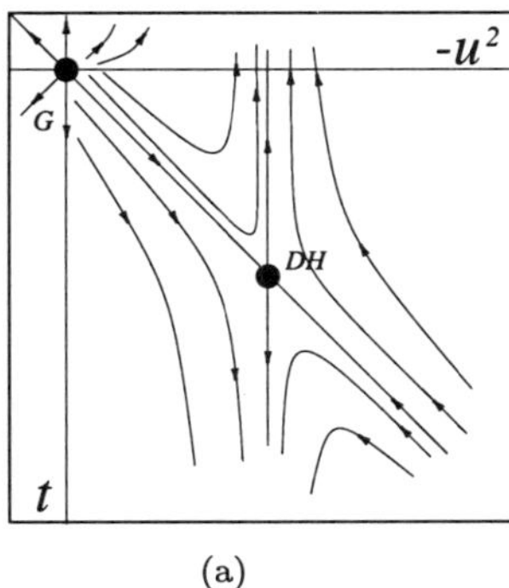
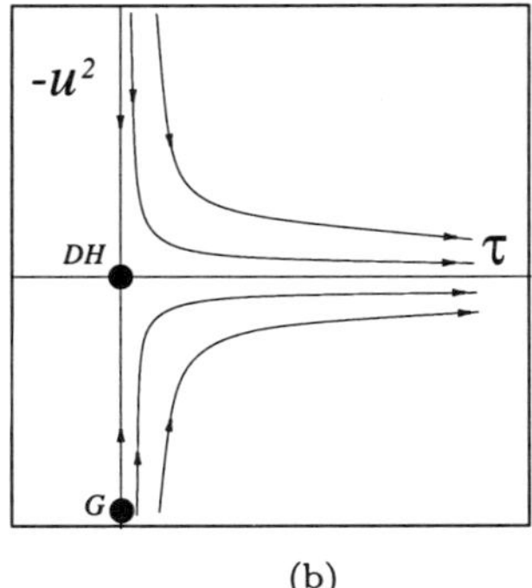

(a)   (b)

Fig. 17. The cross-section of the parameter space showing the flow of parameters (a) $t(\ell)$ when $\tau = \tau^*$ and (b) $\tau(\ell)$ when $t = t^*$ respectively versus $-u^2(\ell)$. DH refers to the nontrivial fixed point and G corresponds to the Gaussian fixed point.

becomes evident here as the eigenvectors are *nontrivial* combinations of the parameters. We obtain

$$\begin{pmatrix} t' \\ \tau' \end{pmatrix} = b^{y_t} \begin{pmatrix} t + \tau \\ 0 \end{pmatrix} + b^{y_\tau} \begin{pmatrix} -\tau \\ \tau \end{pmatrix}, \tag{100}$$

where $y_\tau = 2$, $y_t = 2 - \epsilon/2 = 1/\nu_\phi$, and

$$u' = b^{y_u} u, \tag{101}$$

where $y_u = -\epsilon/2$; primed "$'$" values are the values of the parameters after the RG transformation.

To calculate the scaling behavior of the end-to-end length we must calculate the scaling behavior of the two-point correlation function $\tilde{G}_2(q, t, \tau)$ defined by

$$\delta^d(\mathbf{q} + \mathbf{q}')(2\pi)^d \tilde{G}_2(q, t, \tau) \equiv \langle \phi_1(\mathbf{q})\phi_1(\mathbf{q}') \rangle. \tag{102}$$

Using Eqs. (86) and (102), the behavior of the two-point correlation function under rescaling by b is given by

$$\tilde{G}_2(q, t, \tau) = \zeta_\phi^2 b^{-d} \tilde{G}_2(bq, b^{y_t}(t + \tau) - b^{y_\tau}\tau, b^{y_\tau}\tau)$$

$$= b^{2 - \eta_\phi} \tilde{G}_2(bq, b^{y_t}(t + \tau) - b^{y_\tau}\tau, b^{y_\tau}\tau). \tag{103}$$

We can iterate the RG transformation p times until $b^p \tau = \tau_0$ (τ_0 is nonuniversal and depends on the PE system being studied) to obtain the scaling behavior of the correlation function.

$$\tilde{G}_2(q,t,\tau) = \left|\frac{\tau}{\tau_0}\right|^{-(2-\eta_\phi)\nu_\psi} \Phi\left(q\left|\frac{\tau}{\tau_0}\right|^{-\nu_\psi}, \frac{|\tau/\tau_0|^{\nu_\psi/\nu_\phi}}{(t+\tau-\tau_0|\tau/\tau_0|^{-\nu_\psi/\nu_\phi})}\right), \quad (104)$$

where $\Phi(x,y)$ is an unknown scaling function and we have by convention defined

$$\nu_\phi = \frac{1}{y_t} = \frac{2}{(d-2)}, \quad (105)$$

$$\nu_\psi = \frac{1}{y_\tau} = \frac{1}{2}. \quad (106)$$

Performing the inverse Laplace transform, we obtain

$$P_2(q,t,\tau) = \mathcal{T}_L^{-1}(\tilde{G}_2(q,t,\tau)) \propto \tau^{-(2-\eta_\phi)\nu_\psi}\Psi(\bar{q}\bar{\tau}^{-\nu_\psi}, \bar{\tau}^{\nu_\psi/\nu_\phi}N), \quad (107)$$

where $\Psi(x,y)$ is a dimensionless scaling function and we have defined the dimensionless quantities $\bar{q} = aq$ and $\bar{\tau} = a^2\tau$.

From Eqs. (82) and (107) we obtain the scaling behavior of the end-to-end distance where $1/\bar{\kappa}$ is the dimensionless screening length in Kuhn length units ($\bar{\kappa}^2 = \tau/a^2$). A similar calculation gives the radius of gyration, R_g.

$$\langle R^2 \rangle \sim R_g^2 \sim \frac{a^2}{\bar{\kappa}^{4\nu_\psi}}\mathcal{F}_1(N\bar{\kappa}^{2\nu_\psi/\nu_\phi}). \quad (108)$$

We can check the scaling behavior of $\langle R^2 \rangle$ in the regimes $\bar{\kappa}^2 N \to \infty$ and $\bar{\kappa}^2 N \to 0$. In the limit $\kappa \to 0$ we obtain $\langle R^2 \rangle \sim a^2 N^{2/(d-2)}$ in agreement with the unscreened result[36] whilst for $\kappa \to \infty$ we find

$$\langle R^2 \rangle \sim a^2 N\bar{\kappa}^{2\nu_\psi(1/\nu_\phi-2)}.$$

Micka and Kremer performed extensive simulations on a bead-spring model of polymers with Debye–Hückel interactions between monomers. They obtained "cross-over" results which could not be explained by any of the existing theories.[32] Using Eq. (108) we identified the relevant scaling variables and used them to analyze the simulation results of Micka and Kremer.[44] As Fig. 18 shows, this procedure allows us to collapse their data on a single master curve.

We can also extend the approach to many chain systems (see Ref. 41) as has been done for neutral chains using the scaling approach of Sec. 2.

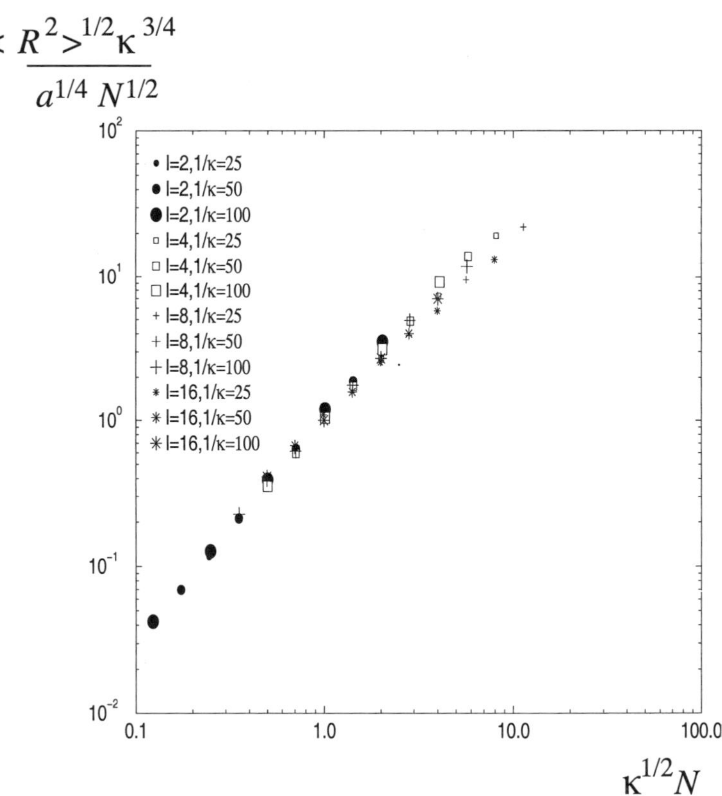

Fig. 18. The scaling of $\langle R^2 \rangle$ with $\bar{\kappa}$ and N.

5. Summary

You have come to the end of a biased summary of some old and new polymer physics given the author's own limited knowledge and research interests. The focus of the new work was on the effect of rigidity and electrostatics on the properties of polymers. For rigid polymers we found that one can obtain interesting new structure on intermediate scales. For charged screened polymers we obtained a new scaling regime when the screening length and chain size are of both large in comparison with the microscopic structure and of the same order.

We have focused on equilibrium properties and have ignored dynamic properties of flexible polymers. There is a huge body of work[1] on this topic. We also did not mention recent work on the dynamics of semiflexible polymers, e.g. Ref. 45, which is particularly interesting for biological applications.

Acknowledgments

These notes are based on a course of lectures given at the Summer School on Scaling and Disordered Systems at the Institute of Advanced Studies in Basic Sciences (IASBS) in Zanjan, Iran, from 3–16 July 1999. I would like to thank the organizers M. R. H. Khajehpour, M. R. Kolahchi, and M. Sahimi for inviting me and all the members of the IASBS for their warmth and hospitality.

It is a pleasure to acknowledge discussions with A. Ajdari, M. E. Cates, S. F. Edwards, R. Everaers, E. Frey, G. Grest, H.-K. Janssen, J.-F. Joanny, M. Kardar, A. Liu, A. Maggs, U. Micka, K. Müller-Nedebock, M. Pützt, J. Rudnick, L. Schäfer, M. Schmidt, B. Söderberg, D. Thirumalai, T. Vilgis, and T. Witten and fruitful collaborations with R. Golestanian, K. Kremer, and M. Stapper on some of the work described here.

Appendix A. Mapping to Field Theory

A.1. *Single chain*

Beginning with Eqs. (73)–(75) we can perform a Hubbard–Stratonovich transformation to get

$$Z_{DH} = \int \prod_{j=1}^{m} \mathcal{D}[\mathbf{r}_j]\mathcal{D}[\psi] \exp\{-\beta\mathcal{H}_P[\mathbf{r}_j] - \mathcal{S}_0[\psi]\}$$

$$\times \exp\left\{-\frac{i}{2}f(\ell_B^{(d-2)}\Omega_d)^{1/2}\sum_j \int \mathrm{d}s\psi[\mathbf{r}_j(s)]\right\}, \tag{109}$$

where

$$\mathcal{S}_0[\psi(\mathbf{x})] = \frac{1}{2}\int \mathrm{d}^d x(|\nabla\psi|^2 + \kappa^2|\psi|^2). \tag{110}$$

Performing the functional integral over ψ produces Eqs. (73)–(75) with $\bar{u} = f^2\ell_B^{d-2}\Omega_d$.

The two point distribution function defined in Eq. (77) of a Debye–Hückel PE is given by

$$P_2(\mathbf{R}, L, \kappa^2) = Z_{DH}^{-1} \int \mathcal{D}[\mathbf{r}]\mathcal{D}[\psi]\delta^d[\mathbf{r}(N) - \mathbf{r}(0) - \mathbf{R}]$$

$$\times \exp\left\{-\beta\mathcal{H}_P[\mathbf{r}] - \frac{i}{2}\sqrt{\bar{u}}\int ds\psi[\mathbf{r}(s)] - \mathcal{S}_0[\psi]\right\}. \quad (111)$$

The functional integral over $\mathbf{r}(s)$ is the evolution operator of a d-dimensional quantum mechanical system from 0 to $\mathbf{R}$ in "imaginary time" N so that

$$P_2(\mathbf{R}, L, \kappa^2) = \int \mathcal{D}[\psi]\exp\{-\mathcal{S}_0[\psi]\}\langle\mathbf{R}|e^{-LH}|\mathbf{0}\rangle \quad (112)$$

where H is the quantum Hamiltonian

$$H = -\frac{1}{2}\nabla^2 + \frac{i\sqrt{\bar{u}}}{2}\psi(\mathbf{r}) \quad (113)$$

and $|\mathbf{r}\rangle$ are position eigenstates ($\langle\mathbf{r}|\mathbf{r}'\rangle = \delta^d(\mathbf{r} - \mathbf{r}')$). Assuming that a complete set of eigenstates $\{\Phi_i\} \to H\Phi_i = E_i\Phi_i$ of the Hamiltonian exists,

$$\langle\mathbf{R}|e^{-LH}|\mathbf{0}\rangle = \sum_j \Phi_j^*(\mathbf{R})\Phi_j(\mathbf{0})\exp\{-E_jL\}.$$

Performing the Laplace transform in Eq. (78), we obtain

$$\tilde{G}_2(\mathbf{R}, t, \kappa^2) = \int \mathcal{D}[\psi]\exp\{-\mathcal{S}_0[\psi]\}\langle\mathbf{R}|(H + t_0)^{-1}|\mathbf{0}\rangle. \quad (114)$$

The matrix element can be expressed as

$$\langle\mathbf{R}|(H + t)^{-1}|\mathbf{0}\rangle = (\text{Det}[H + t_0])^{-n/2}\int \mathcal{D}[\phi]\phi_1(\mathbf{R})\phi_1(\mathbf{0})$$

$$\times \exp\left\{-\frac{1}{2}\int d^dr[|\nabla\phi|^2 + t_0|\phi|^2 + i\sqrt{\bar{u}}|\phi|^2\psi]\right\}, \quad (115)$$

where ϕ is an n component field. Choosing $n = 0$ so the normalizing determinant is unity and using Eqs. (114) and (115), we obtain Eq. (79).

References

1. M. Doi and S. F. Edwards, *Theory of Polymer Dynamics* (Oxford University Press, Oxford, 1986).

2. P. G. de Gennes, *Scaling Concepts in Polymer Physics* (Cornell University Press, Ithaca, New York, 1970).

3. P. J. Flory, *Principles of Polymer Chemistry* (Cornell University Press, Ithaca, New York, 1953).

4. M. Doi, *Introduction to Polymer Physics* (Oxford University Press, Oxford, 1995).

5. B. Alberts, D. Bray, J. Lewis, M. Raff, K. Roberts, and J. D. Watson, *Molecular Biology of the Cell* (Garland, New York, 1994).

6. J. des Cloizeaux and G. Jannink, *Polymers in Solution* (Oxford University Press, Oxford, 1989).

7. I. Noda, N. Kato, T. Kitano, and M. Nagasawa, *Macromolecules* **16**, 668 (1981).

8. K. Binder and D. W. Heermann, *Monte Carlo Simulation in Statistical Physics* (Springer-Verlag, Berlin, 1988).

9. N. Metropolis, A. W. Rosenbluth, M. N. Rosenbluth, A. H. Teller, and E. Teller, *J. Chem. Phys.* **21**, 1087 (1953).

10. K. Binder, in *Computational Modelling of Polymers*, ed. J. Bicerano (Marcel Dekker, New York, 1992).

11. K. Kremer, in *Monte Carlo and Molecular Dynamics Simulations in Polymer Science*, ed. K. Binder (Oxford University Press, Oxford, 1995).

12. J. des Cloizeaux, *J. Phys. (Paris)* **36**, 281 (1975).

13. P. Pincus, *Macromolecules* **9**, 386 (1976).

14. S. F. Edwards, *Proc. Phys. Soc. (London)* **85**, 613 (1965).

15. P. G. de Gennes, *Phys. Lett.* **38A**, 339 (1972).

16. K. Freed, *Renormalization Group Theory of Macromolecules* (John Wiley and Sons, New York, 1987); E. Eisenriegler, *Polymers Near Surfaces* (World Scientific, Singapore, 1993).

17. J. des Cloizeaux, *J. Phys. (Paris)* **43**, 635 (1981).

18. O. Kratky and G. Porod, *Rec. Trav. Chim.* **68**, 1106 (1949).

19. L. D. Landau and E. M. Lifshitz, *Theory of Elasticity* (Pergamon Press, Oxford, 1986).

20. B. A. Dubrovin, A. T. Fomenko, and S. P. Novikov, *Modern Geometry — Methods and Applications Part I: The Geometry of Surfaces, Transformation Groups and Fields* (GTM Springer-Verlag, New York, 1984).

21. C. Bouchiat and M. Mezard, *Phys. Rev. Lett.* **80**, 1556 (1998); D. Moroz and P. Nelson, *Macromolecules* **31**, 6333 (1998).

22. S. Smith, L. Finzi, and C. Bustamante, *Science* **258**, 1122 (1992).

23. J. F. Marko and E. D. Siggia, *Macromolecules* **27**, 981 (1994); J. F. Marko and E. D. Siggia, *Macromolecules* **28**, 8759 (1995).

24. P. Cluzel, A. Lebrun, C. Heller, R. Lavery, J.-L. Viovy, D. Chatenay, and F. Caron, *Science* **271**, 792 (1996); S. B. Smith, Y. Cui, and C. Bustamante, *Science* **271**, 795 (1996); T. R. Strick, J.-F. Allemand, D. Bensimon, A. Bensimon, and V. Croquette, *Science* **271**, 1835 (1996).

25. R. Everaers, R. Bundschuh, and K. Kremer, *Europhys. Lett.* **29**, 263 (1995).

26. T. B. Liverpool, R. Golestanian, and K. Kremer, *Phys. Rev. Lett.* **80**, 405 (1998).

27. M. G. Bawendi and K. F. Freed, *J. Chem. Phys.* **83**, 2491 (1985); J. B. Lagowski, J. Noolandi, and B. Nickel, *J. Chem. Phys.* **95**, 1266 (1991); A. M. Gupta and S. F. Edwards, *J: Chem. Phys.* **98**, 1588 (1993); T. B. Liverpool and S. F. Edwards, *J. Chem. Phys.* **103**, 6716 (1995).

28. T. Burkhardt, *J. Phys. A: Math. Gen.* **30**, L167 (1997); T. Odijk, *Macromolecules* **19**, 2313 (1986); W. Helfrich and W. Harbich, *Chem. Scr.* **25**, 32 (1985); A. R. Khoklov and A. N. Semenov, *Physica* **112A**, 605 (1982).

29. R. Golestanian and T. B. Liverpool, *Phys. Rev. E* (submitted).

30. R. M. Hornreich, R. Liebmann, H. G. Schuster, and W. Selke, *Z. Phys.* **B35**, 91 (1979).

31. F. Oosawa, *Polyelectrolytes* (Marcel Dekker, New York, 1971).

32. J.-L. Barrat and J.-F. Joanny, *Adv. Chem. Phys.* **94**, 1 (1996); S. Forster and M. Schmidt, *Adv. Poly. Sci.* **120**, 51 (1995).

33. R. Golestanian and M. Kardar (this volume).

34. R. R. Netz and H. Orland, *Eur. Phys. J.* **B8**, 81 (1999).

35. T. Odjik, *J. Poly. Sci, Poly. Phys. Ed.* **15**, 477 (1977); J. Skolnick and M. Fixman, *Macromolecules* **10**, 944 (1977).

36. P. Pfeuty, R. M. Velasco and P. G. de Gennes, *J. Phys. Lett. (Paris)* **37** (1977); P. Pfeuty, *J. Phys. Colloque. (Paris)* **C2**, 39 (1978).

37. G. Jug and G. Rickayzen, *J. Phys. A: Math. Gen.* **14**, 1357 (1981); G. Jug, *Ann. Phys. (N.Y.)* **142**, 140 (1982).

38. J.-P. Bouchaud, M. Mezard, G. Parisi, and J. S. Yedidia, *J. Phys. A: Math. Gen.* **24**, L1025 (1991).

39. Y. Kantor and M. Kardar, *Europhys. Lett.* **9**, 53 (1989); Y. Kantor and M. Kardar, *Phys. Rev. Lett.* **83**, 745 (1999).

40. T. B. Liverpool and M. Stapper, *Europhys. Lett.* **40**, 485 (1997).

41. T. B. Liverpool and M. Stapper, *Eur. Phys. J.* **B** (submitted).

42. N. Goldenfeld, *Lectures on Phase Transitions and the Renormalization Group* (Addison-Wesley, Reading Mass, 1992); P. Chaikin and T. Lubensky, *Principles of Condensed Matter Physics* (CUP, Cambridge, 1995).

43. J. Zinn-Justin, *Quantum Field Theory and Critical Phenomena* (Oxford University Press, Oxford, 1996); D. Amit, *Field Theory, the Renormalization Group and Critical Phenomena* (World Scientific, Singapore, 1984).

44. U. Micka and K. Kremer, *Phys. Rev. E.* **54**, 2653 (1996); *Europhys. Lett.* **38**, 279 (1997).

45. R. Everaers, F. Julicher, A. Ajdari, and A. Maggs, *Phys. Rev. Lett.* **82**, 3717 (1999).

Annual Reviews of Computational Physics VIII (pp. 205–227)
Edited by Dietrich Stauffer

THE DISSIPATIVE DYNAMICS AND RELAXATION BEHAVIOR OF A GENERIC MODEL FOR HYDROPHOBIC COLLAPSE

ERKAN TÜZEL and AYŞE ERZAN*

*Department of Physics, Faculty of Sciences and Letters,
Istanbul Technical University, Maslak 80626, Istanbul, Turkey*
Feza Gürsey Institute, P.O. Box 6, Çengelköy 81220, Istanbul, Turkey

A generic model of a random polypeptide chain, with discrete torsional degrees of freedom and Hookean springs connecting pairs of hydrophobic residues, reproduces the energy probability distribution of real proteins over a very large range of energies. We show that this system with harmonic interactions, under dissipative dynamics driven by random noise, leads to a distribution of energy states obeying a one-dimensional Ornstein–Uhlenbeck process and giving rise to the so-called Wigner distribution. We find stretched exponential relaxation under Metropolis dynamics at low temperatures with the exponent $\beta \simeq 1/4$, in agreement with the best experimental results on proteins. The time dependent correlation functions for fluctuations about the native state, computed in the Gaussian approximation for real proteins, have also been found to have the same functional form. Our results indicate that the energy landscape of polypeptide chains exhibits universal features over a very large range of energies and is relatively independent of the specific dynamics.

1. Introduction

A huge amount of effort has recently been invested in modeling the interactions responsible for yielding the native states of proteins as their thermodynamic equilibrium state.[1,2] An outstanding problem which so far has defied satisfactory analysis is that the time scales over which this native state can be selected via a random sampling of phase space differ by orders of magnitude from those that are actually observed, and which must be obtained if the protein is to be able to perform its biological functions. Therefore an understanding of the dynamical processes involved in the folding of proteins, and the structure of the energy (or free energy) landscape continues to be of vital interest.

Recently such features of real proteins have begun to be appreciated,[3-7] as the density of vibrational energy states[8] may be reproduced by coarse-grained model Hamiltonians which capture the essential mechanism driving the folding process, namely hydrophobic interactions.[5-7] At least for relatively short polypeptide chains, it may not be unreasonable to assume that a kind of self-similarity holds over the entire energy landscape, such that not only vibrational but conformational energy states obey the same overall statistics. This self-similarity of the energy spectrum is indeed encountered in other complex systems such as large nuclei.[9,10]

We take the view here that the protein in its native state must essentially correspond to a self-organized system, that is, the "native state" should be conceived of as the attractor of a dynamics. This typically corresponds not to a unique conformation but to a set of conformations to which the trajectory of the phase point representing the molecule is confined after asymptotically long times. Therefore, the aim should eventually be to model the dynamics of the molecule that places it on the correct attractor, and not to find the computer algorithm which zeroes in on the "native conformation" with the greatest efficiency.

As a starting point, we introduce and study a model of N coupled, over-damped torsional oscillators with discrete allowed states. We first map out the statistics of the energy landscape, under a dynamics based on relaxing pairs of rotational degrees of freedom (dihedral angles) sampled with a probability which is a function of the conjugate torques τ_i,

$$P(i) = \frac{|\tau_i|^\eta}{\sum_i |\tau_i|^\eta} \, . \tag{1}$$

The distribution of energy states $P(E)$ visited along a given trajectory depends on the dynamics. The energy landscape is effectively coarse- or fine-grained by tuning the parameter η. We find that the energy histograms obtained in this way, may be very well represented by a Wigner distribution,[9] or "Wigner surmise" for the statistics of level spacings S encountered[10] in the study of large nuclei,

$$P(S) \sim S \exp\left(-\frac{\pi}{4}S^2\right) \, . \tag{2}$$

while the coarse-grained energy level distributions are comparable with the statistics of the nth ($n = 1, 2, 3 \ldots$) neighbor energy level spacing. The distribution of energy steps encountered along different trajectories obey a stretched exponential form, with the power in the exponential depending on η.

We find[11] that preferentially relaxing the maximal torques ($\eta > 0$) drives the system to less stable, high energy states, whereas choosing the dihedral angles either with uniform probability or preferring minimal torques ($\eta \leq 0$) give rise to more successful strategies for reaching low-lying energy states.

On the other hand under low-temperature Metropolis Monte Carlo dynamics,[12] the chain exhibits anomalous, stretched exponential relaxation. The power in the exponent can be related to the distribution of energy steps along relaxation paths, which is also a stretched exponential, albeit with a different exponent.

The paper is organized as follows. In Sec. 2 we define our model, in Sec. 3 we present our simulation results for the "η-dynamics", in Sec. 4 we show that the energy obeys an OU process. In Sec. 5 we present our Metropolis Monte Carlo results for the relaxation of the chain, and relate them to the energy gap distribution. In Sec. 6 we discuss the relationship between this model and other complex systems.

2. The Model

We consider a model[11] consisting of N residues, treated as point vertices, interacting via Hookean potentials. We have been motivated by the model proposed by Haliloğlu, Bahar, and Erman[5] where all interactions between the different residues are governed by confining to square-law potentials.[5-7] In our model, however, the covalent bonds between residues are treated as fixed rods of equal length. The residues located at the vertices may be polar P or hydrophobic H. All the hydrophobic vertices are to be connected to each other with springs of equal stiffness. This feature mimics the effective pressure that is exerted on the hydrophobic residues by the ambient water molecules, and results in these molecules being driven to the relatively less exposed center in the low-lying energy states, whereas the polar residues are closer to the surface (see Fig. 1). It is important to note that we treat all H–H pairs on an equal footing, that is, there is no "teleological" information that is fed into the system by connecting only those H–H pairs which are close to each other in the native configuration for a particular sequence. It is known that real proteins are distinguished by H–P sequences that lead to unique ground states while a randomly chosen H–P sequence will typically give rise to a highly degenerate ground state. In the absence of detailed knowledge regarding the rules singling out the realistic H–P sequences we considered a generic H–P sequence obtained by choosing 50% of the residues to be hydrophobic and distributing them randomly along

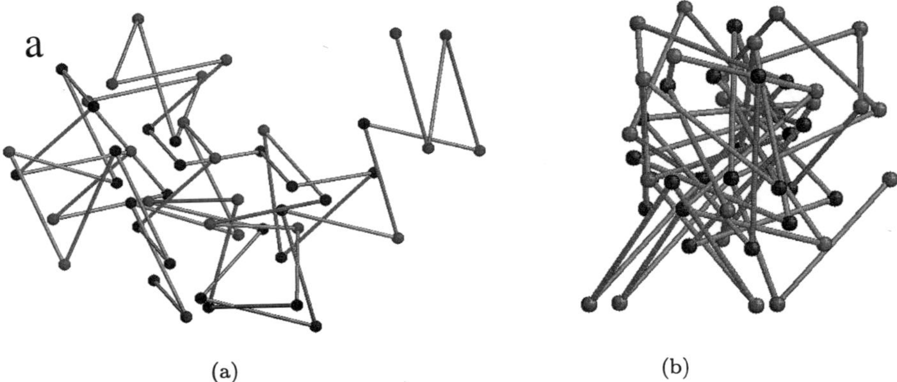

Fig. 1. A chain of $N = 48$ residues, half of which are randomly chosen to be hydrophobic, (darker beads) shown (a) in a random initial configuration and (b) in a folded state reached under Metropolis dynamics. The chain has folded in such a way as to leave the polar residues on the outside. (Generated using RasMol V2.6).

the chain. We have checked that our results were quite robust with respect to changing the sequence of hydrophobic or hydrophilic residues, or even taking all of them to be hydrophobic.

The energy of the molecule is

$$E = \frac{K}{2} \sum_{i,j} c_{i,j} |\mathbf{r}_i - \mathbf{r}_j|^2 = K \sum_{i,j} \mathbf{r}_i^\dagger V_{ij} \mathbf{r}_j \,. \tag{3}$$

If we define $Q_i = 1$ for the ith vertex occupied by a hydrophobic residue, and $Q_i = 0$ otherwise, we may write $c_{i,j} = Q_i Q_j$ and

$$V_{ij} = [(N_H - 1)c_{i,i} - c_{i,j-1} - c_{i,j+1}]\delta_{i,j}$$

$$- (1 - \delta_{i,j})(1 - \delta_{i,j-1} - \delta_{i,j+1})c_{i,j} \,. \tag{4}$$

We take the bond angles $\alpha_i, i = 1, \ldots, N-1$, to have the alternating values of $(-1)^i \alpha$, with $\alpha = 68°$. The dihedral angles ϕ_i can take on the values of 0 and $\pm 2\pi/3$. The state (conformation) of the system is uniquely specified once the numbers $\{\phi_i\}$ are given. The constraints placed on the conformations due to the rigid chemical bond lengths and by restricting the chemical and dihedral angles to discrete values prevent the molecule from trivially collapsing to a point. The residues effectively reside on the vertices of a tetrahedral lattice. The position vectors $\mathbf{r}_i$ of each of the vertices in the chain can be expressed in terms of a sum over the directors $\mathbf{R}_i$ of unit length representing the chemical bonds, which may be obtained from $\mathbf{R}_1$ by successive rotations $\mathbf{M}_k(\alpha_k)$ and $\mathbf{T}_k(\phi_k)$ through the bond and the dihedral angles,[13] viz.

$$\mathbf{r}_i = \sum_{j=1}^{i-1} \prod_{k=j}^{2} \mathbf{T}_k(\phi_k)\mathbf{M}_k(\alpha_k)\mathbf{R}_1 \,, \tag{5}$$

where we may choose $\mathbf{R}_1$ to lie along any of the Cartesian directions in our laboratory frame without loss of generality. We obtain the torques that act at each of the vertices i by substituting this in Eq. (3) and taking the partial derivative with respect to ϕ_i, viz.

$$\tau_i = -\partial E/\partial \phi_i \,. \tag{6}$$

The system is assumed to evolve within a viscous environment, with friction coefficient ζ_r, subject to random kicks from the ambient molecules. The dynamics is overdamped, so that there is no acceleration, and the impulse received with each kick dies immediately. We may write the Langevin equation for the positions of the vertices as

$$\frac{d\mathbf{r}_i(t)}{dt} = \frac{1}{\zeta_r}\mathbf{F}_i + \boldsymbol{\xi}_r(i,t) \,, \tag{7}$$

where $\boldsymbol{\xi}_r(i,t)$ is an independent Gaussian r. v. for each i and t. Equivalently, for the state vector $\boldsymbol{\phi} = (\phi_1, \ldots, \phi_N)$, we have the Langevin equation

$$\frac{d\phi_i(t)}{dt} = \frac{1}{\zeta_\tau}\tau_i + \xi_\tau(i,t) \,, \tag{8}$$

where the torque τ_i is a function of all the angles $\{\phi\}$, ζ_τ is the appropriate friction coefficient and ξ is again a Gaussian random "force" delta correlated in i and time. Viewed in this way the dynamics is similar to a pinned interface[14,15] or a charge density wave system[16–19] in $1+1$ dimensions, where the phase angles of the charge density wave are also cyclic variables. On the other hand, in the present system the interactions are long ranged.

For the discrete, sequential numerical simulation of the evolution of this system, we postulate the following set of rules:

(1) Form the self-similar probability distribution in Eq. (1),

$$P(i) = |\tau_i|^\eta / \sum_i |\tau_i|^\eta \,,$$

(2) Choose a pair of vertices (i, i') according to this probability distribution over $\{\tau_i > 0\}$ and $\{\tau_i < 0\}$,

(3) Set $\phi_i(t+1) = \phi_i(t) + \text{sign}(\tau_i)(2\pi/3)$.

Here η is a tunable parameter defining the dynamics. For large positive values of η, those angles ϕ_i with the maximal conjugate torques are incremented; for negative values of η the small values of the torque are preferred. For $\eta = 0$ the chain evolves completely randomly. If one choses η to be very large, then we find that there is a large probability that the most recently updated ϕ_i still carries a very large torque, resulting in a jamming of the dynamics. After applying the search strategy based on changing the torques according to a distribution, we found that updating the maximal torques ($\eta > 0$) drives the system to a state with relatively high energies, whereas a random search ($\eta = 0$) or preferentially choosing the minimal torques ($\eta < 0$) gives rise to more successful strategies for reaching low-lying energy states.

It can be said that η here plays the role of a coarse-graining parameter in the exploration of the energy landscape. We would like to recall a recent paper[20] (also see references therein) where an effective inertial effect was introduced into the evolution of a pinned interface, by giving an advantage to that point along the interface which moved last. This led to the coarse graining of the interface, by introducing a persistence time and an associated length scale into the system, whose long-time and large-scale scaling behavior, however, was not altered. (In contrast, discouraging the same point from moving at the next time step led to no appreciable change, since this occurred rarely, to start with.) In the present case, changing η has a similar effect; for large η we get very big persistence effects, while values of $\eta \leq 0$ seem to be qualitatively similar to each other.

We have also studied the case where, after picking a pair (i, j) of vertices according to the above distribution, we changed them randomly, that is, not following rule 3 above, but making a totally random change. This is in the spirit of the "Extremal Optimization" method proposed recently by Boettcher,[21] and indeed leads to, distributions peaked at relatively low energies, similar to the $\eta = 0$ dynamics, for all positive η values that we have tried ($0 < \eta \leq 8$). Using the energy stored in a particular spring as the "fitness parameter" (or cost function) rather than the torques on a particular pair of vertices, and then changing a randomly chosen vertex $i < k < j$, in a random way, was even more successful as far as yielding small total energies for the molecule, and work in this direction is in progress.

3. Distribution of Energy States and Level Spacings

The distribution of the energies of the discrete configurational states explored by the chain of $N = 48$ residues shown in Fig. 1, as it evolves under the above dynamics, is shown in Figs. 2–5, for both positive and negative η. After the first 5000 steps were discarded, the statistics were taken over 5000 steps of the trajectory. It can be seen that the shape of the curve does not essentially change with η, while for positive η the peak shifts to successively higher values of the energy, and the distribution is distorted towards a Gaussian, indicating that the states explored are less correlated. These figures should be compared with those reported by ben-Avraham[8] for the density of vibrational states and by Mach *et al.*[22] for the ultraviolet absorption spectra, and also with the energy histograms obtained by Socci and Onuchic[23] for a Monte Carlo simulation on a lattice model of a protein-like heteropolymer. Our model seems to be very successful in producing realistic distributions of energy states over the whole range of relevant energies.

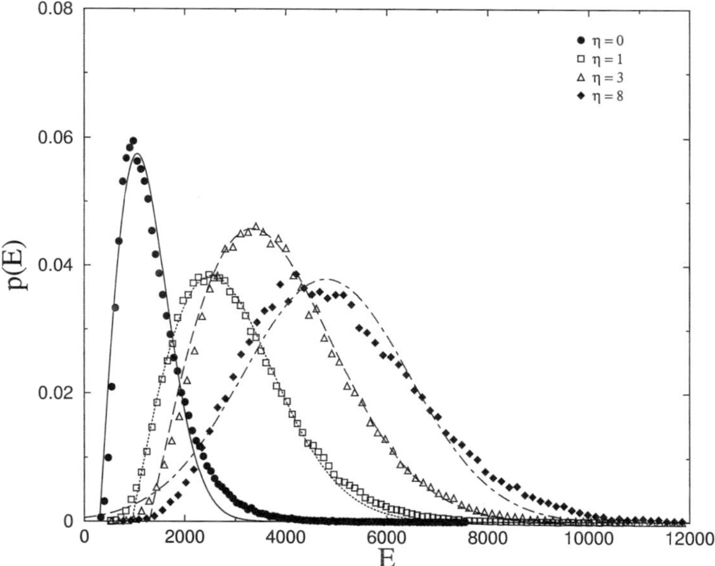

Fig. 2. The normalized energy histograms, averaged over 10 random initial states for chains of $N = 48$, for different $\eta \geq 0$, along paths of 10^4 steps, with the first 5000 steps discarded. The fits are to the Wigner distribution for $\eta = 0, 1, 3$ and Gaussian distribution for $\eta = 8$.

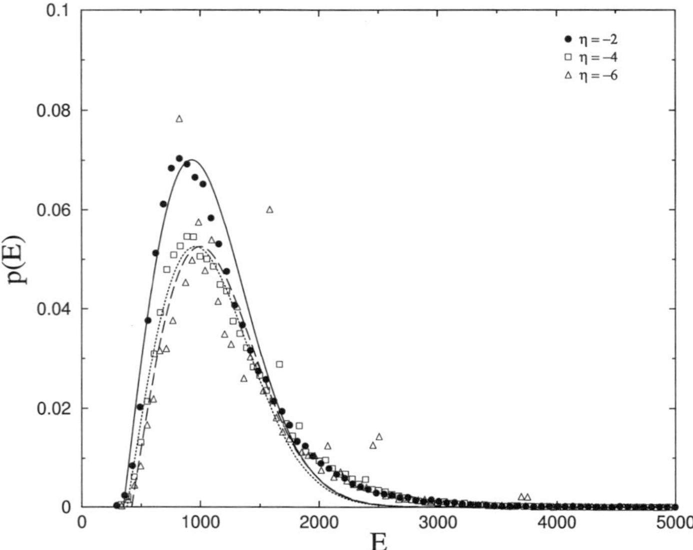

Fig. 3. The normalized energy histograms, for chains of $N = 48$, for different $\eta < 0$ (see Fig. 2). The fits are to the Wigner distribution.

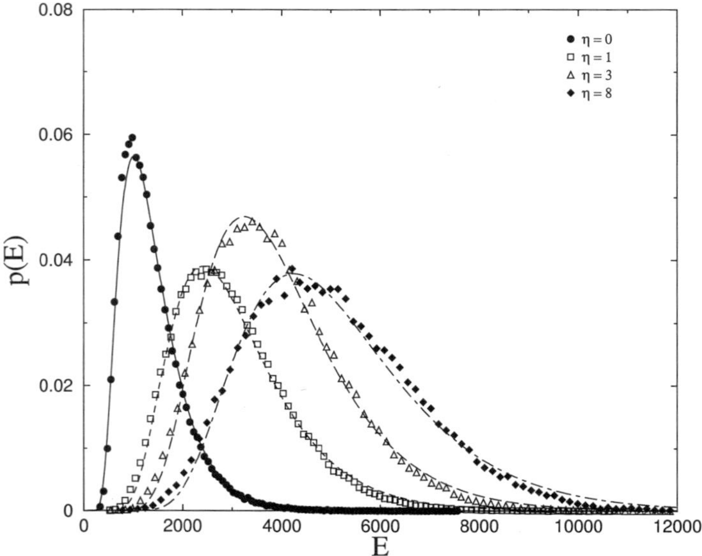

Fig. 4. The normalized energy histograms along trajectories in phase space for the $N = 48$ chain, for $\eta \geq 0$ as in Fig. 2, fitted with the "inverse Gaussian" distribution given in Eq. (10).

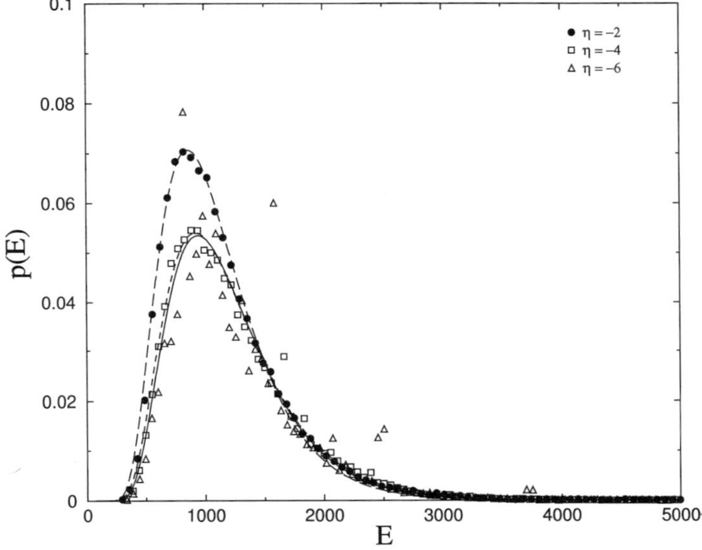

Fig. 5. Energy histograms for $\eta < 0$ as in Fig. 4, fitted with the "inverse Gaussian" distribution given in Eq. (10), for the $N = 48$ chain.

We have been able to fit the simulation results very successfully with a distribution of the Wigner form (Figs. 2 and 3)

$$f_W(E) = a(E - E_0)e^{-b(E-E_0)^2} , \tag{9}$$

for $\eta = -6$ to $\eta = 3$, and the parameters for the fit are given in Table 1. Here E_0 corresponds to the offset due to the lowest energy state attained for different η, and it can be seen that the distribution is shifted to higher values of the energy for higher values of η. The curves become Gaussian for $\eta = 6$ and $\eta = 8$.

It should be mentioned that the same energy distributions may be fitted equally well (see Figs. 4 and 5) by the "inverse Gaussian",[24]

$$f_{IG}(E) = \sqrt{\frac{A}{2\pi E^3}} \, \exp\left[-\frac{A(E - B)^2}{2B^2 E} \right] . \tag{10}$$

It will be noted that this has the same functional form as the distribution of first passage times over a distance d for an Ornstein Uhlenbeck process[25] with diffusion coefficient $D = \sigma^2/2$ and initial drift velocity v, in the regime of small

Table 1. The parameters a, b and E_0 used for fitting the energy histograms to the Wigner distribution $P(E) = a(E - E_0)\exp(-b(E - E_0)^2)$.

η	$a(10^{-4})$	$b(10^{-7})$	E_0
−6	1.50	15.0	420
−4	1.50	15.0	380
−2	2.00	15.0	350
0	1.25	8.7	300
1	0.40	2.0	950
3	0.37	1.2	1300

Table 2. The parameters A, B used for fitting the energy histograms to the inverse Gaussian distribution. The estimated errors are also provided. (Calculated using Levenberg–Marquart algorithm).

η	$A(\times 10^3)$	$\Delta A(\times 10)$	$B(\times 10^3)$	$\Delta B(\times 10)$
−6	7.4	48.0	1.2	2.3
−4	6.8	10.0	1.2	0.5
−2	6.6	6.4	1.1	0.3
0	6.3	4.9	1.4	0.4
1	18.4	7.9	3.1	0.8
3	28.3	32.2	4.0	1.2
6	33.7	39.5	4.6	1.4
8	38.3	59.2	5.2	2.0

times, if one makes the further identifications $A = d^2/(2D)$ and $B = d/v$. We postpone until Sec. 4 a discussion of this result. The results for the fits to parameters A and B are given in Table 2, together with the estimated errors. We find that both the "diffusion constant (mobility)" and the "drift velocity" of the phase point along its trajectory in phase space depend on η, being maximum for $\eta = 0$ and decreasing for positive values of η. For $\eta < 0$ they essentially stay the same.

We have also considered the statistics of energy differences between successive energy states visited along a trajectory obeying the above dynamics.

We found that the distributions were symmetric for ΔE negative or positive, and that they obeyed a stretched exponential distribution (for positive ΔE),

$$P(\Delta E) \sim \exp[-(\Delta E)^c]. \tag{11}$$

The distribution of energy steps for different η are given in Fig. 6(a). The plots of $\ln(-\ln(P(\Delta E)))$ versus $\ln(\Delta E)$ for $\eta = 0$ and $\eta = 8$ are shown in Fig. 6(b). The values of the exponent c range from 0.5 to 0.8 (see Table 3), tending to 1 as η becomes large and positive, again exhibiting a decorrelation effect as the energy landscape is probed with larger and larger η. As $c \to 1$, the distribution tends to a Poissonian, indicating that the system has no cooperativity.

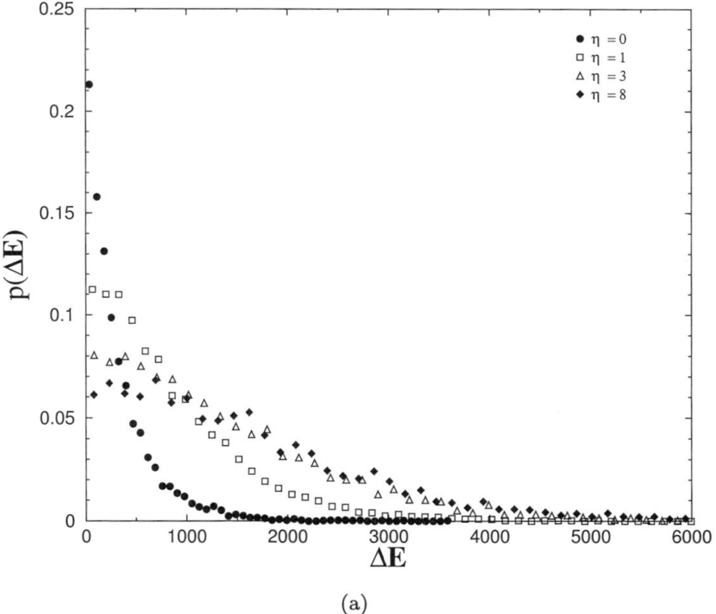

(a)

Fig. 6. (a) The distribution of energy steps along a trajectory in phase space according to the η-dynamics of the $N = 48$ chain, for different η. The last 5000 steps along a 10 000 steps trajectory were considered and (b) The fits, for $\eta = 0$ (left) and for $\eta = 8$ (right) to the stretched exponential form $\sim \exp(-\Delta E^c)$, for $c = 0.58$ and $c = 0.81$ respectively, in the large ΔE limit.

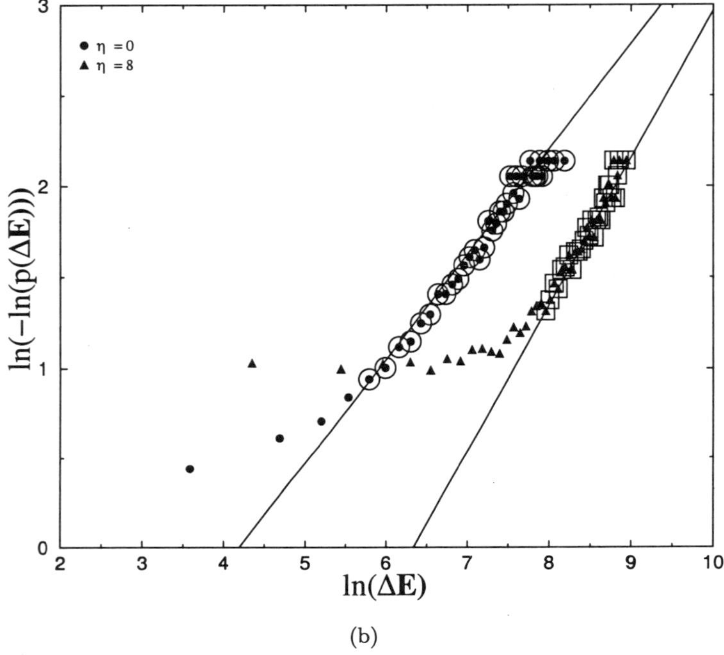

(b)

Fig. 6. (*Continued*).

Table 3. The parameter c used for fitting the distribution of energy steps to a stretched exponential in the form $P(\Delta E) \sim \exp[-(\Delta E)^c]$. The correlation coefficients (r^2) are also provided.

η	c	r^2 (Corr. Coef.)
-8	0.50	0.89
-6	0.49	0.97
-4	0.54	0.98
-2	0.54	0.98
0	0.58	0.97
1	0.74	0.95
2	0.73	0.96
3	0.81	0.95
4	0.73	0.96
6	0.85	0.95
8	0.81	0.95

4. Ornstein-Uhlenbeck Process and the Wigner Distribution

We would now like to show that both the Wigner distribution (9) and the inverse Gaussian distribution (10) arise as limiting forms in an Ornstein–Uhlenbeck (OU) process. We remind the reader that an OU process describes the diffusive motion of a particle subject to a drift velocity proportional to the distance from the origin.[25] It can easily be seen that such a process for a single particle in one dimension would be described by the Langevin equation,

$$\frac{dx}{dt} = -\frac{1}{\zeta} gx + \xi(t) \tag{12}$$

with a Hookean force $F(x) = -gx$ and a delta correlated random force $\xi(t)$, $\langle (\xi(t))^2 \rangle = \sigma^2$. In the absence of the stochastic term which gives rise to diffusive motion, the velocity is simply proportional to the distance from the origin (or the point of equilibrium). For an initial displacement $x(0) = d$, the solution for the distribution of first passage times through the origin is given by

$$f(t) = \frac{2yd}{\pi^{1/2}\sigma} \left(\frac{\rho}{1-y^2} \right)^{3/2} e^{-\rho y^2 d^2/\sigma^2(1-y^2)}, \tag{13}$$

where $\rho = g/\zeta$ and $y = \exp(-\rho t)$. Clearly, without the stochastic term, the solution for (12) is simply $x = d\exp(-\rho t) = dy$. We see that (13) goes over, in the limit of large times, that is $y \ll 1$, to

$$f_{\mathrm{W}}(y) = \frac{2yd\rho^{3/2}}{\pi^{1/2}\sigma} e^{-\rho d^2 y^2/\sigma^2}. \tag{14}$$

On the other hand, for very small times, (13) becomes, to leading order,

$$f_{\mathrm{IG}}(t) = \frac{2\pi d\sigma^2}{(2\pi\sigma^2 t)^{3/2}} e^{-(d-vt)^2/2\sigma^2 t} \tag{15}$$

where we have defined $\rho d = v$.

Recall that $x = dy$ is the "distance remaining to the origin". The distribution function (14) may just as well be considered as a function of x. For late times, we get $f_{\mathrm{W}}(x) \propto x \exp(-\rho x^2/\sigma^2)$ which is in the form of the Wigner surmise (2). On the other hand, for very small times, $x \sim d(1 - \rho t) = d - vt$. The distance from the initial point, $\tilde{x} \equiv (d - x)/v$, becomes simply proportional to the time elapsed and we get the "inverse Gaussian" (10) form,

$$f_{\mathrm{IG}}(\tilde{x}) \sim \left(\frac{\lambda}{2\pi\tilde{x}^3} \right) e^{-(\lambda/2)[(\tilde{x}-\mu)^2/\mu^2\tilde{x}]} \tag{16}$$

where $\lambda = (d/\sigma)^2$ and $\mu = 1/\rho$. It should be noted that both these functions (14) and (15) look extremely alike numerically.

Now we would like to show that the energy E given in Eq. (3) also obeys an OU process, under the dynamics given by (7). Since there is no explicit time dependence of E, we have

$$\frac{dE}{dt} = \sum_i \frac{\partial E}{\partial \mathbf{r}_i} \cdot \frac{\partial \mathbf{r}_i}{\partial t} \, . \tag{17}$$

Substituting from (7) we get,

$$\frac{dE}{dt} = -\frac{1}{\zeta_r} \sum_i \left(\frac{\partial E}{\partial \mathbf{r}_i} \right)^2 + \sum_i \frac{\partial E}{\partial \mathbf{r}_i} \cdot \xi_i(t) \, . \tag{18}$$

From (3) we may compute that

$$\sum_i \left(\frac{\partial E}{\partial \mathbf{r}_i} \right)^2 = \frac{NE}{\zeta_r} + \sum_{\substack{i,j,k \\ i \neq j}} c_{ik} c_{jk} (\mathbf{r}_i - \mathbf{r}_k) \cdot (\mathbf{r}_j - \mathbf{r}_k) \, . \tag{19}$$

We see that the second term is like an average of the products $(\mathbf{r}_i - \mathbf{r}_k) \cdot (\mathbf{r}_j - \mathbf{r}_k)$ over (i, j) pairs $(i \neq j)$, and for a reasonably isotropic configuration, it vanishes. To the same approximation, we may assume that the second term in Eq. (18) is itself equal to a Gaussian stochastic noise, that is, set

$$\xi_E(t) = K \sum_{ij} c_{ij} (\mathbf{r}_i - \mathbf{r}_j) \cdot \boldsymbol{\xi}_i(t) \, .$$

This yields the required result, namely,

$$\frac{dE}{dt} = -\frac{NE}{\zeta_r} + \xi_E \, , \tag{20}$$

which should be compared with (12).

Now, we see that if under the given dynamics, the E distribution obeys one of these limiting forms (14) and (15), then the first passage time distribution for the attainment of the lowest energy state must obey, in turn, Eq. (13). This explains the similarity between the distributions of first passage times for rather general global optimization problems with quadratic cost functions reported by Li et al.[24] and the form of the distribution of energy states which we find from our simulations.

5. Relaxation Behavior

In order to investigate the relaxation properties[12] of the present model, we have employed Metropolis Monte Carlo dynamics. This consists of

(a) choosing a pair (i, i') of dihedral angles randomly on the chain, and updating the $(\phi_i, \phi_{i'})$ in a way that preserves angular momentum, incrementing them in opposite directions by $\Delta\phi = \pm 2\pi/3$,

(b) accepting the move with unit probability if $\Delta E \leq 0$ and with probability $p = \exp(-\gamma \Delta E))$ for $\Delta E > 0$,

(c) repeating the second step once before discarding the pair altogether and going to the first step.

Here γ serves as an effective inverse temperature. We monitor the relaxation of the total energy as a function of "time" measured in the number of MC steps, (i.e. the number of pairs (i, i') sampled) until a steady state is reached, typically in about 10 000 steps. The results for chains of $N = 100$ averaged over 20 randomly chosen initial configurations at zero temperature $(\gamma = \infty)$ are shown in Fig. 7. Defining $\epsilon \equiv (E - E_0)/E_I$, where E_0 is the (time-averaged) equilibrium energy and E_I, the initial value, we find that it obeys a power law, $\epsilon(t) \sim t^{-\sigma}$ with $\sigma = 0.49 \pm 0.01$ for the initial stages of the

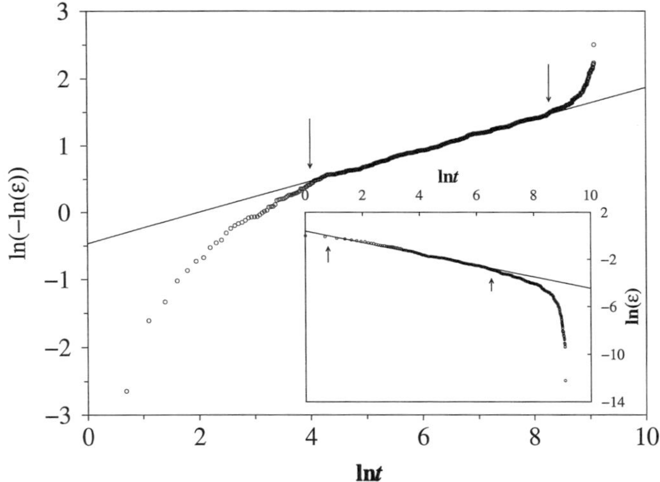

Fig. 7. The decay of the energy of an $N = 100$ chain from a random initial configuration, along a zero temperature Metropolis trajectory, of 10 000 steps, averaged over 20 runs. The later stages fit on a stretched exponential curve $\epsilon(t) \sim \exp(-t^\beta)$ with $\beta = 0.234 \pm 0.003$. The initial stage (inset) is fit by a power law $\epsilon(t) \sim t^{-\sigma}$ with $\sigma = 0.49 \pm 0.01$.

decay, while later stages can be fitted by a stretched exponential $\epsilon(t) \sim e^{-t^{\beta}}$ with $\beta = 0.234 \pm 0.003$.

We also performed simulations for different values of γ, for chains of $N = 48$, averaging over 100 runs with random initial configurations (see Fig. 8). For $\gamma \to \infty$, $\gamma = 0.5$ and $\gamma = 0.3$, the above relaxation behavior continues to hold and the exponents do not seem to depend on γ, with $\beta \simeq 1/4$ and $\sigma \simeq 1/2$ as given in Table 4.

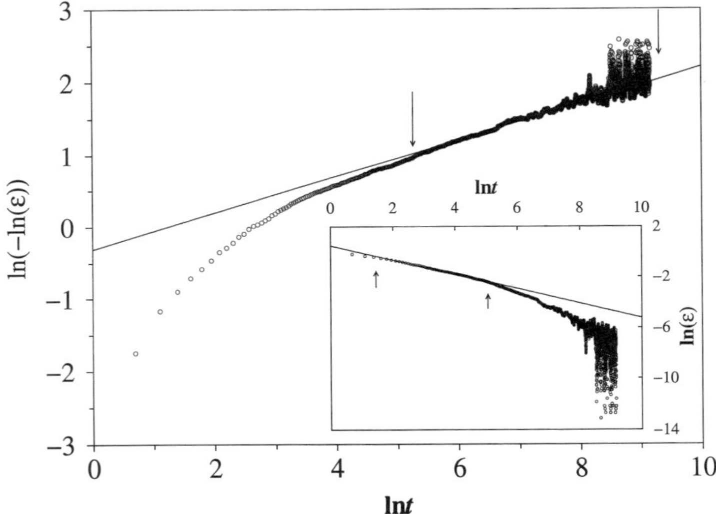

Fig. 8. The decay of the energy of an $N = 48$ chain, along a Metropolis trajectory, from a random initial configuration averaged over 100 runs for $\gamma = 0.3$. The initial stage (inset) is fit by a power law $\epsilon(t) \sim t^{-\sigma}$ with $\sigma = 0.57 \pm 0.01$, and the late stage to a stretched exponential with $\beta = 0.25 \pm 0.03$.

Table 4. The exponent σ and β found for the power law and stretched exponential decay of the total energy with time, for different chain lengths N and inverse temperatures γ. The fits were obtained from a weighted least-squares computation.

N	γ	σ	$\Delta\sigma$	β	$\Delta\beta$
48	∞	0.57	0.01	0.281	0.004
	0.5	0.56	0.01	0.30	0.04
	0.3	0.57	0.01	0.25	0.03
100	∞	0.49	0.01	0.234	0.003

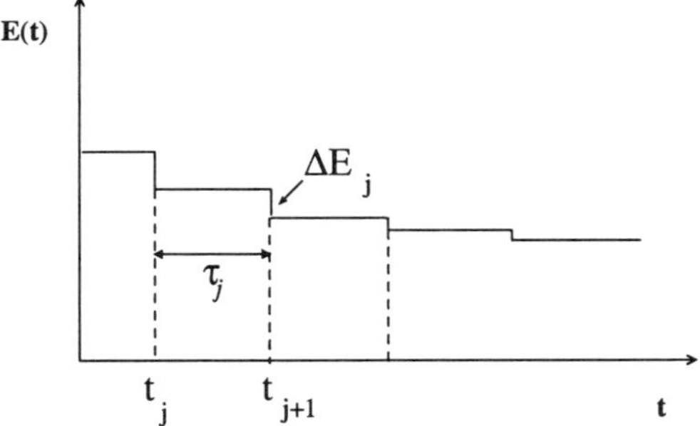

Fig. 9. A schematic plot of the variation of the total energy with time.

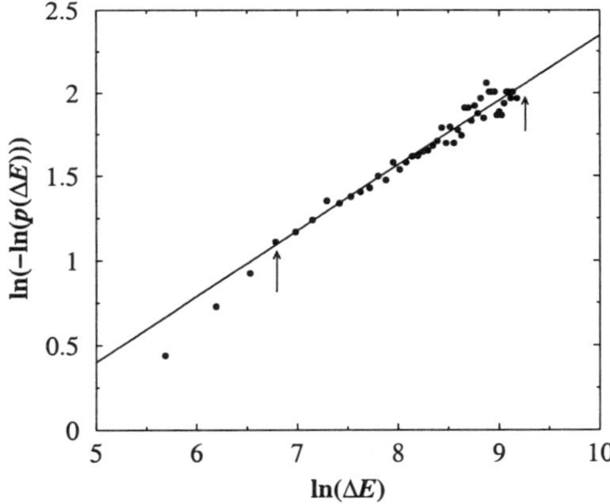

Fig. 10. The distribution of energy differences encountered along the relaxation path are fit to a stretched exponential. Level spacing histograms were formed for chains of $N = 48$ and averaged over 100 runs for the zero-temperature Metropolis relaxation. The exponent α of the stretched exponential is found to be 0.39 ± 0.02.

The variation of the total energy in time is sketched in Fig. 9 over a short sequence of relaxation events. Clearly one may write $E(t)$, averaged over many independent runs, as $\langle E(t) \rangle = \langle E(0) - \sum_{i=1}^{M} \Delta E_i \Theta(t - t_i) \rangle$ where Θ is the Heaviside step function and $t_i = \sum_{k=0}^{i-1} \tau_k$. Taking the time derivative one gets,

$$\langle \dot{E}(t) \rangle = \left\langle -\sum_{i=1}^{M} \Delta E_i \delta \left(t - \sum_{k=0}^{i-1} \tau_k \right) \right\rangle. \tag{21}$$

At zero temperature, the expectation value of $\dot{E}(t)$ can be calculated by carrying out an integration over the distribution of waiting times $\{\tau_k\}$, and the distribution of energy steps encountered along the relaxation path. The expectation value, $\langle \dot{E}(t) \rangle$ is then,

$$\langle \dot{E}(t) \rangle = -\left\langle \sum_{j=1}^{M} \Delta E_j \delta\left(t - \sum_{k=0}^{i-1} \tau_k\right) \right\rangle_{\Delta E, \tau} . \qquad (22)$$

The distribution of waiting times τ_k is dependent only on the configuration of the chain at the kth step and independent of the previous waiting times. Since the dynamics is just changing a pair of dihedral angles in opposite directions, for each conformation $\{\phi_i\}$ one may define an associated chain of $N(N-1)/2$ sites, with each site corresponding to a pair (i, i') on the original chain. On the associated chain, a site will be assigned the value 1 if the corresponding pair has at least one "allowed" move, and the value 0 if both moves are "blocked". Now the probabilities of encountering allowed or blocked moves as one implements the Metropolis dynamics outlined above are simply given by the density of 1' s or 0' s on the associated chain at a given relaxation step, namely, p_k and $q_k = 1 - p_k$. Therefore, in the kth conformation, the probability of making a transition after τ_k blocked moves simply obeys the first passage time distribution,[25]

$$P_k(\tau_k) = \mu_k e^{-\mu_k \tau_k} , \quad \mu_k \equiv |\ln q_k| . \qquad (23)$$

If we were to assume that the τ were distributed identically at each step of the relaxation process (and therefore independently of the ΔE), then we would be able to do the averages in (22) immediately, and we end up with a simple linear decay for $E(t)$! Clearly, to get nontrivial relaxation behavior, one needs to incorporate a dependence of the relevant distributions on the particular conformation reached at that stage of the relaxation.

To see how the distribution of relaxation times depends on the state of the system, one may argue that the larger the energy loss in a relaxation event, the longer it will take for the phase point to make a transition out of this state. Since μ_k is roughly the expectation for τ_k, we assume that $\mu_k \sim 1/\Delta E_k$. With the (rather strong!) assumption that the energy steps encountered along a relaxation path are independently distributed, that is, $P(\Delta E_1 \ldots \Delta E_M) = \prod_{s=1}^{M} P(\Delta E_s)$ for a process of M steps, one finds,

$$\langle \dot{E}(t) \rangle = -\frac{1}{2\pi} \sum_{j=1}^{M} \langle \Delta E_j \rangle \sum_{\ell=1}^{j-1} I_{j\ell}(t) , \qquad (24)$$

where $I_{j,\ell}(t)$ is

$$I_{j,\ell}(t) \equiv \int_0^\infty d(\Delta E_\ell)e^{-t/\Delta E_\ell}P(\Delta E_\ell)\left[\prod_{\substack{k=0\\k\neq\ell}}^{j-1}\left\langle\frac{\Delta E_\ell}{\Delta E_\ell-\Delta E_k}\right\rangle_{\Delta E_k}\right]. \qquad (25)$$

We have determined from our simulations that the distribution $P(\Delta E_\ell)$ (see Fig. 10) is also a stretched exponential $P(\Delta E_\ell) = P_o e^{-(\Delta E_\ell)^\alpha}$. The angular brackets then take the form

$$\Delta E_\ell \int_0^\infty (\Delta E_\ell - \Delta E_k)^{-1}\exp(-(\Delta E_k)^\alpha)d\Delta E_k \qquad (26)$$

which we approximate by $\Delta E_\ell \exp(-(\Delta E_k)^\alpha)$. The integration in Eq. (25) is then straightforward, leading, upon substitution in (24), to

$$E(t) \sim t\sum_{j=1}^{M}\left(\frac{j-1}{j}\right)\exp(-a_j t^\beta) \qquad (27)$$

where $a_j = j(1-\alpha)(\alpha j)^{-\beta}(1+\beta)^{-1}$ and

$$\beta = \frac{\alpha}{\alpha+1}. \qquad (28)$$

Substituting the value of α we find from our simulations, namely $\alpha = 0.39 \pm 0.02$, we get $\beta = 0.28 \pm 0.01$ which is the result we obtained from the fits to the MC simulations within our error bounds.

It is important to note that the distribution of ΔE along a trajectory of the η-dynamics is quantitatively different from the distribution of ΔE encountered along a Metropolis Monte Carlo path, and depends on η. This arises from the highly complex nature of the energy landscape, and the extremely important correlations that arise between the energy steps encountered depending upon how the phase space is being sampled. In particular, we have found out that in the limit of extremely large η, where no cooperativity remains, the distribution of ΔE along a trajectory of the dynamics is Poissonian, which would have led to $\beta = 1/2$ here.

A study of the correlations between fluctuations about the native state[26] for the Beads-and-Springs model, with the $H - P$ sequence and the contact map for the native states for seven real proteins (6lyz, 1cd8, 1bet, 1fil, 1bab, 1csq and 1hiv) was performed by Erman.[27] Using the Gaussian approximation[28] to the coherent scattering function and a normal mode analysis,[4,5,29,30] he also finds a stretched exponential relaxation with $\beta = 1/4$. Experiments on real proteins and polymers[1,31–33] yield $0.2 \leq \beta \leq 0.4$. Our results seem to be closer to 1/4 and smaller than the values most commonly found for spin-glasses,[34] namely 1/3. It should also be noted that glassy behavior is obtained here in the absence of quenched randomness, or due to frustration arising from steric hindrances, which we do not take into account.

Comparing the relaxation behavior near the native state with the behavior we observe at relatively high energies for random heteropolymers, we conclude that the relaxation behavior, and therefore the dynamics and the structure of the energy landscape are universal over a very large range of energies, and are relatively independent of the specific sequence or the details of the dynamics.

6. Discussion

The experimentally reported density of vibrational energy states and UV absorption spectra for a large number of real proteins fall on a universal curve,[8,22] which is consistent with the distributions of Wigner or inverse Gaussian form. Our simple coarse-grained model for the hydrophobic interactions reproduces very closely the same distribution of energy states. Moreover, this same curve appears to describe the distribution of first passage times in global optimization problems with quadratic cost functions.[24] We have provided a way to understand this behavior on the basis of an overdamped Hookean model of hydrophobic interactions.

We would like to make a few concluding remarks about a remaining puzzle, which is the connection between the Wigner distribution, which arises as the distribution of eigenvalue spacings for Gaussian orthogonal matrices,[9,10,35,36] and the present limiting form we have found for the distribution of energy states. The puzzle promises to be fruitful, because we believe that the current state of the art of understanding the energy landscapes of proteins is very similar to that of the study of the energy spectra of very large nuclei in the '50s, when it was realized that the problem might be very advantageously treated as a statistical one. For real, large nuclei, the level spacing distribution, properly scaled with the average density of states for different energies, exhibit a remarkable invariance over the entire energy range.[35] It was then found that the energy levels and nearest-neighbor spacings are governed to a large extent by the statistics of eigenvalues and eigenvalue spacings of orthogonal matrices with a Gaussian distribution of matrix elements.[9,10,35,36] The statistics of nuclear energy level spacings approximately obey the so called Wigner "surmise",[9,10,35] which is given in Eq. (2). This distribution of eigenvalue spacings for 2×2 Gaussian orthogonal matrices can also be obtained by considering the square root of the sum of the squares of two independent but identically distributed Gaussian random variables of zero mean.

It is of great interest to note that the so-called Wigner distribution arises also in "quantum chaos"[37–39] and in the energy spectra of large atomic clusters.[40] Thus, there seems to be a universality to the spectral fluctuations of confined systems of sufficient complexity.[35,41]

It is very intriguing to compare the results for $\eta \geq 0$ (Fig. 2) with the numerically obtained nth neighbor spacing distributions of the eigenvalues for Gaussian orthogonal matrices, as reported by Porter,[42] where the identical shift of the peak and tendency to a symmetric Gaussian distribution is found. This we interpret as reinforcing our observation that larger η dynamics results in a more coarse-grained sampling of the energy landscape. Nevertheless, the connection to nth neighbor spacing distributions of eigenvalues of random matrices and the Ornstein–Uhlenbeck process still remains to be understood.

Acknowledgments

We express our gratitude to the organizers of the Summer School on Scaling in Disordered Systems, for inviting us to this exciting conference, and our admiration for the extraordinary atmosphere of excellence in scientific research which they have been able to create in Zanjan.

References

1. H. Frauenfelder, S. G. Sligar, and P. G. Wolynes, *Science* **254**, 1598 (1991).
2. P. G. Wolynes, J. N. Onuchic, and D. Thirumalai, *Science* **276**, 1619 (1995).
3. K. A. Dill, S. Bromberg, K. Yue, K. M. Feibig, D. P. Yee, P. D. Thomas, and H. S. Chan, *Protein Science* **4**, 561 (1995).
4. M. M. Tirion, *Phys. Rev. Lett.* **77**, 1905 (1996).
5. T. Haliloglu, I. Bahar, and B. Erman, *Phys. Rev. Lett.* **79**, 3090 (1997).
6. B. Erman and K. Dill, *J. Chem. Phys.* **112**, 1050 (2000).
7. B. Erman, Hydrophobic collapse of proteins into their near-native configurations, unpublished.
8. D. Ben-Avraham, *Phys. Rev.* **B47**, 14 559 (1993).
9. E. Wigner, *Proc. Cambridge Phil Soc.* **47**, 790 (1951); *Ann. Math.* **62**, 548 (1955).

10. C. E. Porter, *Statistical Theories of Spectra: Fluctuations* (Academic Press, N.Y., 1965).
11. E. Tüzel and A. Erzan, *Phys. Rev.* **E61**, R1040 (2000).
12. E. Tüzel and A. Erzan, *J. Stat. Phys.* **100**(1/2), 405–422 (2000).
13. P. J. Flory, *Statistical Mechanics of Chain Molecules* (Interscience, N.Y., 1969).
14. For an insightful review see T. Halpin-Healy and Y.-C. Zhang, *Phys. Rep.* **254**, 215 (1995).
15. J. Krug and H. Spohn, Kinetic roughening of growing surfaces, in *Solids Far from Equilibrium* ed. C. Godreche (Cambridge University Press, Cambridge, 1992).
16. A. Erzan, E. Veermans, R. Heijungs, and L. Pietronero, *Phys. Rev.* **B41**, 11522 (1990).
17. E. Veermans, A. Erzan, R. Heijungs, and L. Pietronero, *Physica* **A166**, 447 (1990).
18. G. Parisi and L. Pietronero, *Physica* **A179**, 16 (1991).
19. O. Narayan and D. Fisher, *Phys. Rev.* **B46**, 11520 (1992).
20. Ö. Kayalar and A. Erzan, *Phys. Rev. E.*, to appear.
21. S. Boettcher and A. G. Percus, in math.OC/9904056.
22. H. Mach, D. B. Volkin, C. J. Burke, and C. R. Middaugh, Ultraviolet absorption spectroscopy, in *Methods in Molecular Biology, Vol. 40: Protein Stability and Folding* ed. B. A. Shirley (Humana Press, Totowa, N.J. 1995), pp. 91–114.
23. N. D. Socci and J. N. Onuchic, *J. Chem. Phys.* **103**, 4732 (1995).
24. C. N. Chen, C. I. Chou, C. R. Hwang, J. Kang, T. K. Lee, and S. P. Li, *Phys. Rev.* **E60**, 2388 (1999).
25. W. Feller, *An Introduction to Probability Theory and its Applications* (Wiley, N.Y., 1957), Vols. I and II.
26. B. Erman, *J. Comp. Polym. Sci.*, in press.
27. B. Erman, private communication.
28. B. Ewen and D. Richter, in *Elastomeric Polymer Networks*, eds. J. E. Mark and B. Erman (Prentice Hall, 1992), p. 220.
29. N. Go, T. Noguti, and T. Nishikawa, *Proc. Natl. Acad. Sci.* (USA) **80**, 3696 (1983).
30. M. Levitt, C. Sander, and P. S. Stern, *J. Mol. Biol.* **181**, 423 (1985).
31. J. L. Green, J. Fan, and C. Angell, *J. Phys. Chem.* **98**, 13780 (1994).
32. B. Erman and I. Bahar, *Macromol. Symp.* **133**, 33 (1998).
33. J. Colmenero, A. Arbe, and A. Alegria, *Phys. Rev. Lett.* **71**, 2603 (1993).
34. M. Mezard, G. Parisi, and M. A. Virasoro, *Spin Glass Theory and Beyond* (World Scientific, Singapore, 1987).
35. T. Brody, J. Flores, J. B. French, P. A. Mello, A. Pandey, and S. S. S. Wong, *Rev. Mod. Phys.* **53**, 385 (1981).
36. M. L. Mehta, *Random Matricies and the Statistical Theory of Energy Levels* (Academic Press, N.Y., 1967).
37. M. V. Berry and M. Robnik, *J. Phys.* **A17**, 2413 (1984).
38. E. Yurtsever and J. Brickman, *Phys. Rev.* **A38**, 1027 (1988).

39. E. Yurtsever and J. Brickman, *Phys. Rev.* **A41**, 6688 (1990).
40. D. Wales, private communication.
41. D. Bohigas, M. J. Giannoni, and C. Schmidt, *Phys. Rev. Lett.* **52**, 1 (1984).
42. C. E. Porter, *J. Math. Phys.* **4**, 1039 (1963).

Annual Reviews of Computational Physics VIII (pp. 229–260)
Edited by Dietrich Stauffer

FLUCTUATION-INDUCED PHENOMENA: FROM BIOPHYSICS TO CAVITY QED

RAMIN GOLESTANIAN

Institute for Advanced Studies in Basic Sciences, Zanjan 45195-159, Iran
and
Institute for Theoretical Physics, University of California,
Santa Barbara, CA 93106-4030, USA

MEHRAN KARDAR

Department of Physics, Massachusetts Institute of Technology,
Cambridge, MA 02139, USA

The *static* Casimir effect describes an attractive force between two conducting plates, due to *quantum fluctuations* of the electromagnetic (EM) field in the intervening space. *Thermal fluctuations* of correlated fluids (such as critical mixtures, super-fluids, or liquid crystals) are also modified by the boundaries, resulting in finite-size corrections at criticality, and additional forces that effect wetting and layering phenomena. Modified fluctuations of the EM field can also account for the "van der Waals" interaction between conducting spheres, and have analogs in the fluctuation-induced interactions between inclusions on a membrane. We employ a path integral formalism to study these phenomena for boundaries of arbitrary shape. This allows us to examine the many unexpected phenomena of the *dynamic* Casimir effect due to moving boundaries. With the inclusion of quantum fluctuations, the EM vacuum behaves essentially as a complex fluid, and modifies the motion of objects through it. In particular, from the mechanical response function of the EM vacuum, we extract a plethora of interesting results, the most notable being: (i) The effective mass of a plate depends on its shape, and becomes anisotropic. (ii) There is dissipation and damping of the motion, again dependent upon shape, and direction of motion. (iii) There is a continuous spectrum of resonant cavity modes that can be excited by the motion.

1. Fluctuation-Induced Forces

1.1. *Introduction*

Fluctuation-induced forces are ubiquitous in nature, covering many topics from biophysics to cosmology.[1-5] There are two basic ingredients in these phenomena: (i) A fluctuating medium, such as the electromagnetic (EM) field; and (ii) External objects whose presence suppresses (or in some way modifies) the fluctuations, such as dipoles or conductors. The overall strength of the interaction is proportional to the driving energy of fluctuations (k_BT and $\hbar$ for thermal and quantum fluctuations, respectively); its range is related to that of the correlations of the fluctuations. The most interesting cases are when the interactions are long-ranged, corresponding to scale free fluctuations.

The goal of this review is to provide a glimpse of the unity and simplicity of fluctuation-induced forces. While we attempt to describe a wide range of phenomena, this selection is by no means exhaustive, and highly biased by subjective interests. We have tried to avoid technical details, preferring to present general arguments and dimensional estimates whenever possible. The interested reader is referred to various sources for calculational details.

1.2. *Quantum fluctuations*

The standard Casimir effect[1,3] is a macroscopic manifestation of quantum fluctuations of vacuum. In 1948, Casimir considered the electromagnetic field in the cavity formed by two conducting plates at a separation H. Because the electric field must vanish at the boundaries, the normal modes of the cavity are characterized by wave-vectors $\mathbf{k} = (k_x, k_y, \pi n/H)$, with integer n. Once quantized, these normal modes can be regarded as harmonic oscillators of frequencies $\omega(\mathbf{k}) = c|\mathbf{k}|$; each of which in its ground state has energy $\hbar\omega(\mathbf{k})/2$. While adding up all the ground state energies leads to an infinite contribution to the overall energy $\mathcal{E}(H)$, Casimir showed that a finite *attractive* force is obtained from

$$F(H) = -\frac{\partial\mathcal{E}}{\partial H} = -\hbar c \times \frac{A}{H^4} \times \frac{\pi^2}{240}, \tag{1}$$

where A is the area of the plates. Thus, by measuring the mechanical force between macroscopic bodies, it is in principle possible to gain information about the behavior of the quantum vacuum.

The predictions of Casimir where followed by experiments on quartz[7] and aluminum[8] plates at separations $H > 10^3$ Å. However, these experiments, and others reviewed in Ref. 9, provided results that were at best in qualitative agreement with Eq. (1). It is only quite recently that high precision measurements of the force (using a torsion pendulum) between a gold plate, and gold plated sphere, have confirmed the theoretical prediction to a very high accuracy.[10,11]

1.3. *Thermal fluctuations*

While the Casimir interaction is due to the *quantum fluctuations* of the electromagnetic field, there are several examples in *classical statistical mechanics,* where forces are induced by the *thermal fluctuations* of a correlated fluid. One of the best known examples comes from the finite size corrections to the free energy at a critical point.[4] As demonstrated originally by Fisher and de Gennes,[12] there is a contribution to the free energy of a critical film, that varies with its thickness H, as

$$\delta\mathcal{F}(H) = -k_B T \times \frac{A}{H^2} \times \Delta. \tag{2}$$

This is to be expected on dimensional grounds, as the free energy comes from thermal fluctuations, hence proportional to $k_B T$, and must be extensive. (Similar analysis in d-dimensions leads to a dependence as $1/H^{d-1}$.) The dimensionless amplitude Δ is closely related to the so-called central charge of the critical theory, and in two dimensions exact values can be obtained by employing techniques of conformal field theories.[13] In higher dimensions, they can be estimated numerically,[14] and by $\epsilon = 4 - d$ expansions.[15]

Equation (2) results in a force that decays as $1/H^3$. The difference in the power of H from Eq. (1) is that the fluctuation energy in the latter is quantum in origin, hence proportional to $\hbar c$, which has dimensions of *energy times length.*

1.4. *Superfluid films*

In fact, long-range forces are induced by thermal fluctuations of any *correlated medium,* by which we mean any system with fluctuations that have long-range correlations. A critical system is a very particular example; much more common are cases where long-range correlations exist due to Goldstone modes of a broken continuous symmetry, as in superfluids or liquid crystals. A superfluid is characterized by a complex order parameter, whose phase ϕ may vary across the system. The energy cost of such variations is governed by the Hamiltonian

$$\mathcal{H}[\phi] = \frac{K}{2} \int d^3\mathbf{x}(\nabla\phi)^2\,, \tag{3}$$

where the "phonon stiffness" K is related to the superfluid density. In the Casimir geometry, the free energy of interaction resulting from thermal fluctuations of these modes has the form[16]

$$\delta\mathcal{F}(H) = -k_B T \times \frac{A}{H^2} \times \frac{\zeta(3)}{16\pi}\,. \tag{4}$$

Note that the result is universal, that is, independent of the stiffness K. A similar result is obtained for the electromagnetic field at high temperatures $k_B T \gg \hbar c/H$, which is larger by a factor of two,[17] reflecting the two polarizations of the normal modes (photons).

Liquid Helium tends to wet most substrates. The thickness of the wetting layer is controlled by the strength of the attractive forces that bind the helium film to the substrate,[18] mostly due to van der Waals interactions. In the presence of a chemical potential penalty of $\delta\mu$ per unit volume, the energy of a film of thickness H is

$$\frac{E(H)}{k_B T} = A\left[\frac{\delta\mu}{k_B T}H + \frac{C}{H^2}\right], \tag{5}$$

where C is a *positive* numerical constant. Minimizing this expression leads to a thickness

$$H = \left(\frac{2Ck_B T}{\delta\mu}\right)^{1/3}. \tag{6}$$

When the helium film is in the normal phase, the film thickness is determined solely by the strength of the van der Waals (vdW) force. The numerical value of $C_> = C_{\mathrm{vdW}} > 0$ depends on the substrate, and is nonuniversal. However, when the helium film becomes superfluid, there is an additional attractive fluctuation-induced (FI) force due to Eq. (4), and $C_< = C_{\mathrm{vdW}} + C_{\mathrm{FI}}$; where $C_{\mathrm{FI}} = -\zeta(3)/16\pi \approx -0.02391$. In the vicinity of the superfluid transition, there is a different attractive contribution to the force due to finite size scaling (FSS) of the critical fluctuations, as in Eq. (2), and $C_\lambda = C_{\mathrm{vdW}} + C_{\mathrm{FSS}}$. The best estimate for the finite size scaling amplitude for the XY model in $d = 3$ is $C_{\mathrm{FSS}} \approx -0.03$.[19,20]

The parameter C thus takes three different values in the normal fluid, at the λ-point, and in the superfluid phase. From Eq. (6) we then expect *two* jumps in the film thickness, as the temperature is lowered through the superfluid transition. Such an experiment has indeed been performed recently at Pennsylvania State University.[21] The thickness of the film on a Cu substrate is monitored by a capacitance measurement, for a temperature range of 30 mK close to superfluid transition point of the film (about 2.17 K). The thickness profile as a function of temperature shows a dip at the critical point, confirming the presence of the finite size scaling force. Moreover, the saturated thickness below the critical point is measurably *less* than above the critical point, providing the first experimental evidence for the FI force predicted in Ref. 16.

1.5. *Liquid crystals*

Liquid crystals exemplify anisotropic cases of correlated fluids due to broken symmetry, which again lead to fluctuation-induced forces.[16,23,24] They are also easily accessible, as experiments can be performed at room temperature and require no fine tuning to achieve criticality. The order parameter of a *nematic* liquid crystal is a director field $\mathbf{n}(\mathbf{r})$, characterizing the local preferred direction of the long axis of the molecules.[22] The energy cost of fluctuations of the nematic director $\mathbf{n}(\mathbf{r})$ is given by[22]

$$\mathcal{H}_N = \frac{1}{2} \int d^3\mathbf{r}[\kappa_1(\nabla \cdot \mathbf{n})^2 + \kappa_2(\mathbf{n} \cdot \nabla \times \mathbf{n})^2 + \kappa_3(\mathbf{n} \times \nabla \times \mathbf{n})^2]. \tag{7}$$

Integrating over the nematic fluctuations leads to a free energy contribution

$$\delta\mathcal{E}_N = -k_B T \times \frac{A}{H^2} \times \frac{\zeta(3)}{16\pi} \left(\frac{\kappa_3}{\kappa_1} + \frac{\kappa_3}{\kappa_2} \right). \tag{8}$$

Note that the resulting force does depend on the relative strengths of the elastic coupling constants (reflecting the anisotropy of the system).

In a smectic liquid crystal, the molecules segregate into layers which are fluid like. The fluctuations of these layers from perfect stacking are described by a scalar deformation $u(\mathbf{x}, z)$, which is subject to a Hamiltonian

$$\mathcal{H}_S = \frac{1}{2} \int d^3\mathbf{r} \left[B \left(\frac{\partial u}{\partial z} \right)^2 + \kappa(\nabla^2 u)^2 \right]. \tag{9}$$

The resulting interaction energy

$$\delta\mathcal{E}_S = -k_B T \times \frac{A}{H\lambda} \times \frac{\zeta(2)}{16\pi} \quad \text{with} \quad \lambda \equiv \sqrt{\frac{\kappa}{B}}, \tag{10}$$

falls off as $1/H$, reflecting the extreme anisotropy which has introduced an additional length scale λ into the problem.

Liquid crystal films provide an additional interesting feature in that their surfaces are typically more ordered than the bulk. This is because of the surface tension between the film and the surrounding gas which inhibits fluctuations of the molecules close to the surface.[25] Heat capacity experiments on free standing liquid crystal films near smectic-A (smA) to surface stabilized smectic-I (smI), and smA to surface stabilized smectic-B (smB) transitions have revealed a layer by layer ordering that starts at the surface and proceeds into the bulk.[26,27]

The smB and smI phases are characterized by an additional bond-orientational (hexatic) order in the layers, while the molecules in smI are also tilted.

The wetting analogy suggests that, as in Eq. (6), the thickness of the hexatically ordered phases should grow with the reduced temperature $t = (T - T_c)/T_c$ as $t^{-\nu}$, with $\nu = 1/3$. Here T_c is the bulk transition temperature, and we have assumed that the chemical potential difference is proportional to t. However, experiments find $\nu \approx 0.32 \pm 0.01$ for the smA–smB transition,[27] while at the smA–smI transition,[26] $\nu \approx 0.373 \pm 0.015$. The difference between the two sets of experiments is rather surprising, as the underlying direct interactions (e.g., van der Waals), should be quite similar in the two cases. Is it possible that the fluctuation-induced forces are different in the two cases? The molecular tilt in smI provides a field that further pins the fluctuations of the hexatic order at the surface. The effect of such a surface field on the force is explored in Ref. 28, leading to results consistent with the experimental observations.

1.6. *Charged fluids*

Interactions between a collection of *charged macroions* in an aqueous solution of neutralizing *counterions*, with or without added salt, are in general very complex. The macroions may be charged spherical colloidal particles, charged amphiphilic membranes, stiff polyelectrolytes (e.g., microtubules, actin filaments, and DNA), or flexible polyelectrolytes (e.g., polystyrene sulphonate), and the counterions could be mono- or poly-valent. It is known that, under certain conditions, the accumulation (condensation) of counterions around highly charged macroions can turn the repulsive interaction between two like-charged macroions into an attractive one. The attractive interaction is induced by the enhanced charge-fluctuations close to the macroions, due to the condensation of counterions.[29−31]

Since the entropic attraction between charged macroions is seldom related to the general class of fluctuation-induced forces, in Appendix A we present a path integral formulation that makes this analogy more transparent. The interaction between macroions consists of two parts: A Poisson-Boltzmann (PB) free energy, and a fluctuation-induced correction. Specifically, consider two parallel negatively charged 2D plates with densities $-\sigma$, separated by a distance H in $d = 3$, in a solution of neutralizing counterions with valence z. The PB equation can be solved exactly in this geometry, and the corresponding PB free energy, in the limit of highly charged plates, reads[32]

$$F_{\mathrm{PB}} = \frac{\pi}{2} \times \frac{k_B T A}{z^2 \ell_B H} \left[1 + \frac{1}{4\pi^2 \ell_B^2 z^2 \sigma^2 H^2} + \cdots \right], \tag{11}$$

in which $\ell_B \equiv e^2/\epsilon k_B T$ is the *Bjerrum length*. Note that in the limit $\ell_B z \sigma H \gg 1$, the interaction is independent of the charge densities of the plates; that is, it is *universal*. It is also interesting to note that the (repulsive) PB interaction is highly reduced for multivalent counterions.

The fluctuation-induced correction involves calculation of a determinant (see Appendix A), which depends on the local charge compressibilities. The true compressibility profile (and the charge density profile) emerging from the solution of the PB equation, is generally very complicated. It is usual to simplify the problem by assuming that the surface charge density is so high that the counterions are confined to a layer of thickness $\lambda_{\mathrm{GC}} \ll H$, where $\lambda_{\mathrm{GC}} = 1/2\pi z \ell_B \sigma$ is the Gouy–Chapman length. Then we can use an approximate compressibility profile $m^2(x) = (2/\lambda_{\mathrm{GC}})[\delta(x+H/2)+\delta(x-H/2)]$. In the limit $H/\lambda_{\mathrm{GC}} \gg 1$, we obtain[33]

$$F_{\mathrm{FI}} = -k_B T \times \frac{A}{H^2} \times \frac{\zeta(3)}{16\pi} \left[1 + O\left(\frac{\lambda_{\mathrm{GC}}}{H}\right) \right], \tag{12}$$

for the fluctuation-induced part of the interaction. The calculation of the determinant using the true compressibility profile for the parallel plate geometry has indeed been carried out in Ref. 31, with the result

$$F_{\mathrm{FI}} = -k_B T \times \frac{A}{H^2} \times \left[\frac{\zeta(3)}{16\pi} + \frac{\pi}{4}\left(\frac{\pi}{4}+\frac{1}{2}\right) + \frac{\pi}{4}\ln\left(\frac{H}{\pi\lambda_{\mathrm{GC}}}\right) + O\left(\frac{\lambda_{\mathrm{GC}}}{H}\right) \right]. \tag{13}$$

It is interesting to note that the correct result is considerably stronger than that of the simple "Gouy–Chapman" model (Eq. (12)).[34] Note that at large separations, it is asymptotically identical to the Casimir interaction in Eq. (4). In the opposite limit $H/\lambda_{\mathrm{GC}} \ll 1$, however, it yields[35,36]

$$F_{\mathrm{FI}} = k_B T \times \frac{A}{\lambda_{\mathrm{GC}}^2} \times \Delta_c \times \ln\left(\frac{H}{\lambda_{\mathrm{GC}}}\right), \tag{14}$$

where $\Delta_c \simeq 0.0792$.[36] A similar analysis in d-dimensions leads to asymptotic dependencies as $1/H^{d-1}$ for large separations, and $1/H^{d-3}$ for small separations, where the crossover is set by a generalized Gouy–Chapman length.[36] Interestingly, such attractive interactions have been recently suggested to be responsible for DNA bundle formation,[37] and collapse of stiff polyelectrolytes[38] and rigid membranes.[39]

2. Dispersion Forces

2.1. *Van der Waals interactions*

In addition to his work on the force between plates, Casimir also realized[40] that the van der Waals and London[41] forces can also be understood on the same footing: The presence of the atoms modifies the fluctuations of the electromagnetic field, resulting in an attractive interaction. (For a modern perspective, see the discussion by Kleppner in Ref. 42.)

For example, let us consider two conducting spheres of volumes $V_1 = 4\pi a_1^3/3$ and $V_2 = 4\pi a_2^3/3$, at a distance R. The fluctuation-induced interaction is proportional to the product of the volumes, and thus on dimensional grounds we expect a potential

$$\mathcal{V}(R) = -k_B T \times \frac{V_1 V_2}{R^6} \times \Delta_T \,, \tag{15}$$

due to thermal fluctuations. Usually, the van der Waals interaction is obtained from the instantaneous induced dipoles on the spheres. However, it can be equivalently obtained by examining the electromagnetic fluctuations of the remaining space. When the fluctuations are of quantum origin, Eq. (15) is modified to

$$\mathcal{V}(R) = -\hbar c \times \frac{V_1 V_2}{R^7} \times \Delta_Q \,, \tag{16}$$

with $\Delta_Q = 1287/(256\pi^3)$.[3] The cross-over between the two forms of the interaction occurs at a distance $\lambda_T \sim \hbar c/k_B T$.

Let us compare the above result with the more standard approach to calculating the London force between two neutral *atoms*: While the average dipole for each atom is zero, an instantaneous dipole fluctuation in one can induce a parallel instantaneous dipole on the other, leading to an attraction. Since the direct dipole–dipole interaction decays as $1/R^3$, the induced effect scales as the square, that is, $1/R^6$. In this regard, it is similar to the result in Eq. (15), except that the characteristic energy is set by a typical atomic excitation energy of $\hbar\omega_0$ rather than $k_B T$. The retardation effects are then obtained by taking into account the finite speed of light. We can imagine that for large enough distances when the signal goes from atom-1 to atom-2 to induce the dipole, and return to atom-1 to induce an attraction, it finds the dipole at atom-1 somewhat misaligned; resulting in a weaker attraction. The characteristic time for electron movements can be estimated from the frequency of the orbit as $\tau = 2\pi/\omega_0$. The crossover occurs when the travel time for the signal

independent of κ and $\bar{\kappa}$; its energy scale set by $k_B T$. A generalization of this result that includes the quantum fluctuations of the membranes is given in Ref. 57.

The form of the interaction depends sensitively on the shapes of the inclusions, as demonstrated by the calculation of the fluctuation-induced interaction between rod-like objects.[56] The rods are assumed to be sufficiently rigid so that they do not deform coherently with the underlying membrane. They can thus only perform rigid translations and rotations while remaining attached to the surface. As a result, the fluctuations of the membrane are constrained, having to vanish at the boundaries of the rods. Consider the situation depicted in Fig. 1, with two rods of lengths L_1 and L_2 at a separation $R \gg L_i$. For fluctuating films ($\sigma \neq 0$), there is an attractive fluctuation-induced interaction given by,

$$V_F^T(R, \theta_1, \theta_2) = -\frac{k_B T}{128} \times \frac{L_1^2 L_2^2}{R^4} \times \cos^2[\theta_1 + \theta_2] + O(1/R^6), \qquad (20)$$

where θ_1 and θ_2 are the angles between the rods and the line adjoining their centers, as indicated in Fig. 1. This angular dependence is actually the *square* of that of a dipole–dipole interaction in two dimensions, with L_1 and L_2 as the dipole strengths. The fluctuation-induced interaction on a membrane ($\sigma = 0$) is very similar and given by

$$V_M^T(R, \theta_1, \theta_2) = -\frac{k_B T}{128} \times \frac{L_1^2 L_2^2}{R^4} \times \cos^2[2(\theta_1 + \theta_2)] + O(1/R^6). \qquad (21)$$

The orientational dependence is the *square* of a quadrupole–quadrupole interaction, with the unusual property of being minimized for both parallel and

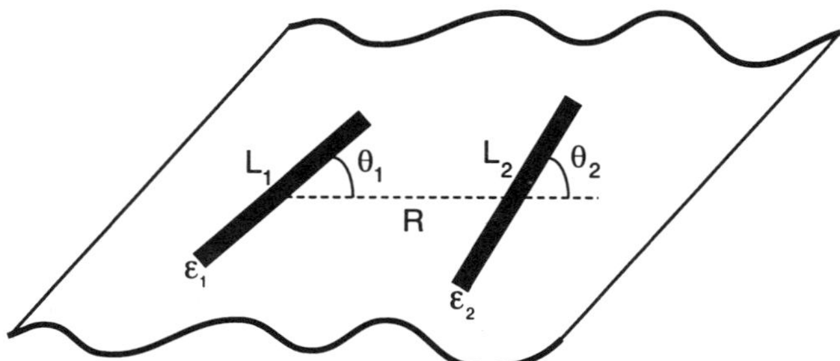

Fig. 1. Two rod-shaped inclusions embedded in a membrane. The rods are separated by a distance R. The ith rod has length L_i, width ϵ_i, and makes an angle θ_i with the line joining the centers of the two rods.

perpendicular orientations of the rods. Note that the strength of the interaction is the same in both cases. The above fluctuation-induced interactions decay less rapidly at large distances than van der Waals forces and may play an important role in aligning asymmetric inclusions in biomembranes. Since orientational correlations are often easier to measure than forces, this result may also be useful as a probe of the fluctuation-induced interaction. Finally, this interaction could give rise to novel two-dimensional structures for collections of rod-like molecules. In particular, the resemblance of the orientational part of the interaction to dipolar forces suggests that a suitable way to minimize the energy of a collection of rods is to form them into chains. (If the rods are not colinear, the interactions cannot be minimized simultaneously.)

Finally, we note that these interactions cannot be obtained by adding two-body potentials on the rods: To find the orientational dependence of additive forces, let us consider an interaction $U(|r_1 - r_2|)du_1 du_2$, between any two infinitesimal segments of two rods of length L at a distance $R \gg L$. Expanding $|r_1 - r_2|$, and integrating over the two rods, leads to the interaction

$$V(R, \theta_1, \theta_2) = L^2 U(R) + \frac{L^4}{6}\left(\frac{U'(R)}{R} + U''(R)\right)$$

$$- \frac{L^4}{12}\left(\frac{U'(R)}{R} - U''(R)\right)(\cos 2\theta_1 + \cos 2\theta_2). \qquad (22)$$

The angular dependence is now completely different, and minimized when the two rods are parallel to their axis of separation. Presumably both interactions are present for rods of finite thickness; the additive interaction is proportional to $L^2(L\epsilon/R)^2$, where ϵ is the thickness. The previously calculated interactions are thus larger by a factor proportional to $(R/\epsilon)^2$ and should dominate at large separations.

2.3. *Polymers on membranes*

The effects of thermal fluctuations of membranes and films on embedded *directed semiflexible polymers* are examined in Ref. 58, using the path integral formulation of Ref. 16. A semiflexible polymer with bending stiffness κ_p is rigid only at distances less than the bare persistence length $\ell_p^0 = \kappa_p/k_B T$. Upon embedding in a fluctuating surface, it is found that the induced interactions soften the rigidity of the polymer. While the reduction in persistence length ℓ_p is not appreciable for polymers embedded in a film, there is a logarithmic reduction upon embedding in a membrane. The softening is more dramatic for two parallel rods, which due to their mutual attraction are inclined to bend towards each other. This leads to an instability in the modes describing their relative *in-plane* fluctuations. The stiffness of the polymer prevents such instabilities only if the length of the polymer is less than a characteristic size $L_p(R) \sim \ell_p^0(R/\ell_p^0)^{3/4}$, that depends on their separation R and is much lower than ℓ_p^0.

The *out-of-plane* fluctuations of the semiflexible polymers have a dramatic effect on their "Casimir" attraction. Consider two polymers embedded in the surface which are, on average, parallel to each other at separations $a \ll R < L$, where a is a microscopic length scale such as membrane thickness. The membrane and polymers have thermal fluctuations but are constrained to remain attached at all times. In the absence of *out-of-plane* polymer fluctuations, there is a fluctuation-induced interaction that falls off as $k_B T(L/R)$. The *out-of-plane* fluctuations actually screen out this interaction, resulting in a *short-ranged* attraction that can be approximately fitted to

$$\frac{F^{\perp}(R)}{L} = -\frac{k_B T}{2\pi} \times \frac{A}{R} \times \exp\left[-b\left(\frac{R}{\lambda}\right)^{\alpha}\right], \tag{23}$$

where $A_F = \pi^2/12$, $A_M = 2.90514$, $b_F = 3.32$, $b_M = 4.17$, $\alpha_F = 0.679$ and $\alpha_M = 0.397$ are parameters obtained from numerical fits, corresponding to films and membranes respectively. The "screening" lengths that determine the range of the interactions are defined as $\lambda_F = (\kappa_p/\sigma)^{1/3}$ for films and $\lambda_M = \kappa_p/\kappa$ for membranes.

3. Path Integral Formulation and Deformed Surfaces

Most computations of Casimir forces are for simple geometries, for example, between two parallel plates. It is natural to consider how these forces are modified by the roughness that is present in most "random" surfaces. There are a number of calculations that go beyond the simple planar geometry. For example, in Ref. 59 a multiple scattering approach is used to compute the interactions for arbitrary geometry in a perturbation series in the curvature. A phenomenological approach is introduced in Ref. 60 in which small deviations from plane parallel geometry are treated by using an additive summation of Casimir potentials. However, as demonstrated in the previous section, fluctuation-induced forces are not additive, and additional steps are necessary to correct the result.[60] Another perturbative approach is introduced in Ref. 61, which is quite general, but suffers from rather cumbersome treatment of the boundary conditions. In Ref. 16, we introduced a path integral approach that made possible relatively simple computations of the fluctuation-induced force. This approach has a number of advantages. First, different manifolds (with arbitrary intrinsic and embedding dimensions) in various correlated fluids can be treated in a similar fashion. Second, the boundary conditions are quite easily implemented, and the corrections can be computed perturbatively in the deformations.

Consider n manifolds embedded in a d-dimensional correlated fluid, with an energy cost appropriately generalized from Eq. (3). The manifolds are described by the functions $R_\alpha(x_\alpha)$, where x_α is a D_α-dimensional internal coordinate ($D_\alpha = 1$ for a polymer and $D_\alpha = 2$ for a membrane), while R_α indicates a position in the d-dimensional fluid. The fluctuation-induced interactions between the manifolds are obtained by integrating over all configurations of the field ϕ, i.e.

$$\exp\left(-\frac{\mathcal{H}_{\text{eff}}[R_\alpha(x_\alpha)]}{k_B T}\right) = \frac{1}{\mathcal{Z}_0}\int \mathcal{D}'\phi(r)\exp(-\mathcal{H}_0[\phi])\,. \tag{24}$$

Here $\mathcal{D}'\phi(r)$ denotes the functional integral with the constraints imposed by the external manifolds, and $\mathcal{Z}_0$ is the partition function for the unperturbed fluid. The boundary conditions ($\phi(R_\alpha(x_\alpha)) = 0$, for $\alpha = 1, 2, \ldots n$) are imposed by inserting delta functions. Using the integral representation of the delta function, Eq. (24) can be rewritten as

$$\exp\left(-\frac{\mathcal{H}_{\text{eff}}}{k_B T}\right) = \frac{1}{\mathcal{Z}_0}\int \mathcal{D}\phi(r)\prod_{\alpha=1}^{n}\mathcal{D}\psi_\alpha(x_\alpha)$$

$$\times \exp\left(-\mathcal{H}_0[\phi] + i\int \mathrm{d}x_\alpha \psi_\alpha(x_\alpha)\phi(R_\alpha(x_\alpha))\right), \tag{25}$$

where $\psi_\alpha(x_\alpha)$ are the auxiliary fields defined on the n manifolds, acting as sources coupled to ϕ. For the quadratic Hamiltonian of Eq. (3), it is easy to integrate over the field ϕ, and obtain the long-range interactions between the sources as

$$\exp\left(-\frac{\mathcal{H}_{\text{eff}}}{k_B T}\right) = \int \prod_{\alpha=1}^{n}\mathcal{D}\psi_\alpha(x_\alpha)\exp(-\mathcal{H}_1[\psi_\alpha(x_\alpha)])\,. \tag{26}$$

The action $\mathcal{H}_1[\psi_\alpha(x_\alpha)]$ is a quadratic form for the n component field $\Psi \equiv (\psi_1, \psi_2, \ldots, \psi_n)$, given by

$$\mathcal{H}_1[\Psi] \equiv \Psi M \Psi^T = \sum_{\alpha=1}^{n} \sum_{\beta=1}^{n} \int dx_\alpha dy_\beta \psi_\alpha(x_\alpha) G^d(R_\alpha(x_\alpha) - R_\beta(y_\beta)) \psi_\beta(y_\beta),$$

$$(27)$$

where $G^d(r) \equiv \langle \phi(r)\phi(0)\rangle_0$ is the two-point correlation function of the field ϕ in free space. Finally, the effective interaction between the manifolds is obtained by performing the Gaussian integrations over the field Ψ as

$$\mathcal{H}_{\text{eff}}[R_\alpha(x_\alpha)] = \frac{k_B T}{2} \ln \det(M[R_\alpha(x_\alpha)]). \qquad (28)$$

The matrix M is a functional of $R_\alpha(x_\alpha)$ and its determinant is in general difficult to evaluate for arbitrary configurations. It is possible, however, to perturbatively calculate the corrections due to small deformations around simple base configurations. Consider two surfaces in $d = 3$, with average separation H. One plate has small deformations described by $h(\mathbf{x})$, that is, $R_1(\mathbf{x}) = (x_1, x_2, 0)$, while $R_2(\mathbf{x}) = (x_1, x_2, H + h(\mathbf{x}))$. The effective Hamiltonian can now be written as $\mathcal{H}_{\text{eff}} = \mathcal{H}_{\text{flat}} + \mathcal{H}_{\text{corr}}$, where $\mathcal{H}_{\text{flat}}$ is the Casimir interaction for two flat plates, and $\mathcal{H}_{\text{corr}}$ is the additional cost of deformations. For flat plates of area A, the interaction energy is

$$\frac{\mathcal{H}_{\text{flat}}}{A} = k_B T \int \frac{d^2\mathbf{p}}{(2\pi)^2} \ln\left(\frac{1}{2\alpha p}\right) - \frac{\zeta(3)}{16\pi} \frac{k_B T}{H^2}. \qquad (29)$$

The first term in Eq. (29) is a contribution to the surface tension which depends on a lattice cut-off. The second term is the usual Casimir interaction, decaying as $1/H^2$ and with a universal amplitude $-\zeta(3)/16\pi \approx -0.02391$. The energy cost of deformations is given by

$$\mathcal{H}_{\text{corr}} = -k_B T \times \frac{3\zeta(3)}{16\pi H^4} \int d^2\mathbf{x}\, h^2(\mathbf{x}) + \frac{k_B T}{4} \int d^2\mathbf{x}\, d^2\mathbf{y}\, [h(\mathbf{x}) - h(\mathbf{y})]^2$$

$$\times \left\{ \frac{1}{8\pi^2 |\mathbf{x} - \mathbf{y}|^6} + \frac{1}{2\pi |\mathbf{x} - \mathbf{y}|^3 H^3} K_1\left(\frac{|\mathbf{x} - \mathbf{y}|}{H}\right) \right.$$

$$\left. + \frac{1}{H^6}\left[K_1\left(\frac{|\mathbf{x} - \mathbf{y}|}{H}\right)\right]^2 + \frac{1}{H^6}\left[K_2\left(\frac{|\mathbf{x} - \mathbf{y}|}{H}\right)\right]^2 \right\}, \qquad (30)$$

with two kernel functions defined by

$$K_1(t) \equiv \int_0^\infty \frac{du}{2\pi} u^2 (e^{2u} - 1)^{-1} J_0(tu), \qquad (31)$$

$$K_2(t) \equiv \int_0^\infty \frac{du}{2\pi} u^2 e^u (e^{2u} - 1)^{-1} J_0(tu) \,. \tag{32}$$

The first term in Eq. (30) represents an instability to deformations that is related to the attraction between the plates. Remarkably, this term can be obtained intuitively by replacing $1/H^2$ with $1/(H + h(\mathbf{x}))^2$ in Eq. (29) and averaging over the position $\mathbf{x}$. The second term represents long-range interactions between the deformations, induced by the fluctuations of the field. The first term in the curly brackets is the conformation energy of the deformed surface in the absence of the second plane, and is independent of H. The remaining terms represent correlations due to the presence of the second plane. Both $K_1(t)$ and $K_2(t)$ approach a constant as $t \to 0$. As $t \to \infty$, $K_1(t) \sim 1/t^3$, while $K_2(t) \sim \exp(-bt)$, with $b \approx 3.3$. The large t behaviors of $K_1(t)$ and $K_2(t)$ determine the long-range interactions between height fluctuations.

Equation (30) can be used to calculate the Casimir force between a flat and a quenched rough surface. Many solid surfaces produced by rapid growth or deposition processes are characterized by self-similar fluctuations.[62] The fluctuations of a self-affine surface grow as

$$\overline{[h(\mathbf{x}) - h(\mathbf{y})]^2} = A_S |\mathbf{x} - \mathbf{y}|^{2\zeta_S} \,, \tag{33}$$

where the overbar denotes quenched average, and ζ_S is a characteristic roughness exponent. Using Eq. (33) we can take the average of the H dependent terms in $\mathcal{H}_{\text{eff}}$, to obtain a free energy per unit area

$$\mathcal{F}(H) = -\frac{k_B T}{H^2} \frac{\zeta(3)}{16\pi} - \frac{k_B T A_S L^{2\zeta_S}}{H^4} \frac{3\zeta(3)}{16\pi} + \frac{k_B T A_S}{H^{4-2\zeta_S}} \frac{C_1}{4} \,, \tag{34}$$

where L is the extent (upper cut-off) of the self-affine structure, satisfying $\Delta H \equiv A_S^{\frac{1}{2}} L^{\zeta_S} \ll H$. (This is the condition that the total width due to roughness, ΔH is less than the average separation H, so that the plates are not in contact.) The coefficient C_1 in Eq. (34) is given by

$$C_1 = \int_{a/H}^{L/H} dt \{ -t^{2\zeta_S - 2} K_1(t) + 2\pi t^{2\zeta_S + 1} [K_1(t)]^2 + 2\pi t^{2\zeta_S + 1} [K_2(t)]^2 \} \,, \tag{35}$$

and does weakly depend on the ratio L/H, but since the functions K_1 and K_2 decay rapidly at large distance, it is quite insensitive to L as long as $L \gg H$. As long as $L \gg H \gg \Delta H$, the interactions in Eq. (34) are arranged in order of increasing strength. The largest effect of randomness is to increase the

Casimir attraction by an amount proportional to $(\Delta H/H)^2$. There is also another correction term, of the opposite sign, that decays as $1/H^{4-2\zeta_S}$, and in principle can be used to indirectly measure the roughness exponent ζ_S. In Eq. (33), if all lengths are measured in units of an atomic scale a_0 (e.g., the diameter of a surface atom), A_S becomes dimensionless. Using a reasonable set of parameters: $\zeta_S \approx 0.35$, $a_0 \approx 5$ Å, $A_S \approx 1$ and $L \approx 300$ Å, we estimate that for surfaces of 1 mm size, and 100 Å apart, the forces generated by the three terms in Eq. (34) are 1.9×10^{-4}, 4.9×10^{-5}, and 3.7×10^{-6} N respectively, (with appropriate lower cut-off ~ 20 Å), which is measurable with the current force apparatus.[10,11]

The calculation of the effective partition function in Eq. (25) is easily extended to four-dimensional space–time. In a relativistic theory, the space and time coordinates are related by Poincaré symmetries, and the action for a free scalar field takes the simple form

$$S = \frac{1}{2} \int d^4 X \partial_\mu \phi(X) \partial_\mu \phi(X) \,, \tag{36}$$

where summation over $\mu = 0, \ldots, 3$ is implicit. Following a Wick rotation, imaginary time appears as another coordinate $X_4 = ict$ in the four-dimensional space–time. In a path integral formulation, the different configurations of the field ϕ are now weighted by $\exp[-S/\hbar]$. The boundary conditions on the field are implemented by inserting delta-functions as before. However, because of the symmetry between space and time coordinates, we are now at liberty to impose the boundary condition on a manifold $\mathbf{r}(\mathbf{x}, t)$, that is, we can treat deformations in space and time on the same footing. This allows us to address the problem of the dynamic Casimir effect, which is the topic of the next section.

4. The Dynamic Casimir Effect

4.1. *Background*

Although less well known than its static counterpart, the dynamical Casimir effect, describing the force and radiation from moving mirrors has also garnered much attention.[63–69] This is partly due to connections to Hawking and Unruh effects (radiation from black holes and accelerating bodies, respectively), suggesting a deeper link between quantum mechanics, relativity and cosmology.[70]

The creation of photons by moving mirrors was first obtained by Moore[63] for a one-dimensional cavity. Fulling and Davis[64] demonstrated that there is a corresponding force, even for a single mirror, which depends on the third time derivative of its displacement. These computations take advantage of conformal symmetries of the (1+1)-dimensional space time, and cannot be easily generalized to higher dimension. Furthermore, the calculated force has causality problems reminiscent of the radiation reaction forces in classical electron theory.[65] It has been shown that this problem is an artifact of the unphysical assumption of perfect reflectivity of the mirror, and can be resolved by considering realistic frequency dependent reflection and transmission from the mirror.[65]

Another approach to the problem starts with fluctuations in the force on a single plate. The fluctuation–dissipation theorem is then used to obtain the mechanical response function,[66] whose imaginary part is related to the dissipation. This method does not have any causality problems, and can also be extended to higher dimensions. (The force in (1+3)-dimensional space–time depends on the fifth power of the motional frequency.) The emission of photons by a perfect cavity, and the observability of this energy, have been studied by different approaches.[67–69] The most promising candidate is the resonant production of photons when the mirrors vibrate at the optical resonance frequency of the cavity.[70] A review, and more extensive references are found in Ref. 71. More recently, the radiation due to vacuum fluctuations of a collapsing bubble has been proposed as a possible explanation for the intriguing phenomenon of sonoluminescence.[72,73]

A number of authors have further discussed the notion of *frictional forces*: Using conformal methods in (1+1)-dimensions, Ref. 74 finds a friction term

$$F_{\text{friction}}(H) = \alpha \, F_{\text{static}}(H) \left(\frac{\dot{H}}{c} \right)^2 , \tag{37}$$

for slowly moving boundaries, where α is a numerical constant that only depends on dimensionality. The additional factor of $(v/c)^2$ would make detection of this force yet more difficult. There have been a few attempts to calculate forces (in higher dimensions) for walls that move *laterally*, i.e. parallel to each other[75–78]: It is found that boundaries that are not ideal conductors, experience a friction as if the plates are moving in a viscous fluid. The friction has a complicated dependence on the frequency dependent resistivity of the plates, and vanishes in the cases of ideal (nondissipating) conductors or dielectrics. The "dissipation" mechanism for this "friction" is by inducing eddy currents in the nonideal conductors, and are thus distinct from the Casimir effect. Experimental evidence of such a contribution to friction has been recently reported in Ref. 79. The experiment employs a quartz crystal microbalance technique to measure the friction associated with sliding of solid nitrogen along a lead surface, above and below the lead's superconducting transition temperature. There is an abrupt drop in the friction at the transition point as the substrate enters the superconducting state.[79]

An interesting analogue of the dynamic Casimir effect is found for the moving interface between two different vacuum states of superfluid 3He.[80] In this system, the Andreev reflection of the massless "relativistic" fermions which live on the A-phase side of the interface provides the corresponding mechanism for friction: The interface is analogous to a perfectly reflecting wall moving in the quantum vacuum.

4.2. *Path integral formulation*

Here, we follow the path integral formulation,[81] outlined in the previous section, for the problem of perfectly reflecting mirrors that undergo arbitrary dynamic deformations. We apply the path integral quantization formalism to a scalar field ϕ with the action in Eq. (36). In principle, we should use the electromagnetic vector potential $A_\mu(X)$, but requirements of gauge fixing complicate the calculations, while the final results only change by a numerical prefactor. (We have explicitly checked that we reproduce the known answer for flat plates by this method.) We would like to quantize the field subject to the constraints of its vanishing on a set of n manifolds (objects) defined by $X = X_\alpha(y_\alpha)$, where y_α parametrize the αth manifold. We implement the constraints using delta functions, and write the partition function as

$$\mathcal{Z} = \int \mathcal{D}\phi(X) \prod_{\alpha=1}^{n} \prod_{y_\alpha} \delta(\phi(X_\alpha(y_\alpha))) \exp\left\{-\frac{S[\phi]}{\hbar}\right\}. \tag{38}$$

The delta functions are next represented by integrals over Lagrange multiplier fields. Performing the Gaussian integrations over $\phi(X)$ then leads to an effective action for the Lagrange multipliers which is again Gaussian.[16] Evaluating $\mathcal{Z}$ is thus reduced to calculating the logarithm of the determinant of a kernel. Since the Lagrange multipliers are defined on a set of manifolds with nontrivial geometry, this calculation is generally complicated. To be specific, we focus on two parallel two-dimensional plates embedded in 3+1 space–time, and separated by an average distance H along the x_3-direction. Deformations of the plates are parametrized by the height functions $h_1(\mathbf{x}, t)$ and $h_2(\mathbf{x}, t)$, where $\mathbf{x} \equiv (x_1, x_2)$ denotes the two lateral space coordinates, while t is the time variable. Following Ref. 16, $\ln \mathcal{Z}$ is calculated by a perturbative series in powers of the height functions. The resulting expression for the effective action (in real time), defined by $S_{\text{eff}} \equiv -i\hbar \ln \mathcal{Z}$, after eliminating h independent terms, is

$$S_{\text{eff}} = \frac{\hbar c}{2} \int \frac{d\omega d^2\mathbf{q}}{(2\pi)^3} [A_+(q,\omega)(|h_1(\mathbf{q},\omega)|^2 + |h_2(\mathbf{q},\omega)|^2)$$

$$- A_-(q,\omega)(h_1(\mathbf{q},\omega)h_2(-\mathbf{q},-\omega) + h_1(-\mathbf{q},-\omega)h_2(\mathbf{q},\omega))] + O(h^3)\,. \tag{39}$$

4.3. *The response function*

The kernels $A_\pm(q,\omega)$, that are closely related to the mechanical response of the system (see below), are functions of the separation H, but depend on $\mathbf{q}$ and ω only through the combination $Q^2 = q^2 - \omega^2/c^2$. The closed forms for these kernels involve cumbersome integrals, and are not very illuminating. Instead of exhibiting these formulas, we shall describe their behavior in various regions of the parameter space. In the limit $H \to \infty$, $A_-^\infty(q,\omega) = 0$, and

$$A_+^\infty(q,\omega) = \frac{1}{360\pi^2 c^5} \begin{cases} -(c^2 q^2 - \omega^2)^{5/2} & \text{for } \omega < cq\,, \\ i\,\text{sgn}(\omega)(\omega^2 - c^2 q^2)^{5/2} & \text{for } \omega > cq\,, \end{cases} \tag{40}$$

where $\text{sgn}(\omega)$ is the sign function. While the effective action is real for $Q^2 > 0$, it becomes purely imaginary for $Q^2 < 0$. The latter signifies dissipation of energy,[66] presumably by generation of photons.[69] It agrees precisely with the results obtained previously[66] for the special case of flat mirrors ($\mathbf{q} = 0$). (Note that dissipation is already present for a single mirror.)

In the presence of a second plate (i.e. for finite H), the parameter space of the kernels subdivides into three different regions as depicted in Fig. 2. In region I ($Q^2 > 0$ for any H), the kernels are finite and real, and hence there is no dissipation. In region IIa where $-\pi^2/H^2 \leq Q^2 < 0$, the H-independent part of A_+ is imaginary, while the H-dependent parts of both kernels are real and finite. (This is also the case at the boundary $Q^2 = -\pi^2/H^2$.) The dissipation in this regime is simply the sum of what would have been observed if the individual plates were decoupled, and unrelated to the separation H. By contrast, in region IIb where $Q^2 < -\pi^2/H^2$, both kernels diverge with infinite real and imaginary parts.[82] This H-dependent divergence extends all the way to the negative Q^2 axis, where it is switched off by a $1/H^5$ prefactor.

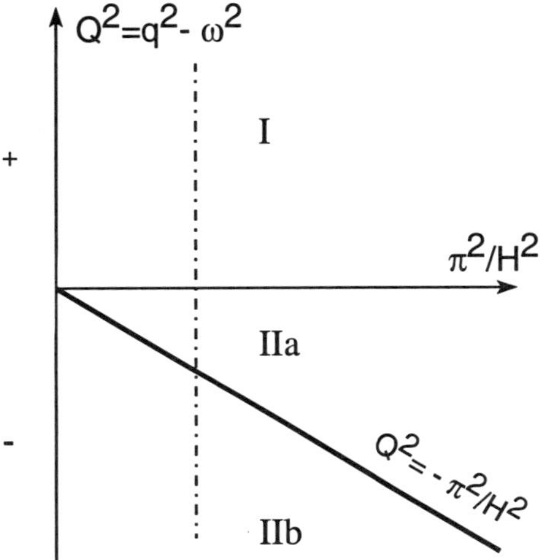

Fig. 2. Different regions of the $(\mathbf{q}, \omega)$ plane.

As a concrete example, let us examine the lateral vibration of plates with fixed roughness, such as two corrugated mirrors.[83] The motion of the plates enters through the time dependencies $h_1(\mathbf{x}, t) = h_1(\mathbf{x} - \mathbf{r}(t))$ and $h_2(\mathbf{x}, t) = h_2(\mathbf{x})$; that is, the first plate undergoes lateral motion described by $\mathbf{r}(t)$, while the second plate is stationary. The lateral force exerted on the first plate is obtained from $f_i(t) = \delta S_{\text{eff}}/\delta r_i(t)$. Within linear response, it is given by

$$f_i(\omega) = \chi_{ij}(\omega)r_j(\omega) + f_i^0(\omega)\,, \tag{41}$$

where the "mechanical response tensor" is

$$\chi_{ij}(\omega) = \hbar c \int \frac{d^2q}{(2\pi)^2} q_i q_j \left\{ [A_+(q,\omega) - A_+(q,0)]|h_1(\mathbf{q})|^2 \right.$$

$$\left. + \frac{1}{2} A_-(q,0)(h_1(\mathbf{q})h_2(-\mathbf{q}) + h_1(-\mathbf{q})h_2(\mathbf{q})) \right\}\,, \tag{42}$$

and there is a residual force

$$f_i^0(\omega) = -\frac{\hbar c}{2} 2\pi\delta(\omega) \int \frac{d^2q}{(2\pi)^2} iq_i A_-(q,0)(h_1(\mathbf{q})h_2(-\mathbf{q}) - h_1(-\mathbf{q})h_2(\mathbf{q}))\,. \tag{43}$$

Let us now consider a corrugated plate with a deformation $h_1(\mathbf{x}) = d\cos(\mathbf{k} \cdot \mathbf{x})$. From the frequency–wavevector dependence of the mechanical response function in Eq. (42), we extract a plethora of interesting results, some of which we list below:

1. *Mass corrections.* For a single plate ($H \to \infty$), we can easily calculate the response tensor using the explicit formulas in Eq. (40). In the limit of $\omega \ll ck$, expanding the result in powers of ω gives $\chi_{ij} = \delta m_{ij}\omega^2 + O(\omega^4)$, where $\delta m_{ij} = A\hbar d^2 k^3 k_i k_j/(288\pi^2 c)$, can be regarded as corrections to the mass of the plate. (Cut-off dependent mass corrections also appear, as in Ref. 71.) Note that these mass corrections are *anisotropic* with

$$\delta m_{\parallel} = \frac{1}{288\pi^2}\frac{\hbar}{c}Ad^2k^5 \,,$$

$$\delta m_{\perp} = 0 \,. \tag{44}$$

Parallel and perpendicular components are defined with respect to $\mathbf{k}$, and A denotes the area of the plates. The mass correction is inherently very small: For a macroscopic sample with $d \approx \lambda = 2\pi/k \approx 1$ mm, density ≈ 15 gr/cm^3, and thickness $t \approx 1$ mm, we find $\delta m/m \sim 10^{-34}$. Even for deformations of a microscopic sample of atomic dimensions (close to the limits of the applicability of our continuum representations of the boundaries), $\delta m/m$ can only be reduced to around 10^{-10}. With the second plate at a separation H, the mass renormalization becomes a function of both k and H, with a crossover from the single plate behavior for $kH \sim 1$. In the limit of $kH \ll 1$, we obtain $\delta m_{\parallel} = \hbar ABk^2d^2/48cH^3$ and $\delta m_{\perp} = 0$, with $B = -0.453$. Compared to the single plate, there is an enhancement by a factor of $(kH)^{-3}$ in $\delta m_{\parallel}$. While the actual changes in mass are immeasurably small, the hope is that its *anisotropy* may be more accessible, say by comparing oscillation frequencies of a plate in two orthogonal directions.

2. *Dissipation.* For $\omega \gg ck$ the response function is imaginary, and we define a frequency dependent effective shear viscosity by $\chi_{ij}(\omega) = -i\omega\eta_{ij}(\omega)$. This viscosity is also anisotropic, with

$$\eta_{\parallel}(\omega) = \frac{1}{720\pi^2}\frac{\hbar}{c^4}Ad^2k^2\omega^4 \,,$$

$$\eta_{\perp}(\omega) = 0 \,. \tag{45}$$

Note that the dissipation is proportional to the fifth time derivative of displacement, and there is no dissipation for a uniformly accelerating plate. However, a freely oscillating plate will undergo a damping of its motion. The characteristic decay time for a plate of mass M is $\tau \approx 2M/\eta$. For the macroscopic plate of the previous paragraph, vibrating at a frequency of $\omega \approx 2ck$ (in the 10^{12} Hz range), the decay time is enormous, $\tau \sim 10^{18}$ s. However, since the decay time scales as the fifth power of the dimension, it can be reduced to 10^{-12} s, for plates of order of 10 atoms. However, the required frequencies in this case (in the 10^{18} Hz range) are very large. Also note that for the linearized forms to remain valid in this high frequency regime, we must require very small amplitudes, so that the typical velocities involved $v \sim r_0\omega$, are smaller than the speed of light. The effective dissipation in region IIa of Fig. 2 is simply the sum of those due to individual plates, and contains no H dependence.

3. *Resonant emission.* The cavity formed between the two plates supports continuous spectrum of normal modes for frequencies $\omega^2 > c^2(k^2 + \pi^2/H^2)$. We find that both real and imaginary parts of $A_\pm(k,\omega)$, diverge in this regime which we interpret as resonant dissipation due to excitation of photons in the cavity. Resonant dissipation has profound consequences for motion of plates. It implies that due to quantum fluctuations of vacuum, *components of motion with frequencies in the range of divergences cannot be generated by any finite external force*! The imaginary parts of the kernels are proportional to the total number of excited photons.[69] Exciting these degrees of motion must be accompanied by the generation of an infinite number of photons; requiring an infinite amount of energy, and thus impossible. However, as pointed out in Ref. 69, the divergence is rounded off by assuming finite reflectivity and transmissivity for the mirrors. Hence, in practice, the restriction is softened and controlled by the degree of ideality of the mirrors in the frequency region of interest.

Related effects have been reported in the literature for $(1+1)$-dimensions,[67–70] but occurring at a *discrete* set of frequencies $\omega_n = n\pi c/H$ with integer $n \geq 2$. These resonances occur when the frequency of the external perturbation matches the natural normal modes of the cavity, thus exciting quanta of such modes. In one space dimension, such modes are characterized by a discrete set of wavevectors that are integer multiples of π/H. The restriction to $n \geq 2$ is a consequence of quantum electrodynamics being a "free" theory (quadratic action): only two-photon states can be excited subject to conservation of energy. Thus the sum of the frequencies of the two photons should add up to the external frequency.[69] In higher dimensions, the appropriate parameter is the combination $\omega^2/c^2 - q^2$. From the perspective of the excited photons, conservation of momentum requires that their two momenta add up to q, while energy conservation restricts the sum of their frequencies to ω. The in-plane momentum q, introduces a continuous degree of freedom: the resonance condition can now be satisfied for a continuous spectrum, in analogy with optical resonators. In Ref. 69, the lowest resonance frequency is found to be $2\pi c/H$ which seems to contradict our prediction. However, the absence of $\omega_1 = \pi c/H$ in 1+1 D is due to a vanishing prefactor,[69] which is also present in our calculations. However, in exploring the continuous frequency spectrum in higher dimensions, this single point is easily bypassed, and there is a divergence for all frequencies satisfying $\omega^2/c^2 > q^2 + \pi^2/H^2$, where the inequality holds in its strict sense.

4. *"Josephson"-like effects.* The constant term in Eq. (43) in sensitive to the shapes of the plates. For two plates corrugated at the same wavelength, with deformations $h_1(\mathbf{x}) = d_1 \cos(\mathbf{k} \cdot \mathbf{x})$ and $h_2(\mathbf{x}) = d_2 \cos(\mathbf{k} \cdot \mathbf{x} + \alpha)$, there is a (time-independent) lateral force

$$\mathbf{F}_{\text{dc}} = \frac{\hbar c A}{2} A_-(k,0)\mathbf{k}d_1 d_2 \sin\alpha \,, \tag{46}$$

which tends to keep the plates 180 degrees out of phase, that is, mirror symmetric with respect to their mid-plane. The dependence on the sine of the phase mismatch is reminiscent of the DC Josephson current in superconductor junctions, the force playing a role analogous to the current in SIS junctions. There is also an analog for the AC Josephson effect, with velocity (the variable conjugate to force) playing the role of voltage: Consider two corrugated plates separated at a distance H, described by $h_1(\mathbf{x},t) = d_1 \cos[\mathbf{k} \cdot (\mathbf{x} - \mathbf{r}(t))]$ and $h_2(\mathbf{x},t) = d_2 \cos[\mathbf{k} \cdot \mathbf{x}]$. The resulting force at a constant velocity ($\mathbf{r}(t) = \mathbf{v}t$),

$$\mathbf{F}_{\text{ac}} = \frac{\hbar c A}{2} A_-(k,0)\mathbf{k}d_1 d_2 \sin[(\mathbf{k} \cdot \mathbf{v})t] \,, \tag{47}$$

oscillates at a frequency $\omega = \mathbf{k} \cdot \mathbf{v}$. Actually both effects are a consequence of the attractive nature of the Casimir force. Experiments are underway at UC-Riverside to try to observe these effects.[84]

5. *Surface tension corrections.* Finally, let us consider the capillary waves on the surface of mercury, with a conducting plate placed at a separation H above the surface. The low frequency–wavevector expansion of the kernel due to quantum fluctuations in the intervening vacuum, starts with quadratic forms q^2 and ω^2. These terms result in corrections to the (surface) mass density by $\delta\rho = \hbar B/48cH^3$, and to the surface tension by $\delta\sigma = \hbar c B/48H^3$. The

latter correction is larger by a factor of $(c/c_s)^2$, and changes the velocity c_s, of capillary waves by $\delta c_s/c_s^0 = \hbar c B/96\sigma H^3$, where σ is the bare surface tension of mercury. Taking $H \sim 1$ mm and $\sigma \sim 500$ dynes/cm, we find another very small correction of $\delta c_s/c_s^0 \sim 10^{-19}$.

4.4. *Radiation spectra*

Where does the energy go when the plates experience viscous dissipation? When the viscosity is a result of losses in the dispersive boundaries,[75-78] the energy is used up in heating the plates. Since we have examined perfect mirrors, the dissipated energy can only be accounted for by the emission of photons into the cavity. The path integral methods can be further exploited to calculate the spectrum of the emitted radiation.[85] The basic idea is to relate the transition amplitude from an empty vacuum (at $t \to -\infty$) to a state with two photons (at $t \to +\infty$), to a two-point correlation function of the field, which is then calculated perturbatively in the deformations. From the transition amplitude (after integrating over the states of one photon) we obtain the probability that an emitted photon is observed at a frequency Ω and with a particular orientation.

Specifically, calculations of the angular distribution and spectrum of radiation were performed in Ref. 85 for a single perfectly reflecting plate, that

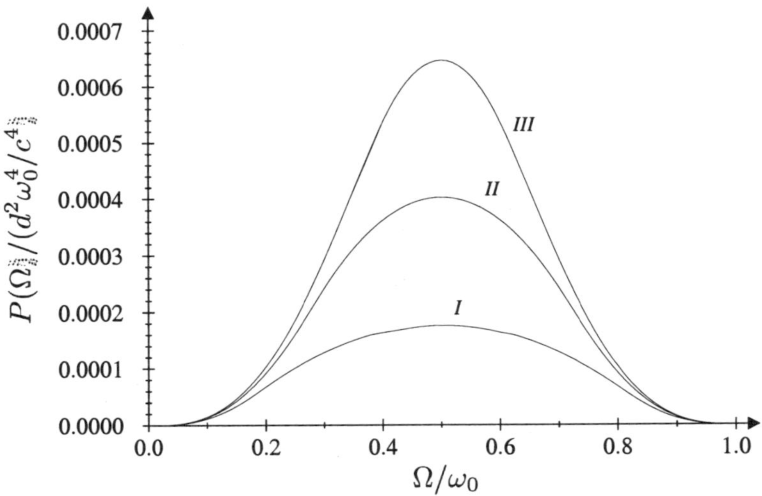

Fig. 3. Spectrum of radiation for different classes. Plot *I* corresponds to $\omega_0/ck_0 = 5/3$, plot *II* corresponds to $\omega_0/ck_0 = 5/2$, and plot *III* corresponds to $\omega_0/ck_0 = 5$.

undergoes harmonic deformations characterized by a height function $h(\mathbf{x}, t) = d\cos(\mathbf{k}_0 \cdot \mathbf{x} - \omega_0 t)$. Depending on the ratio ω_0/ck_0, it is found that radiation at a frequency Ω is restricted to a particular window in solid angle. The total spectrum of radiation $P(\Omega)$, that is the number of photons with frequencies between Ω and $\Omega + d\Omega$ radiated per unit time per unit area of the plate, is found by integrating the angular distribution over the unit sphere, and is shown in Fig. 3. Note that it is a symmetric function with respect to $\omega_0/2$, where it is peaked, which is a characteristic behavior of two-photon processes. The peak sharpens as the parameter $\omega_0/ck_0 \to 0$, and saturates for $k_0 = 0$.

The connection between the dissipative dynamic Casimir force and radiation of photons is made explicit by calculating the total number of photons radiated per unit time and per unit area of the plate. The result is identical to the energy dissipation rate calculated from mechanical response considerations.[81] No radiation is observed at frequencies higher than ω_0, due to conservation of energy, and also for $\omega_0/ck_0 < 1$, in agreement with Sec. B above, where no dissipative forces are found in this regime.

5. Conclusion

In this article we presented various examples of fluctuation-induced phenomena, taken from diverse contexts ranging from biophysics to cavity QED. We hope to have illustrated some of the underlying principles that govern these different manifestations. The basic idea is that if fluctuations of a field (whether of quantum or thermal origin) are hampered by the presence of external objects, there is a back reaction felt by these external objects. Due to their entropic nature, the resulting interactions usually depend on geometrical properties of the objects, with scales set by $\hbar$ (quantum fluctuations) or $k_B T$ (thermal fluctuations). There is a great deal of *universality*, in that the asymptotic limits of the interactions at large distances depend only on the shape of the object, and are independent of microscopic details or energy scales. The dependence on shape, however, can be unexpected and nontrivial; for example, the interaction between extended objects cannot be obtained from a pairwise summation of pair potentials.

Thermal fluctuation-induced interactions are starting to be probed more systematically by experiments on wetting of complex fluids, and colloidal particles. Such interactions may also account for attractions between macroions of like charge, in turn explaining of bundle formation and collapse of DNA filaments and other stiff polyelectrolytes. In fact, the very existence of a fluid phase (condensing from a gas) is due to the first historically recognized fluctuation-induced interactions, the van der Waals force.

The nonintuitive world of frictional Casimir forces and radiation from a perturbed vacuum brings about a deeper understanding of a fundamental physical entity, the quantum vacuum. As any complex fluid the quantum vacuum interacts with objects moving through it, hindering and modifying their motion. While the dynamic Casimir effect appears more as an academic problem at the moment, it has nevertheless raised the hope among some quantum opticians of making an optical resonator from a mechanical one; that is, a laser with mechanical pumping!

A. Path Integral Formulation of Charged Fluids

Here, we introduce a systematic path integral formulation to study fluctuation-induced interactions in a charged fluid. Consider n charged manifolds embedded in a d-dimensional aqueous solution of neutralizing counterions, interacting through Coulomb potentials. The manifolds have charge densities $-\sigma_\alpha$ (all assumed to be negatively charged for simplicity), and are described by the functions $R_\alpha(x_\alpha)$, where x_α is a D_α-dimensional internal coordinate, while R_α indicates a position in the d-dimensional solution. There are N_c positively charged counterions of valence z, each described by a position vector R_i in the d-dimensional solution. The Coulomb Hamiltonian can be written as

$$H_C = \frac{1}{2} \int \mathrm{d}^d X \mathrm{d}^d X' \rho(X) \frac{e^2}{\epsilon |X - X'|^{d-2}} \rho(X') \,, \tag{48}$$

where

$$\rho(X) = -\sum_{\alpha=1}^{n} \int \mathrm{d}x_\alpha \sigma_\alpha \delta^d(X - R_\alpha(x_\alpha))$$

$$+ \sum_{i=1}^{N_c} z \delta^d(X - R_i) \,, \tag{49}$$

is the number density of the charges. Charge neutrality requires $-\sum_{\alpha=1}^{n} \sigma_\alpha A_\alpha + z N_c = 0$, where A_α is the D_α-dimensional area of the αth manifold.

A restricted partition function of the Coulomb system, depending upon the shapes and locations of the macroions, is now given by

$$\mathcal{Z}_{N_c}[R_\alpha(x_\alpha)] = \int \prod_{i=1}^{N_c} \frac{\mathrm{d}^d R_i}{a^d} e^{-H_C/k_B T} \,, \tag{50}$$

in which a is a short-distance cut-off. Using the Hubbard–Stratanovich transformation of the Coulomb interaction,

$$e^{-H_C/k_BT} = \int \mathcal{D}\phi(X) \exp\left\{ -\frac{\epsilon k_BT}{2S_d e^2} \int \mathrm{d}^d X (\nabla\phi)^2 + i \int \mathrm{d}^d X \rho(X)\phi(X) \right\},$$
(51)

we can rewrite the partition function as

$$\mathcal{Z}_{N_c}[R_\alpha(x_\alpha)] = \int \mathcal{D}\phi(X)$$

$$\times \exp\left\{ -\frac{\epsilon k_BT}{2S_d e^2} \int \mathrm{d}^d X (\nabla\phi)^2 - i \sum_{\alpha=1}^{n} \int \mathrm{d}x_\alpha \sigma_\alpha \phi(R_\alpha(x_\alpha)) \right\}$$

$$\times \left(\int \frac{\mathrm{d}^d R}{a^d} e^{iz\phi(R)} \right)^{N_c},$$
(52)

where S_d is the area of the d-dimensional unit sphere. We can introduce a fugacity y, and a rescaled partition function

$$\mathcal{Z}[R_\alpha(x_\alpha)] = \frac{y^{N_c}}{N_c!} \mathcal{Z}_{N_c}[R_\alpha(x_\alpha)],$$
(53)

that can be rewritten as

$$\mathcal{Z} = \sum_{N=0}^{\infty} \delta_{N,N_c} \frac{y^N}{N!} \mathcal{Z}_N[R_\alpha(x_\alpha)]$$

$$= \sum_{N=0}^{\infty} \int_0^{2\pi} \frac{\mathrm{d}\theta}{2\pi} e^{i\theta(N_c-N)} \int \mathcal{D}\phi(X)$$

$$\times \exp\left\{ -\frac{\epsilon k_BT}{2S_d e^2} \int \mathrm{d}^d X (\nabla\phi)^2 - i \sum_{\alpha=1}^{n} \int \mathrm{d}x_\alpha \sigma_\alpha \phi(R_\alpha(x_\alpha)) \right\}$$

$$\times \frac{1}{N!} \left(y \int \frac{\mathrm{d}^d R}{a^d} e^{iz\phi(R)} \right)^{N}.$$
(54)

A shift in the field ϕ by $-\theta$, and use of the neutrality condition renders the θ-integration trivial. We can then sum up the exponential series, and obtain

$$\mathcal{Z}[R_\alpha(x_\alpha)] = \int \mathcal{D}\phi(X) e^{-\mathcal{H}[\phi]},$$
(55)

in which

$$\mathcal{H}[\phi] = \frac{\epsilon k_B T}{2 S_d e^2} \int \mathrm{d}^d X (\nabla \phi)^2 + i \sum_{\alpha=1}^{n} \int \mathrm{d}x_\alpha \sigma_\alpha \phi(R_\alpha(x_\alpha)) - \frac{y}{a^d} \int \mathrm{d}^d X e^{iz\phi(X)}.$$

(56)

Note that the fugacity y can be eliminated using the identity

$$N_c = y \frac{\partial \ln \mathcal{Z}}{\partial y},$$

(57)

which follows from Eq. (53).

We next evaluate the path integral using a saddle point approximation. The extremum of Eq. (55), obtained from $\delta \mathcal{H}/\delta \phi = 0$, is the solution of the Poisson–Boltzmann (PB) equation

$$-\nabla^2(z\psi(X)) - \kappa^2 e^{-z\psi(X)} = -\sum_{\alpha=1}^{n} \int \mathrm{d}x_\alpha \frac{S_d e^2 z \sigma_\alpha}{\epsilon k_B T} \delta^d(X - R_\alpha(x_\alpha)),$$

(58)

for the (real) field $\psi(X) = -i\bar{\phi}(X)$, in which $\kappa^2 = S_d e^2 y z^2 / \epsilon k_B T a^d$ defines the inverse square of the Debye screening length. To study the fluctuations on top of this saddle point, we can set $\phi = \bar{\phi} + \delta \phi$, and expand the Hamiltonian up to quadratic order, to get

$$\mathcal{H}[\phi] = \mathcal{H}[\bar{\phi}] + \frac{\epsilon k_B T}{2 S_d e^2} \int \mathrm{d}^d X [(\nabla \delta \phi)^2 + \kappa^2 e^{-z\psi(X)} \delta \phi^2].$$

(59)

The free energy of the system of charged manifolds in the presence of fluctuating counterions now reads

$$F = F_{\mathrm{PB}} + \frac{k_B T}{2} \ln \det[-\nabla^2 + m^2(X)],$$

(60)

where $F_{\mathrm{PB}} = \mathcal{H}[i\psi(X)]$ is the Poisson–Boltzmann free energy, and $m^2(X) = \kappa^2 e^{-z\psi(X)}$ is a "mass (or charge compressibility) profile". The PB free energy is known to be generically repulsive.[29,48] The fluctuation-induced correction, however, is attractive. For highly charged manifolds, it is indeed reminiscent of the Casimir interactions, but with the boundary conditions being smoothed out. To see this, one should note that the mass profile is indeed identical to the density profile of the counterions. Highly charged manifolds accumulate counterions in their vicinity, and consequently the fluctuations of the "potential" field ϕ are suppressed in a region close to the manifolds, and free in other regions of the solution; hence leading to a Casimir-type fluctuation-induced interaction.

Acknowledgments

This review follows closely an article by the authors in *Reviews of Modern Physics*.[6] We have benefited from collaborations with M. Goulian, H. Li, M. Lyra, and F. Miri on these problems. We are indebted to M. Chan and R. Garcia for supplying us with their experimental details and data prior to publication. Ramin Golestanian acknowledges many helpful discussions with J. Indekeu, and T. Liverpool, and support by the NSF grant PHY94-07194 and the Institute for Advanced Studies in Basic Sciences, Gava Zang, Zanjan, Iran. Mehran Kardar is supported by the NSF grant DMR-98-05833. Many thanks are due to M.R.H. Khajehpour, M. Kolahchi, and M. Sahimi for their efforts in organizing this conference, and their unfailing hospitality.

References

1. H. B. G. Casimir, *Proc. K. Ned. Akad. Wet.* **51**, 793 (1948).
2. I. E. Dzyaloshinskii, E. M. Lifshitz, and L. P. Pitaevskii, *Advan. Phys.* **10**, 165 (1961).
3. V. M. Mostepanenko and N. N. Trunov, *The Casimir Effect and Its Applications* (Clarendon Press, Oxford, 1997).
4. M. Krech, *The Casimir Effect in Critical Systems* (World Scientific, Singapore, 1994).
5. S. Weinberg, *Rev. Mod. Phys.* **61**, 1 (1989).
6. M. Kardar and R. Golestanian, *Rev. Mod. Phys.* **71**, 1233 (1999).
7. I. I. Abricossova and B. V. Deryaguin, *Dokl. Akad. Nauk. SSSR* **90**, 1055 (1953).
8. M. J. Sparnaay, *Physica* **24**, 751 (1958).
9. See, e.g., J. N. Israelachvili and P. M. McGuigan, *Science* **241**, 6546 (1990).
10. S. K. Lamoreaux, *Phys. Rev. Lett.* **78**, 5 (1997).
11. U. Mohideen and A. Roy, *Phys. Rev. Lett.* **81**, 4549 (1998).
12. M. E. Fisher and P.-G. de Gennes, *C. R. Acad. Sci. Ser.* **B287**, 207 (1978); V. Privman and M. E. Fisher, *Phys. Rev.* **B30**, 322 (1984).
13. H. W. J. Blöte, J. L. Cardy, and M. P. Nightingale, *Phys. Rev. Lett.* **56**, 742 (1986).
14. M. P. Nightingale and J. O. Indekeu, *Phys. Rev. Lett.* **54**, 1824 (1985).
15. M. Krech and S. Dietrich, *Phys. Rev. Lett.* **66**, 345 (1991).
16. H. Li and M. Kardar, *Phys. Rev. Lett.* **67**, 3275 (1991); *Phys. Rev.* **A46**, 6490 (1992).
17. J. Schwinger, L. L. DeRaad, and K. A. Milton, *Ann. Phys.* (*N.Y.*) **115**, 1 (1978).
18. For a review see, e.g., S. Dietrich, in *Phase Transitions and Critical Phenomena*, eds. C. Domb and J. L. Lebowitz (Academic, N.Y., 1988), Vol. 12.
19. J. Indekeu, *J. Chem. Soc. Faraday Trans.* **II82**, 1835 (1986).
20. K. K. Mon and M. P. Nightingale, *Phys. Rev.* **B35**, 3560 (1987).
21. R. Garcia and M. Chan, *Phys. Rev. Lett.* **83**, 1187 (1999).

22. P.-G. de Gennes, *The Physics of Liquid Crystals* (Oxford Univ. Press, Oxford, 1974).
23. L. V. Mikheev, *Sov. Phys. JETP* **69**, 358 (1989).
24. A. Ajdari, L. Peliti, and J. Prost, *Phys. Rev. Lett.* **66**, 1481 (1991).
25. J. V. Selinger and D. R. Nelson, *Phys. Rev.* **A37**, 1736 (1988); R. Holyst, D. J. Tweet, and L. B. Sorensen, *Phys. Rev. Lett.* **65**, 2153 (1990).
26. B. D. Swanson, H. Straigler, D. J. Tweet, and L. B. Sorensen, *Phys. Rev. Lett.* **62**, 909 (1989).
27. T. Stoebe, R. Geer, C. C. Huang, and J. W. Goodby, *Phys. Rev. Lett.* **69**, 2090 (1992).
28. M. L. Lyra, M. Kardar, and N. F. Svaiter, *Phys. Rev.* **E47**, 3456 (1993).
29. F. Oosawa, *Biopolymers* **6**, 134 (1968); *Polyelectrolytes* (Marcel Dekker, N.Y., 1971).
30. S. Marcelja, *Biophys. J.* **61**, 1117 (1992), and references therein.
31. P. Attard, J. Mitchell, and B. W. Ninham, *J. Chem. Phys.* **88**, 4987 (1988).
32. See for example Ref. 31 and references therein.
33. P. Attard, R. Kjellander, and D. J. Mitchell, *Chem. Phys. Lett.* **139**, 219 (1987); P. Attard, R. Kjellander, D. J. Mitchell, and B. Jönsson, *J. Chem. Phys.* **89**, 1664 (1988).
34. For a numerical calculation of higher loop corrections see, R. D. Coalson and A. Duncan, *J. Chem. Phys.* **97**, 5653 (1992).
35. P. A. Pincus and S. A. Safran, *Europhys. Lett.* **42**, 103 (1998).
36. R. Golestanian and M. Kardar, unpublished (1998).
37. B.-Y. Ha and A. J. Liu, *Phys. Rev. Lett.* **79**, 1289 (1997); *Phys. Rev. Lett.* **81**, 1011 (1998).
38. R. Golestanian, M. Kardar, and T. B. Liverpool, *Phys. Rev. Lett.* **82**, 4456 (1999).
39. A. W. C. Lau and P. A. Pincus, *Phys. Rev. Lett.* **81**, 1338 (1998).
40. H. B. G. Casimir and D. Polder, *Phys. Rev.* **73**, 360 (1948).
41. F. London, *Z. Phys. Chem.* **B11**, 222 (1930).
42. D. Kleppner, *Physics Today*, October 1990, p. 9.
43. P.-G. de Gennes, *C. R. Acad. Sci. Paris* **II292**, 701 (1981).
44. D. Beysens, J.-M. Petit, T. Narayan, A. Kumar, and M. L. Broide, *Ber. Bunsen-Ges. Phys. Chem.* **98**, 382 (1984).
45. T. W. Burkhardt and E. Eisenreigler, *Phys. Rev. Lett.* **74**, 3189 (1995).
46. B. Alberts, J. Lewis, M. Raff, K. Roberts, and J. D. Watson, *Molecular Biology of the Cell* (Garland, N.Y., 1994).
47. R. B. Gennis, *Biomembranes, Molecular Structure and Function* (Springer-Verlag, NY, 1989).
48. J. Israelachvili, *Intermolecular and Surface Forces* (Academic Press, San Diego, 1992).
49. O. G. Mouritsen and M. Bloom, *Annu. Rev. Biophys. Biomol. Struct.* **22**, 145 (1993).
50. M. Goulian, R. Bruinsma, and P. A. Pincus, *Europhys. Lett.* **22**, 145 (1993); Erratum in *Europhys. Lett.* **23**, 155 (1993).
51. N. Dan, P. Pincus, and S. A. Safran, *Langmuir* **9**, 2768 (1993).

52. P. B. Canham, *J. Theor. Biol.* **26**, 61 (1970); W. Helfrich, *Z. Naturforsch.* **28c**, 693 (1973).

53. P.-G. de Gennes and C. Taupin, *J. Phys. Chem.* **86**, 2294 (1982).

54. F. Brochard and J. F. Lennon, *J. de Phys.* **36**, 1035 (1975).

55. F. David and S. Leibler, *J. de Phys.* **II1**, 959 (1991).

56. R. Golestanian, M. Goulian, and M. Kardar, *Europhys. Lett.* **33**, 241 (1996); *Phys. Rev.* **E54**, 6725 (1996).

57. E. D'Hoker, P. Sikivie, and Y. Kanev, *Phys. Lett.* **B347**, 56 (1995).

58. R. Golestanian, *Europhys. Lett.* **36**, 557 (1996).

59. R. Balian and B. Duplantier, *Ann. Phys.* **112**, 165 (1978).

60. M. Bordag, G. L. Klimchitskaya, and V. M. Mostepanenko, *Int. J. Mod. Phys.* **A10**, 2661 (1995).

61. L. H. Ford and A. Vilenkin, *Phys. Rev.* **D25**, 2569 (1982).

62. M. Kardar, in *Proc. 4h Int. Conf. Surface X-Ray and Neutron Scattering*, eds. G. P. Felcher and H. You, *Physica* **B221**, 60 (1996).

63. G. T. Moore, *J. Math. Phys.* **11**, 2679 (1970).

64. S. A. Fulling and P. C. W. Davies, *Proc. R. Soc.* **A348**, 393 (1976).

65. M.-T. Jaekel and S. Reynaud, *Phys. Lett.* **A167**, 227 (1992).

66. P. A. Maia Neto and S. Reynaud, *Phys. Rev.* **A47**, 1639 (1993).

67. G. Calucci, *J. Phys. A: Math. Gen.* **25**, 3873 (1992); C. K. Law, *Phys. Rev.* **A49**, 433 (1994); V. V. Dodonov, *Phys. Lett.* **A207**, 126 (1995).

68. O. Meplan and C. Gignoux, *Phys. Rev. Lett.* **76**, 408 (1996).

69. A. Lambrecht, M.-T. Jaekel, and S. Reynaud, *Phys. Rev. Lett.* **77**, 615 (1996).

70. P. Davis, *Nature* **382**, 761 (1996).

71. G. Barton and C. Eberlein, *Ann. Phys. (N.Y.)* **227**, 222 (1993).

72. C. Eberlein, *Phys. Rev. Lett.* **76**, 3842 (1996); *Phys. Rev.* **A53**, 2772 (1996).

73. P. Knight, *Nature* **381**, 736 (1996).

74. V. V. Dodonov, A. B. Klimov, and V. I. Man'ko, *Phys. Lett.* **A142**, 511 (1989).

75. L. S. Levitov, *Europhys. Lett.* **8**, 499 (1989).

76. V. E. Mkrtchian, *Phys. Lett.* **A207**, 299 (1995).

77. J. B. Pendry, *J. Phys. Condens. Matter* **9**, 10301 (1997).

78. C. Eberlein, *Phys. World* **11**, 27 (1998).

79. A. Dayo, W. Alnasrallah, and J. Krim, *Phys. Rev. Lett.* **80**, 1690 (1998).

80. G. E. Volovik, *Pisma ZhETF* **63**, 457 (1996); *JETP Lett.* **63**, 483 (1996).

81. R. Golestanian and M. Kardar, *Phys. Rev. Lett.* **78**, 3421 (1997); *Phys. Rev.* **A58**, 1713 (1998).

82. The divergence of kernels in IIb comes from integrations over space-time. Given a cut-off L in plate size, and an associated cut-off L/c in time, the kernels diverge as $\exp[(K - 2)L/H]/[K(L/H)^3]$, with $K = 2QH/\pi$. Some care is necessary in the order of limits for $(L, H) \to \infty$.

83. Corrugated boundaries have been recently used for an interesting experimental manifestation of non-additivity of Casimir forces: A. Roy and U. Mohideen, *Phys. Rev. Lett.* **82**, 4380 (1999).

84. U. Mohideen, private communication (1999).

85. F. Miri and R. Golestanian, *Phys. Rev.* **A59**, 2291 (1999).

Annual Reviews of Computational Physics VIII (pp. 261–286)
Edited by Dietrich Stauffer
© 2000 World Scientific Publishing Company

SCALING AS INFLATION SYMMETRY, AND ITS PHYSICAL CONSEQUENCES

NICOLAS RIVIER

*Laboratoire de Dynamique des Fluides Complexes,
Université Louis Pasteur, 67084 Strasbourg cedex, France*
E-mail: nick@ldfc.u-strasbg.fr

In these lectures, scaling is taken as a symmetry, the symmetry of inflation. Quasicrystals are real materials with inflation symmetry. Their generic physical properties (electrical resistivity, nonstick, nonwet surface, etc.) are caused directly by inflation symmetry. Inflation symmetry is not only esthetically pleasing, but is also a group of specific operations enabling us to calculate the physical properties of the material, a real-space renormalization group.

1. Introduction and Outline

These lectures at a school on Scaling and Disorder consider scaling as a symmetry, the symmetry of inflation. A set of Russian dolls (matrioshki) is a familiar illustration of inflation symmetry. Quasicrystals[6] are actual materials based on inflation. Their physical and technological (nonstick cookware[1]) properties are direct consequences of inflation symmetry.

We will first describe inflation symmetry, geometrically and arithmetically. Since it applies to real, three-dimensional materials and their surfaces, inflation must be compatible with rotation symmetry and with space-filling, and this extends the range of crystallography, classically based on translational symmetry and periodic space-filling.

Extend is the proper verb: Quasicrystals are more, rather than less symmetric than crystals. We give three reasons for this statement:

(i) A crystal, cut along an arbitrary plane, loses its translational symmetry on the cleavage plane. By contrast, a cut quasicrystal keeps its inflation symmetry. Cut Russian dolls are Russian (hierarchical, in the geometrical, rather than political sense of the word) circles or ellipses, depending on the direction of the cleavage plane. Accordingly, physical properties caused by inflation symmetry, such as the very low density of electronic states at the Fermi level (the "pseudogap"), are retained at the surface of the material. This fact is responsible for the remarkable nonwet, nonstick properties of quasicrystalline coatings.[1]

(ii) A quasicrystal is a combination of inflation (by the irrational inflation parameter ν) and rotation symmetries (by a (topological) angle $\theta = 2\pi/n$). For these symmetries to be compatible, $\cos\theta \in Q(\nu)$ must belong to the field extension[2] of rationals Q by ν.[3] Furthermore, ν must be a quadratic irrational (solution of a second-degree equation with integer coefficients) for the inflation to be context-free. In crystals, the compatibility condition is reduced to the familiar restriction $\cos\theta \in Q$, so that translation can be regarded as inflation with a rational multiplier (solution of a linear equation), that is, as real-space renormalization (decimation) of a periodic lattice.[4] Crystals are the rational approximants of quasicrystals.

(iii) The cut-and-projection construction of quasicrystals (whereby it is represented by a strip, with an irrational slope, of a crystalline lattice in a space of *higher* dimension[10]), also indicates that quasicrystals are more symmetric than crystals.

A second part of these lectures will deal with the physical consequences of inflation symmetry, chiefly on the electronic (band) structure induced by the group of inflation operations, which is indeed a renormalization group, a natural form of decimation. Since inflation is not broken, this renormalization of the electronic structure, and the physical properties which it implies, are preserved on the free surface of the material. In obtaining the maximum metallic resistance of long, thin wires, D. Thouless[29] gave the first, and simplest example of renormalization of the electronic structure by decimation, actually by concatenation, which is the inverse operation.

The following, relevant topics are not discussed in these written lecture notes:

I also mentioned in Zanjan inflation of entropy, discussing the motion of a ball rolling on a rough, inclined plane.[38] The finite entropy, or complexity of the trajectory of the ball is due to the many bifurcations it has met. Physically, there is degradation of the kinetic energy of the ball at each bifurcation, and the ball has a finite, limiting velocity; it is not accelerated for a range of angles of incline. It is degradation of kinetic energy, as in the Second Law of Thermodynamics,[23] and not dissipation produced by friction. The limiting velocity of the ball is independent of its mass. (To be pedantic, it should depend on the ratio of inertial to gravitational mass, so that Eotvos's experiment could be repeated in Rennes.) The trajectory of the ball, a succession of intervals where it is accelerated, and of bifurcations rectifying (relaxing) its velocity projected along the trajectory, is the experimental illustration of Drude's model for the ohmic conductivity of an electron in a metal. The finite limiting velocity is due to concatenation of bifurcations, that is, to inflation, rather than to disorder. It can be observed in Galton's machine, a perfectly ordered array of pegs (an executive toy displaying Gaussian fluctuations).

If the symmetry of scaling is inflation, the symmetry of disordered materials is not inflation: It is a local, gauge invariance. It is also the symmetry of (maximum) entropy or ignorance. The entropy (through Boltzmann's formula $S = k \ln \Omega$) is the (logarithm of the) number Ω of different configurations corresponding to the same macroscopic state of the material. The fact that S is maximum implies that there is a large number Ω of physically equivalent configurations, and the symmetry of the disordered material reflects this equivalence. Moreover, S maximum means that a small, local modification of configuration does not change the physical state of the system. This local, equivalent modification is a gauge transformation.[39]

These equivalent configurations can be enumerated in the very simple case of two-dimensional tilings generated by points regularly spaced on a spiral.[35] Such tilings can be observed in the arrangement of florets in the core of a daisy or a sunflower. The entropy or complexity S of the generating spiral[34] is indeed equal to the logarithm of the number of topologically different tilings. Although here, neither the tiling, nor the generating spiral have any disorder, $\Omega > 1$ and the entropy is finite. Inflation (growth of the flower or lengthening of the spiral) imposes better convergents (higher i approximations in Eq. (2)) of the irrational governing the spacing of generating points on the spiral. These convergents are arranged on a (Farey) tree, with bifurcations which are again responsible for the entropy of the structure.

2. Symmetry of Order

2.1. *Symmetry of crystals*

It is a combination of translation and rotation symmetries. Compatible rotation and translations constitute the crystallographic (space- or point-) groups.

(One distinguishes between Bravais lattices — 7 point groups ("crystal systems") and 14 space groups ("Bravais lattices") — and crystal structures, where the basis is arbitrary — 32 crystallographic point groups and 230 crystallographic space groups — Ref. 9).

The point groups are a subset of the set of discrete subgroups of $SO(3)$, the group of proper rotations in 3D. This set contains the groups C_n ($n = 1, 2, 3, \ldots$), D'_n ($n = 2, 3, \ldots$), T, W and P, where C_n is the (cyclic) group of rotations by $\theta = 2\pi/n$ about an axis; D'_n contains the same operations, plus any two-fold symmetries about n axes separated by $\theta/2$, in a plane perpendicular to the first axis; the "Platonic" groups T, W and P leave invariant the tetrahedron, the cube (or octahedron) and the dodecahedron (or icosahedron), respectively.

This result can be obtained combinatorially as the set of *regular, topological tilings* of the sphere. Tiles are topologically identical, with n sides and n vertices. Vertices are junctions of z edges and z tiles. A vertex, tile or edge, is a pole of multiplicity z, n, or 2, of the elements of the rotation group which leave it invariant. One obtains the regular polyhedra $\{n, z\}$, with F tiles, E edges and V vertices. They are related by incidence/adjacency, $nF = 2E = zV$, and by Euler's network formula (on the sphere, of Euler characteristic $\chi = 2$), $F - E + V = 2$. Thus,

$$1 \leq 2/n + 2/z = 1 + 2/E, \tag{1}$$

for n, z and E integers ≥ 2. Solutions of (1) are $\{n, z\} = \{2, E\}$ and $\{E, 2\}$, corresponding to groups D'_E (with edge inversion symmetry) and C_E (without). Also, $\{3, 3\}$, $\{3, 4\}$, $\{4, 3\}$, $\{3, 5\}$ and $\{5, 3\}$ (tetrahedron (T), octahedron and cube (W), icosahedron and dodecahedron (P), respectively). C_E and D'_E are essentially two-dimensional (rotation about a privileged axis). Only the Platonic polyhedra (i.e. rotations by $2\pi/3$, 4 or 5) correspond to truly three-dimensional point groups.

Among these discrete subgroups of $SO(3)$, crystallographic point groups must be compatible with translation symmetry in crystals, or with inflation symmetry in quasicrystals. This compatibility can be stated as a crystallographic condition: A tiling has a finite number of tiles differing in shape and sizes, uniformly distributed. In particular, there is a finite, smallest length (smallest distance between atoms), distributed uniformly.

Let this smallest distance OP separate two lattice points O and P. A rotation by $\theta = 2\pi/n$ about O sends P into P'. To have PP' $\geq$ OP requires $n \leq 6$.

Equality (PP$'$ = OP) imposes coincidence. The case $n = 5$ is eliminated since a rotation by $2\pi/5$ about P, sending O into O$'$, results in OP$'$ < OP.

Crystallographic rotation symmetries $\theta = 2\pi/n$, with $n = 1, 2, 3, 4$ and 6, are those for which $\cos\theta \in Q$ is the solution of a linear equation with integer coefficients. Compatibility demands matching projections, as we shall see (Fig. 5) in the more general case of inflation symmetry.

Thus, five-fold rotation symmetry is incompatible with translation symmetry. To include regular pentagons in space-filling tilings demands a great deal of ingenuity. To include also octagons, decagons, dodecagons and heptagons, as on the ceiling of an eivan of the Friday Mosque in Isfahan (Fig. 1) is a supreme geometric and artistic achievement.

The publication in 1984,[7] of a rather sharp diffraction pattern with five-fold (icosahedral) symmetry by a piece of Al-Mn, showed physicists that sharp diffraction peaks did not demand a periodic structure, with translation symmetry, and that a sufficient condition is not necessary. (Bragg's Law is a good example of the prejudices which can be induced by good physics teaching — tell the truth, nothing but the truth, but not all the truth.) It was soon realized, by looking at the radial structure of the pattern, through the work of Levine and Steinhardt, but especially with the example of Penrose tilings,[8,6] that the symmetry responsible for the sharp diffraction peaks is inflation. And

Fig. 1. Isfahan, Friday Mosque, ceiling of the eastern eivan. This tiling incorporates pentagons, heptagons, octagons, and decagons or stars.

inflation symmetry is compatible with five-, eight-, ten- or twelve-fold rotation symmetry.

2.2. *Inflation symmetry*

2.2.1. *The inevitability of inflation*

Imagine a finite, straight boundary between two perfect, regular, but tilted lattices in 2D. This boundary separates k reticular lines of atoms on the left from $m(< k)$ lines on the right. A longer boundary would accommodate $k' > k$ and $m' > m$ reticular lines. If $m/k = m'/k'$, the tangent ν of the angle of tilt between the two lattices is a rational number $1 - m/k$. If ν is irrational, m'/k' is a better approximation than m/k. There is also irrational mismatch in epitaxial growth; $1 - \nu$ is the ratio of interatomic spacings in the material and in the substrate.

Examples are grain boundaries (tilt or twist) between two crystalline domains of the same material, the boundary between the substrate and a single crystal grown by epitaxy (or between worker and drone cells in a bee honeycomb), an irrational cleavage surface (with terraces) of a crystal, and, of course, quasicrystals. In all these examples, the inflatable structure has the lowest energy, because it keeps apart as much as possible structural units (atoms, or atomic surfaces in quasicrystals, defects like dislocations in grain boundaries or jogs and terraces in free surfaces) which repel each other.

Consider a sequence made of two kinds of structural units L and S (atoms, dislocations, jogs, ...), whith repulsive interactions, $E_{LL}, E_{SS} > 0$, tending to keep members of each family as much apart as possible and to prevent long strings of L or S. (For nearest neighbor interactions, the condition is $2E_{LS} < E_{LL} + E_{SS}$. It favors a homogeneous mixture of L and S. The opposite inequality favors clustering of structural units). What is the structure of the sequence, given that the ratio between numbers of S and L structural units should be as close as possible to ν? The answer is given by the cut-and-projection (or strip) method.[10]

The cut-and-projection method is illustrated in Fig. 2, together with its inflation symmetry. The construction of a line of irrational slope ν on a square lattice, approximated by a polygonal funnel with principal convergents as its vertices, is by Felix Klein.[11] Inflation is an area-preserving transformation of the lattice unit cell. The strip is defined by sliding the unit cell on the line. The alternative atomic surface description,[12] and the covering of the square lattice

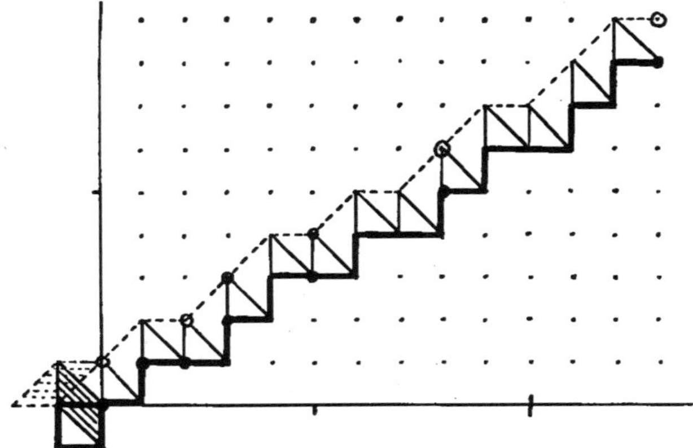

Fig. 2. Illustration of inflation symmetry, cut-and-projection method and rational convergents to an irrational number. Here, the inflation multiplier is the golden mean $1/\tau$. Circles mark its convergents 1/1, 1/2, 2/3, 3/5, etc. The quasiperiodic structure is an irregular staircase (thick line, vertical step S, horizontal step L), its deflated and inflated versions are the dashed and thin lines, respectively. The points of the square lattice which constitute the quasiperiodic structure lie within a strip, defined by translating the lattice unit cell (bottom left) along the line of slope $1/\tau$. The unit cell is deflated according to the substitution rules $L' = LS$, $S' = L$. The convergents are conserved under inflation. The figure also illustrates Rothen's relation (4).

by an oblique tiling[13] are useful to visualize structural defects and matching rules.

The number ν (< 1, say) is the slope of a line $E_\parallel$, drawn through the origin of a lattice Λ, with numbers x of units L on the abscissa, and y of units S on the ordinate. The strip method yields the geometrical structure of the sequence, an irregular staircase in Λ, which is the smoothest alternance of units L and S in a ratio approximating ν. S are the vertical and L, the horizontal steps of the staircase. The units L and S are kept apart as much as is allowed by ν. The one-dimensional problem is thus completely solved by the strip method. One distinguishes three classes of structures, periodic (ν rational), inflatable (ν quadratic irrational), and general. Only the first two classes have a relation between one part of the structure and another, a symmetry. For ν rational, this relation is a repetition (translation symmetry). For ν quadratic irrational, it is the inflation symmetry, which, through the golden ratio $\nu = \tau = [1 + \sqrt{5}]/2$, has inspired architects (Palladio (Fig. 3) and his English followers), garden designers ("Capability" Brown), besides rabbit-breeders.

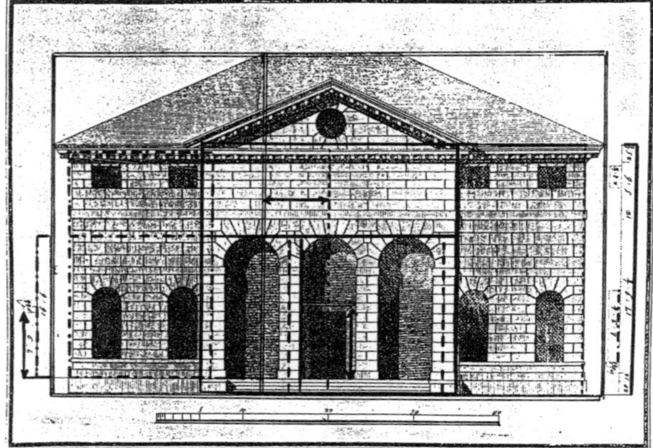

Fig. 3. Villa at Cesalto, designed by Andrea Palladio (1508–1580).[14] It is based on the golden rectangle and its inflation symmetry: A golden rectangle (a, b) can be decomposed into a square (b, b) and a similar rectangle $(b, a - b)$, thus $\tau = a/b = b/(a - b)$, $\tau^2 = \tau + 1$, and this *ad infinitum*. τ is the golden section or Divine proportion — "suavississima sottile e admirabile doctrina" (Luca Pacioli).[15]

The largest golden rectangle (double lines) is based on the horizontal scale. The main golden rectangle (single and dashed lines) is based on the larger right vertical scale. The larger left vertical scale (height of the central arches, horizontal elements of the roof and, deflated, its height) relates to the first two scales. It forms a golden rectangle (dash-dot) on the wings of the villa (the height of the arches is half this scale, the smaller scale is on the right). The smaller left scale (arrow) is the height of the central door; it can also be seen to relate to the larger two golden rectangles. The scales are in the drawing of Palladio. The various lines are by the author and may be complete nonsense: There may be no relation other than esthetic between the three scales.

Numerically, and geometrically,[11] the best approximations are given by the principal convergents A_i/B_i to ν, obtained by truncation of its continued fraction representation,

$$\nu = 1/\{q_1 + 1/[q_2 + \cdots]\} = [0, q_1, q_2, \ldots] \approx 1/[q_1 + 1/(\cdots + 1/q_i)] = A_i/B_i. \quad (2)$$

A finite segment has A_i S units, and B_i L units. A_i and B_i are related to the natural integers q_i by the recursion relation

$$
\begin{aligned}
B_i = q_i B_{i-1} + B_{i-2}, &\quad B_0 = 1, \quad B_{-1} = 0, \\
A_i = q_i A_{i-1} + A_{i-2}, &\quad A_0 = 0, \quad A_{-1} = 1.
\end{aligned}
\quad (3)
$$

We shall see that this recursive procedure corresponds geometrically to inflation of a finite structure by ν. For this inflation to be automatic, that is, context-free, independent of the step number i, the sequence $[\ldots, q_i, q_{i+1}, \ldots]$ must ultimately be periodic, so that ν is a quadratic irrational.[11]

[Figure 2 also illustrates a superposition relation between a Fibonacci sequence (projected staircase, with inflation multiplier $\nu = \tau$) $Q(S_p, L_p) \equiv Q^{(p)}$, with short and long elements of lengths S_p and L_p, respectively, and its inflated, and deflated and shifted counterparts, first noticed (1990) by Rothen,[41]

$$Q^{(p)} + \mathbf{T}(S_p)Q^{(p+1)} = Q^{(p-1)}, \tag{4}$$

where the operator $\mathbf{T}(S_p)$ translates the sequence by S_p. This superposition relation is valid regardless of the projection direction, for sequences projected from the same staircases. Notably for sequences $Q^{(\tau)}$ which are topologically and metrically quasiperiodic, $L_p = S_{p+1} = \tau S_p$ (orthogonal projection of the staircase), or for "metric" sequences with the staircase projected so that $L_p = F_{p+1}a$, $S_p = F_p a$, where F_p are Fibonacci numbers, $F_p = F_{p-1} + F_{p-2}$, $F_1 = F_2 = 1$, $F_3 = 2$, etc. Sequence $Q^{(1)}$ is metrically periodic since $S_1 = L_1 = a$, although still topologically a quasiperiodic sequence (of equal elements). Diffraction is, of course, only a matter of metric spacings, so that the Fourier spectrum of $Q^{(1)}$ is a lattice of points separated by $2\pi/a$. The additional diffraction peaks associated with inflation (the pattern causing the original *furore*[7]) in $Q^{(2)}$ and (phase-shifted) $Q^{(3)}$ all add up to zero (extinction in $Q^{(1)}$). Accordingly, Sibon[41] measured a much sparser diffraction spectrun (only two inflation peaks) for the "metric" sequence $Q^{(2)}$ than for the quasiperiodic sequence $Q^{(\tau)}$ (which has an infinite number of peaks). The diffraction peaks are located at powers of $1/\tau$, since both sequences have inflation symmetry (they originate from the same topological staircase), $x' = \tau x \Rightarrow k' = k/\tau.$]

2.2.2. *Irrationals, infinite descent and inflation*

The golden section, or golden mean τ is also the ratio between the diagonal and the side of a regular pentagon (Fig. 4), which must have pleased the Pythagoricians very much. One simply uses Thales's theorem for similar triangles made of two diagonals and one side (thick lines in Fig. 4). Irrationality of τ follows, *ad absurdum*, from the infinite regression of the geometrical figure. This technique of infinite descent has been invented by Fermat to prove FLT (Fermat's Last Theorem) for exponent 4.[40] FLT for exponent 3 was proved by Euler in 1753, also using infinite descent. But he had to use imaginary numbers,

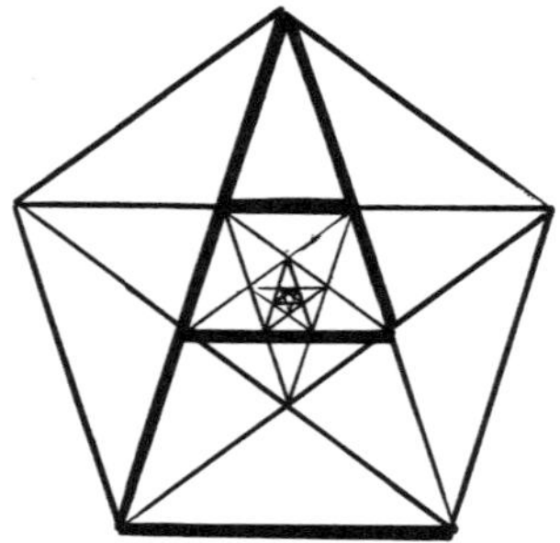

Fig. 4. The golden mean $\tau = [1 + \sqrt{5}]/2$ is the ratio between the diagonal and the side of a regular pentagon. (Use Thales's theorem for the similar triangles made of two diagonals and one side). Irrationality of τ follows *ad abs.* from the infinite regression of the geometrical figure. This technique of infinite descent has been invented by Fermat.

introduced 200 years earlier (1560) by Bombelli and described by Leibniz as "almost an ámphibian between being and nonbeing" (Ref. 40, p. 93).

2.3. *Inflation in 1D*

Inflation of a finite sequence generates a longer one with the same structural properties. It can be done either geometrically (by a different choice of the unit cell of Λ, whose translation along $E_\parallel$ defines the strip), algebraically (by matrix multiplication) or by substitution (replacing by "words" the letters L, S in the sequence). All these methods are equivalent. They are also iterative.

[For example, the Fibonacci sequence LSLLSLSLLSLLS... has substitution rule S $\to$ L, L $\to$ LS, golden quadratic irrational $\nu = \tau = [1 + \sqrt{5}]/2$, all $q_i = 1$, and successive Fibonacci numbers A_i and $B_i = 1, 1, 2, 3, 5, 8, 13, 21, \ldots$. There are interesting examples in biological helices[16]: Collagen is based on the Boerdijk–Coxeter (B–C) helix, characterized by the sequence LLSL LLSL LLS LLSL..., the quadratic irrational $\nu = 1 + \sqrt{3} = [2, 1, 2, 1, 2, 1, \ldots]$ and the substitution rule S $\to$ LLS, L $\to$ LLSL at each successive (double) step. The α-helix is the sequence LLLSL LLLSL LLLS LLLSL..., characterized by the quadratic irrational $2 + \sqrt{3} = [3, 1, 2, 1, 2, 1, \ldots]$; the two-steps substitution rule is a bit tricky because $q_0 = 3$ is not part of the period of the sequence $\{q_i\}$.]

Equation (3) is a particular inflation operation, replacing the finite staircase w_{i-1} from 0 to (B_{i-1}, A_{i-1}) by a longer one, w_i, to (B_i, A_i). Let us approximate line $E_\parallel$ by a broken, continuous line E_B made of segments with rational slope linking principal (and intermediate) convergents. $E_\parallel$ can be deformed into E_B without crossing any lattice point, so that strips on $E_\parallel$ and E_B yield identical staircases. They are obtained by the substitution (concatenation) rules,

$$w_i = w_{i-1}^{q_i} w_{i-2}, \quad w_0 = \mathrm{L}, w_{-1} = \mathrm{S}. \tag{5}$$

(The order of the words in (4) follows from the fact that the last two steps of a segment are LS or SL if its end point is above or below $E_\parallel$. Successive principal convergents lie alternatively above (i odd) and below (i even) $E_\parallel$, so w_i, a sequence of q_i w_{i-1}'s and one w_{i-2} must end with w_{i-2}. Intermediate convergents $w_i^{(s)} = w_{i-1}^s w_{i-2}, 1 < s < q_i$, are constructed in the same way.)

It is the integer q_i which sets the inflation rule (3) or (5) at stage i. Inflation symmetry generates longer structures by repeated application of the same rule. This compels the sequence of integers $\{q_i\}$ to be periodic after some stage, and ν to be quadratic irrational (Lagrange[11]). Conversely, if the sequence $[\ldots, q_i, q_{i+1}, \ldots]$ were not periodic, the rules would depend on the stage of inflation or, through (5), on the position in the structure, and inflation would not be context-free. Quasicrystalline structures with inflation symmetry are characterized by quadratic irrationals. A rational ν has finite sequence $\{q_i\}$ and periodic (crystalline) structure.

For any quadratic irrational ν, there is a 2×2 inflation matrix $\mathbf{S}$ with integer coefficients S_{ij}, $\mathrm{tr}\,\mathbf{S} > 0$ and $|\det\mathbf{S}| = 1$, eigenvalues $s_+ = |1/s_-| > 1$. $\mathbf{S}$ sends lattice point $\mathbf{w} = (x, y)$ into $\mathbf{w}' = (x', y')$, $\mathbf{w}' = \mathbf{S}.\mathbf{w}$ in matrix notation. (Alternatively, $\mathbf{S}^{-1}$ transforms the unit cell of L which defines the strip.[10]) We require that the eigenvector of $\mathbf{S}$ corresponding to s_+, $\mathbf{e}_+ = (1, \nu)/\sqrt{1 + \nu^2}$, is along $E_\parallel$. Then,

$$s_+ = S_{12}\nu + S_{11}. \tag{6}$$

$\mathbf{S}$ maps any point $\mathbf{w}$ of Λ within the strip into another, closer to $E_\parallel$ ($|s_-| < 1$) but further from the origin, and inflates structure $0\mathbf{w}$ into $0\mathbf{w}'$, with multiplier s_+. Quadratic irrationals ν and s_+ belong to the same field $Q(\nu)$.

The physical structure is the staircase $0\mathbf{w}$ projected on $E_\parallel$. Its length w_+ is the component of $\mathbf{w}$ along $\mathbf{e}_+$,

$$w_+ = [x + y(S_{12}/S_{11})\nu]/\sqrt{1 + \nu^2}. \tag{7}$$

It is inflated consistently, $w_+' = s_+ w_+ = [x' + y'(S_{12}/S_{21})\nu]/\sqrt{1 + \nu^2}$. Projection and eigenvectors are orthogonal only if $S_{12} = S_{11}$.

Inflation symmetry fails for algebraic irrationals of degree $d > 2$:

(i) Suitable inflation multipliers s_+ are PV numbers (algebraic integers with all $d-1$ conjugates < 1 in modulus). But s_+ is now an eigenvalue of a $d \times d$ matrix, and the sequence involves d different structural units. The physical setup of structural units repelling each other becomes arbitrary.

(ii) Inflation is no longer context-free for $d > 2$,[3] so that a consistently deflatable structure w_+ (7) cannot be defined. For example, let $q_4 = 2$ pop in at stage 4 of the Fibonacci sequence, that is, $\nu = [0,1,1,1,2,1,1,\ldots]$, instead of $[0,1,1,\ldots]$. The structure (obtained from rule (4)) is almost Fibonacci's, LSLLS LSLLS L<u>SL</u> LSLLS LSLLS L<u>SLL</u>S L<u>SL</u>, with a jog (phason flip, underlined) switching horizontal and vertical steps 12 and 13. This jog is repeated further along, by inflation. Deflating the structure by substitution rule yields, after three steps, the sequence LLSLLLS, which contains the forbidden LLL. After four steps, SLSSL is not deflatable any longer. Inflation is a symmetry and its operations form a group, including an inverse operation, independent of the context (local structure) on which they operate.

A better example is collagen,[16] which is, as mentioned earlier, based on the B–C helix, an array of regular tetrahedra. In Euclidean space, the number of edges per turn of the B–C helix is $2\pi/\arccos(-2/3) = [2, 1, 2, 1, 2, 1, 1, 2, 1, 7, 6,$ etc.$]$, which is not context-free inflatable. Natural collagen, which is eminently inflatable ("from the molecule to the individual animal, here a submarine worm"[17]), is based on an inflatable helix with $1 + \sqrt{3} = [2,1,2,1,2,1,2,1,\ldots]$ edges per turn. The two helices are identical through 112 steps.

2.4. *Compatibility inflation–rotation (Pleasants)*

Quasicrystals in higher dimensions can be represented as tilings or as n-grids. An n-grid is either dual of the tiling (a vertex of the grid corresponds to a tile, a vertex of the tiling to a cell of the grid), or decoration of its tiles (Ammann grid). (See Ref. 6, Sec. 1.4.) The n-grid is made of n identical arrays of parallel lines in the n directions, so that rotation by $2\pi/n$ is a symmetry (up to affine deformation). n-grids can be regular (without intersection of three or more lines through one vertex) or singular (with many coincidences). The parallel lines may be equidistant (periodic) or quasiperiodic. Triangular and kagome tilings are periodic three-grid. The former is singular, the latter is regular. Ammann's decoration of the Penrose tiles is a quasicrystalline, regular five-grid.

Inflation (by the same multiplier ν) and rotation (angle θ) are compatible provided that all $\cos\theta \in Q(\nu)$. This result is due to Pleasants.[3] It generalizes to quasicrystals the crystallographic condition (a tiling with only a finite number of tile shapes and sizes, such that every finite configuration occurs throughout the tiling with uniform frequency and there are only finitely many such configurations). One can construct a quasicrystalline tiling as a direct product (grid) of ν-inflatable arrays along the symmetry directions of a finite symmetry group, provided the inflatable arrays have the same inflation multiplier ν, and $Q(\nu)$ contains the cosines of all the angles between the symmetry directions.

$[Q(\nu)$ is a simple extension of the field of rationals Q, of degree 2 if ν is quadratic irrational. Its elements are $r + s\nu$, r, $s \in Q$. For $Q(\nu)$ to be a field (with an inverse operation), it is sufficient, but not necessary that ν be quadratic irrational. For example, $Q(\sqrt{2}+\sqrt{3})$ is a field, with $x = \sqrt{2}+\sqrt{3}$ satisfying the quartic equation $x^4 - 10x^2 + 1 = 0$.[2]$]$

Here is a geometrical proof (Fig. 5). Consider one axis, 1 say, cut at right angle by one array of parallel lines of the grid, and at an angle $\pi/2 - \theta$ by another array. θ is the angle between the two directions. Only if some intersections coincide is the crystallographic condition satisfied, notably the existence of a smallest length. Then, by inflation, there are infinitely many coincidence points, uniformly distributed through the tiling, amd at finite distances of each other. The distance (7) between any two lines of the array i is $w_{i+} = [x_i + y_i(S_{12}/S_{21})\nu]/\sqrt{1+\nu^2}$. Arrays 1 and 2 intersect on axis 1

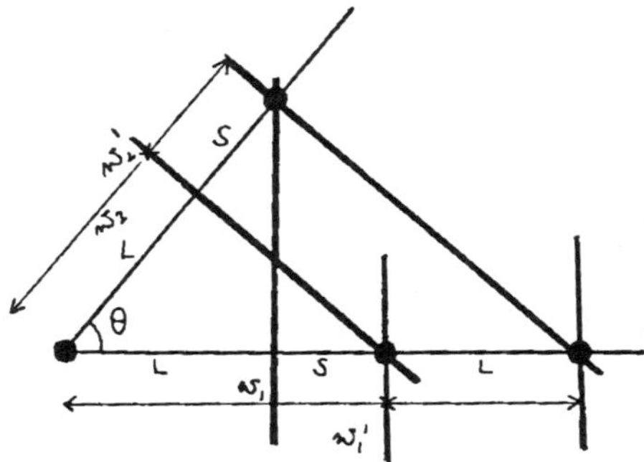

Fig. 5. Pleasants' theorem: Multiple coincidences between inflationary arrays of lines in different directions guarantee a finite minimal distance throughout the network (crystallographic condition). Coincidences are conserved under affine deformations; rotation symmetry is only topological.

if $w_{1+}\cos\theta = w_{2+}$, so that $\cos\theta = w_{2+}/w_{1+} = w'_{2+}/w'_{1+} \in Q(\nu)$, and is invariant by inflation.

Remarks

(a) Coincidence and parallelism are conserved by affine deformations, so that the rotation symmetry is only a topological framework which may not be manifest in the structure (see Fig. 5).

(b) Crystallographic patterns are restricted by the compatibility of translation and rotation symmetries. Quasicrystallographic patterns are similarly restricted, with inflation replacing translation. The direct product comes as a bonus. Inflation and rotation operations commute.

(c) The inflation multiplier ν determines the rotation symmetry. ν, hence $\cos\theta$ are quadratic irrational for context-free inflation. This restricts quasicrystals to those generated with the symmetry of regular n-gons, $n = 5$, 8, 10 and 12, which are all real materials. The corresponding inflation multipliers are $\nu = (1+\sqrt{5})/2 = [1]$ for $n = 5$ and 10, $\nu = (1+\sqrt{2})/2 = [2]$ for $n = 8$, and $\nu = 1+\sqrt{3} = [2,1]$ for $n = 12$. As seen above, only $n = 5$ corresponds to a truly three-dimensional finite subgroup of $SO(3)$.[5] The other patterns $n = 8$, 10 or 12 are periodic in the third dimension.

2.5. *Inflation symmetry is preserved through cut*

Inflation symmetry, as generated by the cut-and-projection method, is not broken if the physical space $E_{\parallel}$ is interrupted. Thus, the physical properties of bulk quasicrystals caused by inflation will continue to hold on its surface.

Any cut of a quasicrystal is also a quasicrystal (modulo affine deformations, but with the same inflation multiplier in all directions). Contrast this with a crystal, whose symmetries are only conserved if it is cut on particuliar cleavage planes. Conversely, a combination of inflationary arrays in $n = 5$, 8, 10 or 12 directions (with the same inflation multiplier ν such that $\cos 2\pi/n \in Q(\nu)$) has inflation symmetry (lift of inflation symmetry by the theorem of Pleasants).

Conservation of symmetries through cut or lift is a general property of the cut-and-projection method. Restriction of the physical space $E_{\parallel}$ reduces automatically the acceptance strip, but the algebraic properties responsible for inflation symmetry are not affected. Foams constitute an interesting example: A cut foam is a two-dimensional foam, conserving its essential symmetry, which is local, gauge invariance. A cut Laguerre–Telley (LT) foam (generalization of Voronoi's) is a LT foam. A cut Voronoi foam is a LT foam.[19]

3. Inflation on Physics

3.1. *Inflation and wetting on quasicrystals*

Quasicrystals make excellent frying pans (they are nonwetting, have low solid friction, etc.). These extraordinary properties, generic of quasicrystalline coatings, have been discovered and patented by Dubois.[1] They are associated with low surface energy, itself a consequence of inflation symmetry of quasicrystals,[4,20,21] which, as we have seen, is not broken by the free surface of the material.

[The discovery of nonstick quasicrystalline coatings by Dubois set a fundamental question in physics (how can a metallic alloy be nonstick?), suggested the answer (its electronic structure is that of a poor metal, actually, more like a dirty, p-type semiconductor), its cause (a fundamental symmetry — inflation — unbroken at the free surface of the material) and the means to demonstrate it (inflation is a technique of renormalization).[22] We are looking at one of the few cases where engineering has been responsible for a discovery in basic physics and suggested its explanation (the others are Carnot with the second law of thermodynamics and Shannon with information theory.[23]) The experimental discovery of a pseudogap at the Fermi level in quasicrystals is due to Belin-Ferre *et al.*.[24] There is, up to now, only indirect evidence that the pseudogap persists to the surface of the material.]

3.2. *Wetting, sticking and electronic contribution to the surface tension*

Sticking is a wetting problem. What wets sticks. Classical wetting theory is a branch of thermodynamics. Macroscopically, what matters is the surface tension of the solid (solid–liquid interface). In quasicrystalline coatings, the characteristic length of the solid (quasicrystalline cluster size ≥ 1 nm) is much smaller than the persistence length of the comestible material (the "liquid"). Nonstickiness of the quasi-frying pan is thus due to a fundamental property of the coating, its morphology and inflation symmetry .

The electrons near the Fermi level and close to the surface of the material, make up the electronic surface tension γ,[20]

$$\gamma A = \int dz \int^{0} d\epsilon\epsilon [g(\epsilon, z) - g_{\text{bulk}}(\epsilon)\theta(z_d - z)]. \tag{8}$$

Here, z is the coordinate normal to the surface (of area A), $z = z_d$ defines Gibbs's dividing surface, θ is the Heaviside (step) function; $g(\epsilon, z)$ is the local density of states (DOS) and $g_{\text{bulk}}(\epsilon)$, the bulk DOS. The quantity in brackets (difference between surface and bulk electronic DOS) is only nonzero near the Fermi energy $\epsilon = 0$. If it is small, the surface tension is small. Thus, the material has small surface tension if its bulk DOS is low at the Fermi energy ("pseudogap") and if this pseudogap persists to the surface of the material. This is the case of quasicrystals.[24,25]

How can a metal have low surface tension? A bad crystalline (transition) metal with low electronic DOS at the Fermi energy in the bulk (neighboring electron orbitals form bonding–antibonding combinations, split in energy) has a high surface tension, because the electronic orbitals at the surface of the material, without neighbors with which to combine, are dangling out. They straddle the Fermi energy half-filled. Tungsten, with a minimum density of electronic states at Fermi energy in the bulk, is used as a catalyst or as tip for scanning electron microscopes.

In a quasicrystal, inflation replaces translation to produce similar bonding–antibonding combinations of electronic orbitals. The combinations persist on the surface of the material, because inflation symmetry is not broken. Thus, low surface tension is due to inflation symmetry, a generic property of quasicrystals and their approximants, absent in crystalline materials.

3.3. *Crystallographic inflation symmetry*

The result of Pleasants[3] (Fig. 5) that quasicrystals are direct products of the same one-dimensional inflation pattern, rotated by $2\pi/n$ (up to affine deformations), has two implications:

(i) Since the bulk material is an inflatable structure, then any surface (a cut in an arbitrary direction) is also inflatable. If the pseudogap in the bulk is a result of inflation, then it will necessarily survive on the surface.

(ii) It is sufficient to show that a pseudogap is formed in a one-dimensional inflatable structure. By convolution of the 1D DOS, it will also occur in 2D (surface) and 3D (bulk). (See Sec. 3.5.)

3.4. *Pseudogap in one dimension*

Take a tight-binding Hamiltonian $\mathbf{H}$ on a 1D structure. For simplicity, consider a Fibonacci sequence and discriminate betweeen two "sites" S and L. Each "site" represents a group of energy levels centered at energy ϵ_S or ϵ_L. For a periodic chain, $\epsilon_0 = 0$. The matrix element of $\mathbf{H}$ between nearest neighbor sites, t, is the same everywhere.

There is a technique called dynamical or energy renormalization, whereby the energy of an electron in a (one-dimensional) crystal is renormalized by decimation of the lattice (elimination of every other atom).[26] In a crystal, this renormalization takes the form of a one-variable, nonlinear map of the energy. If the electron energy lies initially within the band, it remains inside the band, where the dynamical map is elliptic. If it lies outside the band (where the map is hyperbolic), it is sent to infinity by renormalization, and the electronic density of states is duly zero.[4] In a quasicrystal, deflation symmetry $\mathrm{LS} \to \mathrm{L}, \mathrm{L} \to \mathrm{S}$, is a natural decimation. Renormalization of the electron energy is a dynamic map involving three variables.[28]

3.4.1. *Periodic chain*

To calculate the 1D electronic density of states (DOS), renormalize the energy. For a periodic chain, this is done by decimating the chain. At each step of the renormalization procedure, remove every other site and rescale the Schroedinger equation $t(c_{i+1} + c_{i-1}) = \epsilon c_i$, or $\mathbf{X_{i+1}} = \mathbf{M} \cdot \mathbf{X_i}$. Here, $\mathbf{X_{i+1}}$ is the transposed matrix (c_{i+1}, c_i), with c_i, the probability amplitude that the electron is at site i. The transfer matrix $\mathbf{M}$ is a 2×2 real, unimodular matrix with determinant 1 and trace $s = \operatorname{tr} \mathbf{M} = \epsilon/t$ (here in the representation $M_{11} = s$, $M_{21} = 1 = -M_{12}$, $M_{22} = 0$).

The recursion relation, here the logistic map $s' = s^2 - 2$,[26] is a nonlinear trace map, obtained by taking the trace of the Cayley–Hamilton formula for all four matrix elements of $\mathbf{M}$, $\mathbf{M}^2 - s\mathbf{M} + \mathbf{I} = 0$ ($\mathbf{I}$ is the 2×2 unit matrix, and $s' = \operatorname{tr} \mathbf{M}^2$). States with $|s| < 2$ form a band (elliptic region) and remain within the band upon renormalization. States with energies outside the band $|s| > 2$ are mapped to infinity and their density tends to zero (hyperbolic region). Band edges constitute the Euclidean region. We obtain $g(s) = 1/[\pi\sqrt{4 - s^2}]$, $|s| < 2$, the familiar tight-binding DOS, which is an invariant of the logistic map.

3.4.2. *Quasicrystalline chain*

For a quasicrystalline chain, deflation symmetry is also a natural means of decimation. The tight-binding transfer matrices still have determinant 1, but site-dependent trace $\operatorname{tr} \mathbf{M_S} = s_S = (\epsilon - \epsilon_S)/t$ and $\operatorname{tr} \mathbf{M_L} = s_L = (\epsilon - \epsilon_L)/t$.

Moreover, s_S and s_L are no longer single energy levels but *groups* of levels, spanning a range of energies $0 \le |s_S - s_L| \le w$. In a quasicrystal, each tile (the unit on which inflation acts) contains several atoms, and many electronic states.

Inflation operation is the concatenation $w_{i+1} = w_i \cdot w_{i-1}$ of sequences (words) w_i, through substitution $S \to L \to LS$. On the transfer matrix, the inflation operation is $\mathbf{M}_{i+1} = \mathbf{M}_{i-1} \cdot \mathbf{M}_i$, for a particular pair of levels (s_S, s_L) of the groups. $\mathbf{M}$ remains real unimodular.[4,37] The inflation relation is inverted $\mathbf{M}_{i-2} = \mathbf{M}_i \cdot \mathbf{M}_{i-1}^{-1}$, and yields, by taking the trace of $\mathbf{M}_{i+1} + \mathbf{M}_{i-2}$, the three variables $(x, y, z) = (s_{i-2}, s_{i-1}, s_i)$ Kohmoto map,[28]

$$s_{i+1} = s_i s_{i-1} - s_{i-2}, \tag{9}$$

with initial conditions $s_{-1} = s_S$, $s_0 = s_L$, $s_1 = \mathrm{tr}\,\mathbf{M_S} \cdot \mathbf{M_L} = s_L s_S - 2$, $s_i = \mathrm{tr}\,\mathbf{M_i}$. (Specifically, $x' = y$, $y' = z$, $z' = zy - x$, with $x_0 = 2$ (a useful, but not necessary restriction), $y_0 = s_S$, $z_0 = s_L$.)

The states $(x, y, z) = (s_{i-2}, s_{i-1}, s_i)$ remain on the surface $x^2 + y^2 + z^2 - xyz = \kappa$, an invariant of the map. There is one surface with parameter $\kappa = (s_S - s_L)^2 + 4$ for each pair of levels (s_S, s_L). To groups of levels correspond a stack of invariant surfaces, with $4 \le \kappa = (s_S - s_L)^2 + 4 \le w^2 + 4$. The stack of surfaces foliates the space of all possible states. State (x, y, z) has energy $\epsilon/t = f(x^2 + y^2 + z^2)$, where f is a monotonically increasing function of its argument. The full stack of invariant surfaces contributes to the DOS, whereas a Fibonacci chain with one level per site has its electronic states on a single surface.

In addition, map (9) has fixed points, some isolated (e.g. (2, 2, 2) is a fixed point of order 1 for $\kappa = 4$), others belonging to one-dimensional families, such as

- $(-1, x, -1)$, order 4,   and $(1, x, 1)$, order 12;
      (stack $\kappa = x^2 - x + 2$)
- $(-1, -x, -1)$, order 4,   and $(1, x, -1)$, order 12;
      (stack $\kappa = x^2 + x + 2$)
- $(\pm x, \pm x/[x - 1], \pm x)$, order 2 $(+ + +)$ or 6 $(+ - -)$;
      (stack $\kappa = x^2[1 - (x - 1)^{-1} + (x - 1)^{-2}]$)
- $(x, 0, 0)$, order 6;
      (stack $\kappa = x^2$), $\tag{10}$

for any $x \ne 0$. If there are any other fixed points, they have dimension ≤ 1.

Each invariant surface has elliptic and hyperbolic regions, with positive, respectively negative, Gaussian curvature. As with the logistic map, a point (x, y, z) remains in the elliptic regions of the surface when iterated by map (9). The energy density of these states is finite. Conversely, if the point is in the hyperbolic regions, it will be mapped to infinity, and the density of such states is zero, except if (x, y, z) is a fixed point. States in the hyperbolic regions are usually at higher energies than those in the elliptic regions.

Consider all states of energy less than ϵ/t. They are located inside a sphere, intersecting the stack of invariant surfaces. Since the states in the hyperbolic region have zero density, a state contributes to the integrated density of states, either if

(a) it is in the elliptic region of its invariant surface, or if
(b) it is a fixed point (10) within the hyperbolic region.

Families of fixed points (10) give a small contribution to the DOS, above the main band (a). Thus, the integrated DOS is $N(\epsilon) = \int_< d\epsilon' g(\epsilon') +$ contribution of fixed point families (10), where the symbol $<$ restricts the integral to energies less than ϵ/t and within the elliptic region of the stack of invariant surfaces.

The manifold separating the regions of positive and negative curvatures of the stack of invariant surfaces resembles a deformed cube. It is given by the equation (in polar coordinates, $3(\epsilon/t)^2 = r^2$):

$$0 = 16 - 4r^2[1 - \sin^2\theta(\cos^2\theta + \sin^2\phi\cos^2\phi\sin^2\theta)]$$
$$+ 3r^4 \sin^2\phi\cos^2\phi\sin^4\theta\cos^2\theta.$$

The electronic density of states $g(\epsilon)$ looks like that of a heavily p-doped semiconductor: For each value of κ — each pair of levels (s_S, s_L) — there is a 1D tight-binding "valence" band (a), plus the isolated energies of the fixed points. For groups of energy levels — stacks of κ's — the individual valence bands are superposed and the families of fixed points (10) constitute an "impurity" band, a dense set of states (in ϵ, or in $0 \leq \kappa - 4 \leq w^2$). The density of states rises through an inflexion point to a maximum, then falls to the much smaller contribution of the fixed point families (10). This sharp fall is the experimentally observed wall (or "pseudogap").[24] The Fermi level is in the "impurity" band, just above the top of the "valence" band.[4,21]

3.5. *In 3D and 2D*

In two and three dimensions, even though the quasicrystalline structure is a direct product of inflatable sequences,[3] the combinatorics at each vertex or each tile, and thus the precise tight-binding Hamiltonian, are complex. However, inflation is a symmetry operating at a macroscopic scale, on tiles containing many atoms, averaging over the microscopic fluctuations due to the atomic structure of the material. This is why, in 1D, inflation operates (9) on groups of energy levels instead of single atomic orbitals. For the same reason, it is *independently* and *uniformly* in two (in 2D) or three directions (in 3D), that it operates on groups of energy levels. The electronic density of states is obtained by simple (in 2D) or double (in 3D) convolution of the 1D DOS. Conversely, the features of the DOS, like inflation symmetry itself, are preserved under cut, at the surface of the material.

Uniform quasiperiodic patterns of structural units are seen at grain boundaries. The boundary between two perfect crystalline lattices, tilted or twisted, is a quasiperiodic grid of dislocations.[18]

3.6. *Fixed point states are localized*

The electronic density of states at the Fermi level is finite. These states (10) are fixed points, closed orbits of the Kohmoto map (9). They are all localized.

Suppose, *ad absurdo*, that an electron could escape in a finite time Δt from the fixed point state, as it happens for the Friedel–Anderson virtual bound state of a transition metal impurity in a metal. By the uncertainty principle, the state broadens into a resonance, of finite width $\Delta\epsilon = h/\Delta t = \pi g_c V^2$, where g_c is the density of the conducting states into which the electron escapes, with Hamiltonian matrix element V. But these conducting states are in the hyperbolic region of the map, and their energy is sent to infinity by inflation. Their density g_c is zero. Thus, the fixed point states are localized, albeit with a large localization length.[25] Nonexponentially-localized states in quasicrystals lie at a finite energy from the Fermi level.

3.7. *Conductivity of quasicrystals at low temperatures*

With a finite density of localized electronic states at the Fermi level, electric conductivity at low temperatures is by variable-range hopping

$$\sigma = C \exp[-(T_0/T)^{1/4}]. \tag{11}$$

The electron hops through (localized) states, at an average distance which is a compromise between two factors: A Boltzmann factor penalizes nearby states because they are less likely to be at the same energy, and a factor proportional to the square of the overlap of the wavefunctions penalizes long hops.[31] Delahaye *et al.*[25] measured $T_0 = 24/(\pi k)[1/(g_0\xi^3)] \approx 1$ mK. This low temperature implies a large localization length $\xi \approx 300$ nm, as well as a substantial density of localized, fixed point states (10) at the Fermi level in quasicrystals. The remarkable observation of a variable-range hopping conductivity in i-AlPdRe at low temperatures[25] thus establishes the electronic structure of quasicrystals: finite density of localized states at the Fermi level.

3.8. *Variable-range hopping conductivity*

If the density of states at the Fermi energy is finite, but all the states at, and near the Fermi energy are localized, that is, if the eigenstates of the Schroedinger equation are localized, but dense in energy around the Fermi level, electrical transport at low temperature is by variable-range hopping, with the characteristic (Mott) behavior for the conductivity

$$\sigma = C \exp[-B/(kT)^{1/(D+1)}], \tag{12}$$

where D is the dimensionality of the gas of charge carriers (electrons or holes).[31] The conducting states are too far away in energy ($>E_c$), inaccessible at low temperatures, $\exp(-E_c/kT) \ll 1$, and electrons must hop between localized states.

The hopping conductivity from site I to site J at distance R, and energy difference W, is the product of two, competing factors:

- One factor is equal to the modulus square of the overlap between the wave functions of the two (exponentially localized) states, $\exp(-2R/\xi)$, where ξ is the localization length. If R_0 denotes the average interatomic distance, ξ/R_0 is larger than 1.
- The Boltzmann factor $\exp(-W/kT)$.

Here, $\xi/R_0 \gg 1$, and the electron can hop far to a state of nearly the same energy. At low temperatures, hopping between localized states is always preferred to promotion to conducting states.

Electrons will normally (on average) hop to a site J with the lowest energy W, within a distance $\leq R(T)$. For a three-dimensional material $D = 3$, there are $(4/3)\pi(R/R_0)^3$ sites available, with a density of states $g_0 = (dn/d\epsilon)_0$ per unit volume, thus $W = 1/[(4/3)\pi R^3 g_0]$. A site will be found within R, but at a minimal energy cost W proportional to R^{-D}.

Thus, the probabilty of a hop to a site at distance R is proportional to $\exp - f(R)$, with $f(R) = 2\bar{R}/\xi + W(R)/kT$, where $\bar{R} = [D/(D+1)]R$ is the average hopping distance within R. The conductivity is thus proportional to $\exp - f_{\min}$, which is Mott's law (12), with $B^{D+1} = [1/(g_0\xi^D)][(D+1)2^D/\Omega_D]$, where Ω_D is the volume of the D-sphere of unit radius. In $D = 3$, $B = B_0[1/(g_0\xi^3)]^{1/4}$, with $B_0 = 2[3/(2\pi)]^{1/4}$.

Variable-range hopping had been originally introduced as the mechanism for charge transfer in amorphous semiconductors, where an "impurity band" of localized states straddles the Fermi energy, filling the gap between valence and conduction bands in crystalline semiconductors, into a pseudogap. Quasicrystals are, as we have seen, in exactly the same physical situation as heavily p-doped semiconductors (an "impurity band" of localized, fixed point states, straddling the Fermi energy). The measurement of a Mott conductivity by Delahaye *et al.*[25] implies that the states in the pseudogap are all localized.

3.9. *Application: a photon sieve*

Ebbesen *et al.*[32] found that a metallic plate, pierced by a (periodic) lattice of microscopic holes, lets through light of wavelength longer than the size of the holes, with a normalized flux considerably larger than the geometric area of the holes. An explanation is that the pierced metallic sheet behaves like an aerial, with surface plasmons (collective oscillations of the electrons) on both sides coupled through the holes. If the holes are drilled on a quasicrystalline lattice, no transmission should occur, since metallic conduction is destroyed by inflation symmetry.

3.10. *Maximum metallic resistance of thin wires*

Inflation symmetry has been used to calculate the resistance of a thin wire of length L, consisting of a sequence of similar, but not identical metallic elements (tiles), with length L_0 each.[29] Each (mesoscopic), metallic element is described by a group of energy levels, spaced at energy intervals $d\epsilon/dN = 1/(Vg_0)$. (There are N electrons in a volume V, and $g_0 = (1/V)dN/d\epsilon$ is the electronic density of states.)

A voltage difference Φ at the ends of the tile, excites $e\Phi dN/d\epsilon$ electrons, carrying a current e/t each, where $t = L_0^2/D$ is the diffusion time across the tile of length L_0. The total current is $J = (1/r)\Phi$, and the conductance $1/r = (e^2/t)dN/d\epsilon$. Diffusion constant $\mathbf{D}$, conductivity σ (or mobility $\mu = \sigma/[e(N/V)]$) and density of states are related by Einstein's relation for a degenerate gaz, $\sigma = (1/2)e^2 g_0 \mathbf{D}$.

Each level of the group is shifted by $\Delta\epsilon = \hbar/t$, through the uncertainty principle, thus,

$$\Delta\epsilon = \hbar\mathbf{D}/L_0^2 = (\hbar/e^2)(2\sigma/(L_0^2 g_0)) = (\hbar/e^2)(2/r)(d\epsilon/dN). \tag{13}$$

Consider now a cable made of s elements in series. Connecting the elements perturbs the energies of their levels, by an amount $\Delta\epsilon = \hbar/t$. There are two possibilities:

(i) $\hbar/t > d\epsilon/dN$, the spacing between levels in the group. Then, conduction between two consecutive tiles involves all the levels of the group, and the ohmic behavior is conserved upon renormalization (concatenation). Indeed, $L \to 2L$, $\hbar/t \to \hbar/(4t)$, $N \to 2N$, and $r \to 2r$ is the metallic conductance of two elements of wire in series.

(ii) But, if $\hbar/t < d\epsilon/dN$, each energy level of one tile is connected to one single level of the next tile by the matrix element of the Hamiltonian V, which solders one element to the next, so that the resistance of s elements will be proportional to V^s, that is, it increases exponentially with the length $L = sL_0$ of the wire. The cable is no longer metallic. Maximum metallic resistance $r_{\max}$ of a cable is thus, from (13),[29]

$$r_{\max} = 2\hbar/e^2 \approx 8 \ k\Omega \tag{14}$$

($h/(4e^2) = 6453.204 \ \Omega$ is von Klitzing's unit of resistance).

3.11. *Conductivity of DNA*

A recent application: Electrical conductance of a DNA molecule.[36] Fink and Schoenenberger measured, for a chain of $L = 600$ nm, a resistance of 2.5 MΩ, which is comfortably nonmetallic. One can (over-) estimate from this resistance (assuming a radius of 1 nm) a value for the resistivity of DNA. One obtains 1.25 mΩcm, which is of the order of the maximum metallic resistivity (as estimated from the Ioffe–Regel criterion[31]). Thus, DNA is intrinsically just metallic (it can be regarded as consisting of metallic elements), but a chain of $L = 600$ nm is too long and behaves like an insulator. (Note that the value of the resistivity is an overestimate, since it assumes that the resistance of the DNA molecule increases with its length. The assumption is, as we have seen, unwarranted for $L = 600$ nm.)

4. Conclusion

Inflation symmetry has generic physical consequences: The electronic density of states looks like that of a heavily doped p-type semiconductor, with, at the Fermi level, a small but finite density (pseudogap) of localized states. Inflation symmetry is preserved at the surface of the material, as observed through scanning tunneling microscopy.[33] The material (nonstick, nonwet, but also low oxidation, low corrosion, hardness) and physical properties claimed by patents[1] and experiments[24] are comprehensive: They apply to all metals with inflation symmetry (quasicrystals, but also their rational approximants), in the bulk and on the surface. "Generic", "inflation symmetry", and "unbroken" are now concepts in patent Law.

References

1. J. M. Dubois and P. Weinland, CNRS, European Patent 89402187.2 (1989); J. M. Dubois, A. Proner, B. Bucaille, P. Cathonnet, C. Dong, V. Richard, A. Pianelli, Y. Massiani, S. Ait-Yaazza, and E. Belin-Ferre, Quasicrystalline coatings with reduced adhesion for cookware, *Ann. Chim. Fr.* **19**, 3–25 (1994).
2. I. Stewart, *Galois Theory* (Chapman and Hall, London, 1973).
3. P. A. B. Pleasants, Quasicrystallography: Some interesting new patterns, in *Elementary and Analytic Theory of Numbers* (Banach Center Publications, PWN, Warsaw, 1984), pp. 439–461; W. F. Lunnon and P. A. B. Pleasants, Quasicrystallographic tilings, *J. Maths. Pures Appl.* **66**, 217–263 (1987).
4. N. Rivier and D. Boose, Pseudogap on the surface of quasicrystals, in *ICQ5*, eds. C. Janot and R. Mosseri (World Scientific, 1995), pp. 802–806.
5. B. Doubrovnine, S. Novikov, and A. Fomenko, *Geometrie Contemporaine*, 1 ere partie, Mir, Moscou 1982, paragraph 20.
6. C. Janot, *Quasicrystals. A Primer* (Oxford Univ. Press, 1992).
7. D. Shechtman, L. Blech, D. Gratias, and J. W. Cahn, Metallic phase with long-range orientational order and no translational symmetry, *Phys. Rev. Lett.* **53**, 1951–1953 (1984).
8. R. Penrose, The role of aesthetics in pure and applied research, *Bull. Inst. Math. Appl.* **10**, 266–271 (1974); R. Penrose, Pentaplexity, *Eureka* **39**, 16–22 (1978); M. Gardner, Extraordinary nonperiodic tiling that enriches the theory of tiles, *Sciet. Amer.* **Jan.** 110–121 (1977).

9. N. W. Ashcroft and N. D. Mermin, *Solid State Physics* (Holt, Rinehart and Winston, 1976), Chapter 7.

10. R. K. P. Zia and W. J. Dallas, A simple derivation of quasicrystalline spectra, *J. Phys. A: Math. Gen.* **18**, L341–L345 (1985); V. Elser, The diffraction pattern of projected structures, *Acta Cryst.* **A42**, 36–43 (1986); A. Katz and M. Duneau, Quasiperiodic patterns, *Phys. Rev. Lett.* **54**, 181–196 (1985); P. A. Kalugin, A. Y. Kitayev, and L. S. Levitov, $Al_{0.86}Mn_{0.14}$: A six-dimensional crystal, *JETP Lett.* **41**, 145–149 (1985).

11. H. Davenport, *The Higher Arithmetic* (Hutchinson, London, 1952).

12. P. Bak, Icosahedral crystals from cuts in six-dimensional space, *Scripta Met.* **20**, 1199–1204 (1986); T. Janssen, Crystallography of quasicrystals, *Acta Cryst.* **A42**, 261–271 (1986).

13. C. Oguey, M. Duneau, and A. Katz, A geometrical approach to quasiperiodic tilings, *Comm. Math. Phys.* **118**, 99–118 (1988); See also: P. Kramer, *J. Math. Phys.* **29**, 516 (1988).

14. D. Pedoe, *Geometry and the Visual Arts* (Penguin, Harmondsworth, 1976).

15. M. Cleyet-Michaud, *Le Nombre d'Or* (Que Sais-je 1530, PUF Paris, 1973).

16. J. F. Sadoc and N. Rivier, Boerdijk–Coxeter helix and biological helices, *Eur. Phys. J.* **B12**, 309–318 (1999).

17. F. Gaill and Y. Bouligand, Supercoil of collagen fibrils in the integument of *Alvinella*, an abyssal annelid, *Tissue & Cell* **19**, 625–642 (1987).

18. N. Rivier, Quasicrystals and grain boundaries, in *Quasicrystals and Incommensurate Structures in Condensed Matter*, eds. M. J. Yacaman, *et al.* (World Scientific, 1990), pp. 50–68; N. Rivier, The topological structure of grain boundaries, in *Number Theory and Physics*, eds. J.-M. Luck, *et al.* (Springer, 1990), pp. 118-127.

19. H. Telley, *Modelisation et Simulation bidimensionnelle de la Croissance des Polycristaux* (PhD Thesis, EPFL Lausanne, 1989); N. Rivier, Geometry and fluctuations of surfaces, *J. Phys. (Paris)* **51**, C7-309-317 (1990).

20. N. Rivier, Non-stick quasicrystalline coatings, *J. Non-cryst. Solids* **153–154**, 458–462 (1993).

21. C. Janot and M. de Boissieu, Quasicrystals as a hierarchy of clusters, *Phys. Rev. Lett.* **72**, 1674–1677 (1994).

22. N. Rivier, Wetting on quasicrystals, in *New Horizons in Quasicrystals*, eds. A. I. Goldman *et al.* (World Scientific, 1997), pp. 188–199.

23. R. P. Feynman, R. B. Leighton, and M. Sands, *The Feynman Lectures on Physics* (Addison-Wesley, 1963), I.44.1.

24. E. Belin, Z. Dankhazi, A. Sadoc, J. M. Dubois, and Y. Calvayrac, Aluminium electronic distributions in Al-Cu-Fe alloys, *Europhys. Lett.* **26**, 677–682 (1994); C. Berger, Electronic properties of quasicrystals; experimental, in *Lectures on Quasicrystals*, eds. F. Hippert and D. Gratias (Editions de Physique, Les Ulis, 1994), Chapter. 10, pp. 463–504.

25. J. Delahaye, J. P. Brison, and C. Berger, Evidence for variable range hopping conductivity in the ordered quasicrystal i-AlPdRe, *Phys. Rev. Lett.* **81**, 4204–4207 (1998).

26. D. A. Lavis, B. W. Southern, and G. S. Davinson, A real space rescaling treatment of the spectral properties of an adatom-contaminated crystal system, *J. Phys.* **C18**, 1387–1399 (1985).

27. N. Rivier and T. Aste, Curvature and frustration in cellular systems, *Phil. Trans. R. Soc. Lond.* **A354**, 2055–2069 (1996).

28. M. Kohmoto, L. P. Kadanoff, and C. Tang, Localization problem in one dimension: Mapping and escape, *Phys. Rev. Lett.* **50**, 1870–1873 (1983).

29. D. J. Thouless, Maximum metallic resistance in thin wires, *Phys. Rev. Lett.* **39**, 1167–1169 (1977).

30. N. Rivier and M. Durand, Variable-range hopping conductivity in quasicrystals, *Mat. Sci. Eng.* **A** (2000), to appear.

31. N. F. Mott and E. A. Davis, *Electronic Processes in Non-Crystalline Materials* (Oxford Univ. Press, 1971).

32. T. W. Ebbesen, H. J. Lezec, H. F. Ghaemi, T. Thio, and P. A. Wolff, Extraordinary optical transmission through sub-wavelength hole arrays, *Nature* **391**, 667–669 (1998); See also the N&W comment: J. R. Sambles, More than transparent, *ibid.*, pp. 641–642; J.-P. Dufour, Un curieux "tamis à photons" étonne les physiciens, *Le Monde*, 13/2/1998.

33. T. M. Schaub, D. E. Burgler, H. J. Guntherodt, and J. B. Suck, Quasicrystalline structure of icosahedral $Al_{68}Pd_{23}Mn_9$ resolved by scanning tunneling microscopy, *Phys. Rev. Lett.* **73**, 1255–1258 (1994).

34. Y. Dupain, T. Kamae, and M. Mendes-France, Can one measure the temperature of a curve?, *Arch. Rat. Mech. Anal.* **94**, 155–163 (1986); I. Stewart, La physique des courbes, *Visions Géométriques* (Bibliothèque Pour la Science, Paris, 1994), Chapter. 5.

35. N. Rivier and A. Goldar, Entropy of aperiodic crystals generated by spirals, in *Aperiodic '97*, eds. M. de Boissieu, J.-L. Verger-Gaugry, and R. Currat (World Scientific, 1998), pp. 115–119.

36. H. W. Fink and C. Schoenenberger, Electrical conduction through DNA molecules, *Nature* **398**, 407–410 (1999).

37. N. Destainville and J. F. Sadoc, Excitations in one dimension: A geometrical view of the transfer matrix method, *J. Math. Phys.* **38**, 1849–1863 (1997).

38. F. X. Riguidel, A. Hansen, and D. Bideau, *Europhys. Lett.* **28**, 13 (1994); D. Bideau, I. Ippolito, and N. Rivier, Une bille qui roule sur une surface rugueuse (Pour la Science, 2001).

39. N. Rivier, Gauge theory and geometry of condensed matter, in *Geometry in Condensed Matter Physics*, ed. J. F. Sadoc (World Scientific, 1987), pp. 1–88.

40. S. Singh, *Fermat's Last Theorem* (Fourth Estate, London, 1997).

41. M. Sibon, N. Rivier, and F. Rothen, Optical Fourier transform of quasiperiodic sequences, in preparation.

Annual Reviews of Computational Physics VIII (pp. 287–300)
Edited by Dietrich Stauffer

PERCOLATION SIMULATION: LARGE LATTICES, VARYING DIMENSIONS

DIETRICH STAUFFER[*,†] and NAEEM JAN[†]

*Institute for Theoretical Physics, Cologne University,
50923 Köln, Euroland

†Physics Department, StFX University, Box 5000,
Antigonish, N.S. B2G 2W5, Canada

Tricks are explained to simulate percolation lattices with one program for variable dimensionality, and to look at large lattices in two (or three) dimensions.

1. Introduction

The Hoshen–Kopelman algorithm, which to my deep regret was invented in a department of chemical engineering, is an elegant way to analyze existing large lattices in terms of their clusters.[4] In two dimensions, only one line at a time has to be stored in the computer's main memory; in three dimensions we need one plane to be stored. Complete programs are given in the literature.[1] For the convenience of the reader we first give the Leath algorithm.[2] Similar algorithms may be useful for the other percolation problems discussed in this volume by Erzan, Marrink *et al.*, and Stauffer (on markets).

2. Recycling Hoshen–Kopelman

Basically, in percolation each site of a large lattice is occupied for probability p and is empty for probability $1 - p$; clusters are groups of occupied neighboring sites. To find these clusters, each occupied site gets a label in the following way: We set INDEX = 0 and start, on the square or triangular lattice, in the upper left corner and proceed like a typewriter from left to right, and then from top to bottom. If we get to a new occupied site, we check if it has an occupied neighbor which was investigated before, that is, on the left, or in the upper line. If it is not connected in this way, then INDEX increases by 1 and the site gets this new INDEX as its label. If, on the other hand, it has a neighbor which was already given a label, then the new site gets this label. In this way each cluster gets its own label, and at any time of the simulation, we have to store only one lattice line: the left part of the current line, then the current site and its upper neighbors, and the right part of the previous line.

However, what do we do if the site has two or more already investigated neighbors, which have different labels since up to the current situation the computer had "thought" they were different clusters; the presently investigated site joins these disconnected branches. Then the site gets the smallest of the neighbor labels as the proper label, and the larger labels of the other neighbors are marked as "bad". This is accomplished by using a label N such that for a good label we have $N(\text{label})$ positive and equal to the number of sites in this cluster up to now, while for a bad label, $-N(\text{label})$ points to another label which may be better. For example, if at some site of a triangular lattice a cluster with label 2 joins one with label 4 and another with label 5, then $N(2)$ remains at 2, while $N(4)$ and $N(5)$ become -2. If later the labels 1 and 2 are found to join, then $N(2)$ becomes -1. Thus by continuously checking if $N(\text{label})$, $N(-N(\text{label})$, $N(-N(-N(\text{label})))$ etc is positive, we find the proper root label. This loop is achieved in the Fortran program by the labels 13 and 14, and the corresponding back jumps GOTO 13 and GOTO 14. (Empty sites are not labeled with zero but with MAX, the largest label number allowed by the program. In this way, their label is never smaller and thus "better" than that of an occupied site.)

In this way, the array $N(\text{INDEX})$ contains the connection information. Its index increases by one unit every time it seems as if a new cluster will start. In the square lattice this happens with probability $p(1-p)^2$, in the triangular lattice with probability $p(1-p)^3$, both of which are quite small. Nevertheless, for large enough lattices, the memory needed for N may be larger than that to store the line (or plane in three dimensions). Then a recycling of unused labels, like recycling of used paper, helps; this was invented two decades ago by Reynolds and Nakanishi at Boston University. The program contains this recycling at the beginning, between the GOTO 20 statement and the label 20. Basically, when INDEX gets dangerously close to the upper limit MAX of the computer program, one repeats the analysis of the last line and makes a simplified Hoshen–Kopelman label check: Each occupied site becomes the label of its left neighbor if that is occupied; otherwise the label of that occupied site is checked for being positive (good) or negative in the loop given by label 22 and the GOTO 22 command. When the whole line is finished, the cluster sizes as stored in N are added up for the histogram of the cluster size distribution. Basically, in this way we discard all the clusters which are definitely finished above the current line, and keep only those needed for the analysis of the lower parts of the lattice, that is, those that touch the present line.

In this way, MacLeod and Jan on a Dec Alpha workstation at a small university could simulate within two weeks a 2 million × 2 million lattice, four times larger than the largest Ising lattices simulated in two dimensions. (It was for MacLeod an undergraduate project in the thermal physics course.) For the simple cubic lattice, percolation with 10001^3 sites was possible, as large as the largest Ising lattices. (In three dimensions, recycling was needed every two planes.) The higher the dimensionality of the lattice, the larger is the fraction of the lattice belonging to the single hyperplane to be stored; thus the Hoshen–Kopelman is not as memory saving. The two-dimensional

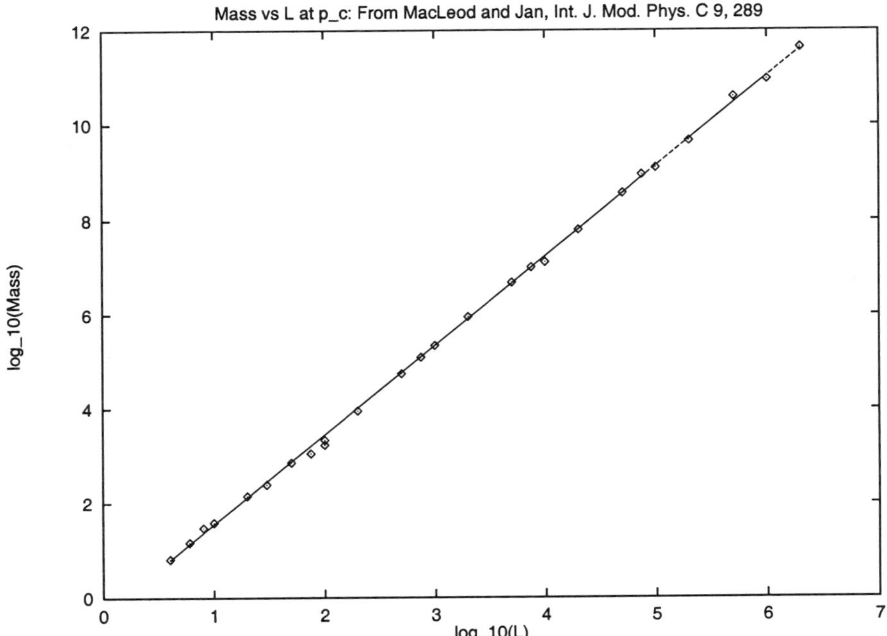

Fig. 1. Fractality of the largest cluster right at the percolation threshold 0.592746. From Ref. 3).

simulations confirmed, for example, that the largest cluster right at the critical point is a fractal with the theoretically predicted fractal dimension $D = 91/48$: Its mass in an $L \times L$ lattice at p $= 0.592746$ increases as L^D for large L, Fig. 1. Present efforts of parallelization by domain decomposition may soon allow even larger lattice sizes.

3. Varying Dimensions

In percolation theory, the upper critical dimension is six, and thus one wants to sometimes write one program for dimensions d varying from two to seven. Writing different programs may easily cause errors: one corrects an error in one of the programs but not in all of them. An innermost loop over $2d$ neighbors in the d-dimensional hypercubic lattice is inefficient for vector computers. The trick, invented originally for immunology,[5] is to use the parameter declaration of Fortran together with IF-conditions for jumps, which are then already evaluated at compilation time.

Thus if you declare at the beginning of the program the dimensionality IDIM to be 3, you can use

$$\texttt{PARAMETER(IDIM} = \texttt{3)},$$

which causes the compiler to insert a 3 wherever it finds the constant IDIM. (Of course, now you are not allowed to change this parameter, e.g., through IDIM=IDIM+1.) If afterwards you use

$$\texttt{IF(IDIM.EQ.3)GOTO 7},$$

then already at the compilation the computer evaluates whether it jumps to 7 or continues to the next line. It may even give you a warning that you never come to the next line. In any case, the evaluation no longer has to be done during the execution of the program, and little speed is lost. The third program counts clusters in IDIM dimensions. While it looks quite long, many statements are just repetitions of the same commands, for the different lattice dimensions. Dietrich Stauffer has been using this type of program since many years, particularly for market simulations. Naeem Jan generalized it such that it checks, in the number ICONN, how many sites of the last hyperplane are still connected to the first hyperplane.

In the exercise during the meeting, we programmed the simpler case to check if on the square lattice a cluster connects top with bottom; this is our last program here.

```fortran
c     First program: Leath
      parameter(L=1001,L2=L*L)
      integer neighb(4),isite(-L+1:L2+L),list(L2)
      data p/0.592746/, iseed/1/
      ibm=2*iseed-1
      ip=(2.0*p-1.0)*2147483647.0
      neighb(1)= 1
      neighb(2)=-1
      neighb(3)= L
      neighb(4)=-L
      init=L2/2
      max=0
      isize=1
      list(isize)=init
      do 6 j=-L+1,L2+L
6         isite(j)=2
1     min=max+1
      isite(init)=0
      max=isize
      do 2 idir=1,4
        nb=neighb(idir)
        do 3 index=min,max
          j=list(index)+nb
          if(isite(j).lt.2) goto 3
          isite(j)=1
          ibm=ibm*16807
          if(ibm.lt.ip) isite(j)=0
c         unclear=2, empty=1, occupied=0
          if(isite(j).eq.0) then
            isize=isize+1
c           new cluster site found
            list(isize)=j
          end if
3       continue
2     continue
      do 4 i=1,L
4       if(isite(i-L)*isite(i+L2).eq.0) goto 8
      if(isize.gt.max) goto 1
8     do 7 i=1,L-2,3
        do 7 k=1,L-2,3
      j=(i-1)*L+k
      j1=j+L
      j2=j1+L
7     if(isite(j)+isite(j+1)+isite(j+2)+isite(j1)+isite(j1+1)+
     1 isite(j1+2)+isite(j2)+isite(j2+1)+isite(j2+2).le.4)print*,k,i
      print 5, p,isize,L
5     format(1x,f9.6,2i9)
      stop
      end
```

```
         PROGRAM PERC
C        SECOND PROGRAM: TRIANGULAR SITE PERCOLATION
         WITH NAKANISHI RECYCLING DIMENSION
         LEVEL( 2501),N(10000 ),NS(64)
         LOGICAL TOP,LEFT,BACK
         P=RAND(3)
C
C
         MAX=10000
         L=2500
         P=0.5
C        L*L LATTICE AT CONCENTRATION P OF OCCUPIED SITES (ONE
         SIMULATION).
C        COMPUTER TIME PROPORTIONAL TO L*L AND ROUGHLY INDEPENDENT
         OF P.
C        HERE L UP TO 2500. LARGER L REQUIRES INCREASE OF DIMENSION
         OF ARRAY
C        LEVEL(L+1) AS WELL AS OF DIMENSION OF N(0.07*L*L). HOWEVER,
         THE RECYCLING
C        ALLOWS, WITH SOME ADDITIONAL TIME, TO SIMULATE LARGER
         SYSTEMS WITHOUT
C        INCREASING N TOO MUCH. NUMBER OF RECYCLINGS (OUTPUT "IREC")
         SHOULD BE
C        MUCH SMALLER THAN NUMBER L OF LINES TO AVOID WASTE OF
         COMPUTER TIME.
C        WORLD RECORD IS L = 2 MILLION TIMES 2 MILLION (NAEEM JAN 1998)
C
C        MAX MUST AGREE WITH DIMENSION FOR ARRAY N. LARGE
         SHOULD BE ENLARGED
C        FOR LARGER SYSTEMS TO AVOID TOO MUCH OUTPUT
C
C
         DO 998 I=1,64
998      NS(I)=0
         LARGE=100000
         TAUM1=96./91.
         ALOG2=1.0000001/ALOG(2.0)
         MAX3=MAX*3
         N(MAX)=MAX
         LP1=L+1
         LIMIT=MAX-L*0.2
         INDEX=0
         IREC=0
         CHI=0.
C           IF(L.GT.1000) STOP 1
         DO 1 I=1,LP1
1        LEVEL(I)=MAX
         DO 3 K=2,LP1
         IF(INDEX.LT.LIMIT) GOTO 20
         IREC=IREC+1
         IF(K.LE.3) STOP 2
         J=INDEX
         DO 21 I=2,LP1
         LEV=LEVEL(I)
         IF(LEV.EQ.MAX) GOTO 21
         IF(LEVEL(I-1).NE.MAX) GOTO 24
         LABEL=LEV
         IF(N(LEV).GE.0) GOTO 27
         MS=N(LEV)
22       LABEL=-MS
         MS=N(LABEL)
```

```
        IF(MS.LT.0) GOTO 22
        N(LEV)=-LABEL
27      IF(LABEL.GT.INDEX) GOTO 25
        J=J+1
        IF(J.GE.MAX) STOP 3
        N(J)=N(LABEL)
        N(LABEL)=-J
        LEVEL(I)=J-INDEX
        GOTO 21
24      LEVEL(I)=LEVEL(I-1)
        GOTO 21
25      LEVEL(I)=LABEL-INDEX
21      CONTINUE
        IF(IREC.EQ.1) PRINT 2, P,J
C       END OF RECYCLING; NOW ANALYSIS OF FINISHED CLUSTERS
        DO 26 IS=1,INDEX
        NIS=N(IS)
        IF(NIS.LE.0) GOTO 26
        FNIS=NIS
        CHI=CHI+FNIS*FNIS
        IF(NIS.GE.LARGE) PRINT 97, NIS
        NIS=ALOG(FNIS)*ALOG2+1.
        NS(NIS)=NS(NIS)+1
26      CONTINUE
        INDEX1=J-INDEX
        IF(INDEX1.LE.0) STOP 5
        DO 23 IND=1,INDEX1
23      N(IND)=N(IND+INDEX)
        INDEX=INDEX1
        IF(IREC.NE.(IREC/100)*100) GOTO 20
        PRINT 2, P,INDEX,J,K
        PRINT 96, NS,J
20      CONTINUE
        MOLD=MAX
        DO 3 I=2,LP1
        MBACK=MOLD
        LBACK=MOLD
        MOLD=LEVEL(I)
        IF(RAND(0).GT.P) GOTO 9
        MLEFT=LEVEL(I-1)
        LTOP=LEVEL(I)
        MTOP=LTOP
        IF(MLEFT+MTOP+MBACK.EQ.MAX3) GOTO 4
        LEFT=MLEFT.LT.MAX
        TOP=MTOP.LT.MAX
        BACK=MBACK.LT.MAX
C       FIRST HOSHEN-KOPELMAN CLASSIFICATION OF TOP NEIGHBORS
        IF(.NOT.TOP.OR.N(LTOP).GE.0) GOTO 12
        MS=N(LTOP)
13      MTOP=-MS
        MS=N(MTOP)
        IF(MS.LT.0) GOTO 13
        N(LTOP)=-MTOP
C       NOW COMES THE BACK NEIGHBOR (LEFT OF TOP)
12      IF(.NOT.BACK.OR.N(LBACK).GE.0) GOTO 11
        MS=N(LBACK)
14      MBACK=-MS
        MS=N(MBACK)
        IF(MS.LT.0) GOTO 14
        N(LBACK)=-MBACK
```

```
C        LEFT NEIGHBOR NEEDS NO RECLASSIFICATION
11       MNEW=MIN0(MTOP,MBACK,MLEFT)
         LEVEL(I)=MNEW
         ICI=1
         IF(TOP) ICI=ICI+N(MTOP)
         IF(LEFT.AND.MTOP.NE.MLEFT) ICI=ICI+N(MLEFT)
         IF(BACK.AND.MBACK.NE.MLEFT.AND.MBACK.NE.MTOP)
        1 ICI=ICI+N(MBACK)
         N(MNEW)=ICI
C        ICI IS THE SIZE OF THE CLUSTER AT THIS STAGE
         IF(TOP .AND.MTOP .NE.MNEW) N(MTOP )=-MNEW
         IF(LEFT.AND.MLEFT.NE.MNEW) N(MLEFT)=-MNEW
         IF(BACK.AND.MBACK.NE.MNEW) N(MBACK)=-MNEW
         GOTO 3
4        INDEX=INDEX+1
C        START OF NEW CLUSTER
         LEVEL(I)=INDEX
         N(INDEX)=1
         GOTO 3
9        LEVEL(I)=MAX
3        CONTINUE
C        NOW FINAL ANALYSIS
         IF(INDEX.EQ.0) GOTO 35
         DO 6 IS=1,INDEX
         NIS=N(IS)
         IF(NIS.LT.0) GOTO 6
         FNIS=NIS
         CHI=CHI+FNIS*FNIS
         IF(NIS.GE.LARGE) PRINT 97, NIS
         NIS=ALOG(FNIS)*ALOG2+1.
         NS(NIS)=NS(NIS)+1
6        CONTINUE
35       PRINT 96, NS
96       FORMAT(" NS:",2I11,5I9,5(/,2I7,10I6))
97       FORMAT(" CLUSTER OF SIZE ",I12)
2        FORMAT(F15.8,3I10,2F20.4)
         CHI=(CHI/L)/L
         PRINT 2, P,L,INDEX,IREC,CHI
         I=2**21
         ISUM=0
         DO 905 INDEX=1,22
         NIS=23-INDEX
         ISUM=ISUM+NS(NIS)
         CHI=(ISUM*I**TAUM1/L)/L
         IF(ISUM.GT.0) PRINT 2, CHI,ISUM,I,NIS
905      I=I/2
         STOP
         END
```

```fortran
c     Third program: Variable dimension
c     random site percolation in d dimensions, counts clusters and connectivity
c     free bound.cond. in 2 directions and helical in the other d-2 directions
      parameter(idim=3,L=101,nplane=L**(idim-1),nc=nplane/L,
     1 Lplane=nplane+nc,max=(nplane*L)/8)
c     denominator for max to be empirically adjusted, also in KLASS
      dimension n(max),level(Lplane),ns(0:30)
      logical left,top,back,four,five,six,seven
      real*8 fnis, fmax, chi, alog2, p
      common level,n
      data p,ibm,nrun/0.311608d0,1,10 /, ns/31*0/
      print *, p,L,idim,ibm,nrun
      if(idim.gt.7.or.idim.lt.2) stop 2
      ibm=2*ibm-1
      ipro=(2.0d0*p-1.0d0)*2147483648.0d0
      alog2=1.0d0/0.69315d0
      L1=0
      L2=0
      L3=0
      L4=0
      L5=0
      if(idim.gt.2) L1=L
      if(idim.gt.3) L2=L*L1
      if(idim.gt.4) L3=L*L2
      if(idim.gt.5) L4=L*L3
      if(idim.gt.6) L5=L*L4
      do 1 irun=1,nrun
      chi=0.0d0
      maxclu=0
      left=.false.
      top =.false.
      back=.false.
      four=.false.
      five=.false.
      six =.false.
      seven=.false.
      do 5 i=1,nc
5        level(i)=max
      do 6 i=nc+1,Lplane
6        level(i)=1
      n(max)=1
      n(1)=0
      index=1
C     INITIALIZATION OF EMPTY TOP PLANE FINISHED, NOW L FURTHER PLANES
      do 2 k=1,L
      do 3 i=nc+1,Lplane
      ibm=ibm*16807
      if(ibm.gt.ipro) then
        level(i)=max
```

```
          goto 3
          end if
          left=level(i- 1).lt.max
          top =level(i-L1).lt.max
          if(idim.eq.2) goto 7
          back=level(i-L2).lt.max
          if(idim.eq.3) goto 7
          four=level(i-L3).lt.max
          if(idim.eq.4) goto 7
          five=level(i-L4).lt.max
          if(idim.eq.5) goto 7
          six =level(i-L5).lt.max
          if(idim.eq.6) goto 7
          seven=level(i ).lt.max
7         continue
          if(.not.(left.or.back.or.top.or.four.or.five.or.six.or.seven))
     1        goto 4
          mleft=max
          mtop =max
          mback=max
          mfour=max
          mfive=max
          msix =max
          mseven=max
C     NO NEED TO RECLASSIFY LEFT NEIGHBOR: JUST DONE
          if(left) mleft=level(i-1)
          if(top ) mtop =klass(i-L1)
          if(idim.eq.2) goto 8
          if(back) mback=klass(i-L2)
          if(idim.eq.3) goto 8
          if(four) mfour=klass(i-L3)
          if(idim.eq.4) goto 8
          if(five) mfive=klass(i-L4)
          if(idim.eq.5) goto 8
          if(six ) msix =klass(i-L5)
          if(idim.eq.6) goto 8
          if(seven) mseven=klass(i)
8         mnew=min0(mleft,mtop,mback,mfour,mfive,msix,mseven)
          ici=1
          if(left) then
             ici=ici+n(mleft)
             if(mleft.ne.mnew) n(mleft)=-mnew
          endif
          if(top ) then
             if(mtop.ne.mleft) ici=ici+n(mtop)
             if(mtop .ne.mnew) n(mtop )=-mnew
          endif
          if(idim.eq.2) goto 9
          if(back) then
```

```
         if(mback.ne.mleft.and.mback.ne.mtop) ici=ici+n(mback)
         if(mback.ne.mnew) n(mback)=-mnew
       endif
       if(idim.eq.3) goto 9
       if(four) then
         if(mfour.ne.mleft.and.mfour.ne.mtop.and.mfour.ne.mback)
     1      ici=ici+n(mfour)
         if(mfour.ne.mnew) n(mfour)=-mnew
       endif
       if(idim.eq.4) goto 9
       if(five) then
         if(mfive.ne.mleft.and.mfive.ne.mtop.and.mfive.ne.mback.and.mfive
     1      .ne.mfour) ici=ici+n(mfive)
         if(mfive.ne.mnew) n(mfive)=-mnew
       endif
       if(idim.eq.5) goto 9
       if(six ) then
         if(msix.ne.mleft.and.msix.ne.mtop.and.msix.ne.mback.and.msix
     1      .ne.mfour.and.msix.ne.mfive) ici=ici+n(msix)
         if(msix .ne.mnew) n(msix )=-mnew
       endif
       if(idim.eq.6) goto 9
       if(seven) then
         if(mseven.ne.mleft.and.mseven.ne.mtop.and.mseven.ne.mback.and.
     1      mseven.ne.mfour.and.mseven.ne.mfive.and.mseven.ne.msix)
     2      ici=ici+n(mseven)
         if(mseven.ne.mnew) n(mseven)=-mnew
       endif
9      level(i)=mnew
       n(mnew)=ici
       goto 3
4      index=index+1
C      A NEW CLUSTER STARTS
       level(i)=index
       n(index)=1
3      continue
       if(index.ge.0.99*max) then
         print *, ' error: ', index,max,L**idim,k
         stop 9
       endif
2      continue
       iconn=0
       do 14 I=nc+1,Lplane
14       if(level(i).eq.1) iconn=iconn+1
         do 11 is = 1, index
           nis = n(is)
           if(nis .lt. 1) goto 11
           if(maxclu .lt. nis) maxclu = nis
           fnis = nis
```

```fortran
                 ibin = dlog(fnis)*alog2+0.000001d0
                 if(ibin.le.30) ns(ibin) = ns(ibin) + 1
                 chi = chi + fnis*fnis
11               continue
                 fmax = maxclu
                 chi = chi - fmax*fmax
                 print 100, irun,index,maxclu,(chi/L)/nplane,iconn
1                continue
                 do 12 ibin = 0,30
12                 if(ns(ibin).gt.0) print *, 2**ibin, ns(ibin)
100              format(1x,i5,2i9,f12.3,i5)
                 stop
                 end
                 function klass(index)
                 parameter(idim=3,L=101,nplane=L**(idim-1),nc=nplane/L
      1          ,Lplane=nplane+nc,max=(nplane*L)/8)
                 dimension level(Lplane),n(max)
                 common level,n
                 ms=n(level(index))
                 if(ms.lt.0) goto 13
                 klass=level(index)
                 return
13               klass=-ms
                 ms=n(klass)
                 if(ms.lt.0) goto 13
                 n(level(index))=-klass
                 return
                 end
```

```fortran
c     Fourth program: simple square lattice
      dimension level(3001), n(1000000)
      data p/0.592746/, L/3000/, ibm/7/, max/1000000/
      print *, p, L, ibm, max
      ip=(2.0*p-1.0)*2147483647
      index = 1
      n(1) = 1
      n(max)= max
      level(1)=max
      do 1 i=2,L+1
c        top line occupied; left column empty = max
1        level(i)=1
      do 2 k=1,L
         line=max
         do 3 i=2,L+1
           ibm=ibm*16807
c          check if site is occupied through integer random number
           if(ibm.gt.ip) goto 9
           mleft=level(i-1)
           mtop =level(i)
           if(mleft+mtop.eq.2*max) goto 4
           ms=n(mtop)
           if(ms.eq.mtop) goto 7
6          ms=n(ms)
c          Hoshen-Kopelman classification: search for root n(ms)=ms
           if(n(ms).ne.ms) goto 6
           n(mtop)=ms
           mtop  =ms
7          mnew=min(mtop,mleft)
           n(mnew) =mnew
           level(i)=mnew
           if(mtop .lt.max.and.mtop .ne.mnew) n(mtop )=mnew
c          relabel the bad label: n(bad)=good
           if(mleft.lt.max.and.mleft.ne.mnew) n(mleft)=mnew
           goto 3
4          index=index+1
c          start new cluster
           n(index)=index
           level(i)=index
           goto 3
c          make site empty
9          level(i)=max
c          check if at least one site has level = 1; stop otherwise
3          line=min(line,level(i))
2        if(line.gt.1) stop 2
      print *, index
      end
```

References

1. D. Stauffer and A. Aharony, *Introduction to Percolation Theory* (Taylor and Francis, London, 1994); A. Bunde and S. Havlin, *Fractals and Disordered Systems* (Springer, Berlin-Heidelberg, 1996); M. Sahimi, *Applications of Percolation Theory* (Taylor and Francis, London, 1994).
2. H. G. Evertz, *J. Stat. Phys.* **70**, 1075 (1993).
3. S. MacLeod and N. Jan, *Int. J. Mod. Phys.* **C9**, 289 (1998); N. Jan and D. Stauffer, *ibid*, p. 341.
4. J. Hoshen and R. Kopelman, *Phys. Rev.* **B14**, 3428 (1976).
5. D. Stauffer and G. Weisbuch, *Physica* **A180**, 42 (1992); M. Sahimi and D. Stauffer, *Phys. Rev. Lett.* **71**, 4271 (1993).
6. D. Tiggemann seems to have developed a parallel code which has produced data on a Cray-T3E consistent with the square-lattice data of Ref. 3; and A. B. MacIsaac is developing a code for multiple workstations coupled in parallel with fast software (Beowulf) between the nodes.

Annual Reviews of Computational Physics VIII (pp. 301–306)
Edited by Dietrich Stauffer

SOME ASPECTS OF DYNAMICS OF JOSEPHSON-JUNCTION ARRAY AT GOLDEN MEAN FRUSTRATION

MOHAMMAD R. KOLAHCHI

*Institute for Advanced Studies in Basic Sciences, Gava Zang,
P. O. Box 45195-159, Zanjan, Iran*

In the array of Josephson-junctions at golden mean frustration there exists time-temperature superposition principle similar to that of supercooled liquids for temperatures close to $0.3J/k_B$. Below T^* such that $0.2 < k_B T^*/J < 0.3$ this scaling is lost. We propose that T^* marks the temperature where well-defined metastable states are developed shaping a rugged landscape and causing the loss of scaling.

An experimental realization of frustrated XY models, is an array of Josephson-junctions in a magnetic field. The frustration is essentially due to competition between flux quantization and energy minimization. The parameter measuring the frustration is the ratio of flux in a plaquette to the quantum of flux. For nonintegral values of this parameter, f, currents must necessarily be excited. The excitations of the model at low temperatures are well described by the array of such current loops: the vortex lattice.[1]

The vortex lattice is the manifestation of long-range phase ordering. As temperature is increased, this phase ordering or coherence across the lattice, is disrupted. The vortices that act somewhat like Coulomb particles in their interaction with each other, start to move. The transition can happen in two steps, that is, the vortex lattice may first depin and then melt.[2] The details of the transition appear to depend crucially on the particular structure of the vortex lattice.[3]

Aside from commensurability effects, the local environment set up by the neighboring vortices can act as if a vortex is in a frozen field, hence hampering its motion.[4] For the vortex lattice to be (globally) commensurate with the underlying periodicity of the grid, and simultaneously having locally stable organization of vortices, is a problem that is more serious for some values of f. Ironically, it may happen that a large number of solutions exist that nearly satisfy both. The system can take a macroscopically long-time to move in its phase space and find the best solution. This is because states that are close in energy may not be so in configuration. Besides, the motion in the phase space becomes exceedingly slow as temperature is lowered because the chance of overcoming the energy barriers is reduced. Such temperature dependent slow processes are generally referred to as structural relaxation processes.[5] For such systems, it is said that the energy landscape is rugged. This ruggedness is the result of many locally (in the phase space sense) stable states, which are also called metastable states.

Empirically, it appears that for metastable states, f is the density of vortices. (In a sense this is true always, due to charge neutrality, or flux conservation, but a well-defined vortex only exists for metastable states.) Due to the symmetry present in the problem (similar to that of the problem of a superconducting ring in a magnetic field), the maximally frustrated system is one with $f = 1/2$. This value presents such a strict condition on the system on a square grid, that a well-separated solution, a deep minimum, is singled out; the problem has an exact solution just as when the field is off. Interestingly, this is a characteristic of complex systems where, as some control parameters are varied, the problem exhibits such easy–hard–easy nature.

What values of f make the problem hard, that is, create a landscape that is exquisitely rugged, and why? These are open questions. Physically, it is plausible that to have locally frozen fields, which inhibit easy relaxation, higher density of vortices is required. Heuristically, this means that at higher densities, vortices get in each other's way.

Now, if we think of the possible ways the lattice can organize itself as it begins to form, when the temperature is decreased, the hierarchical property of some values of f may play a part in increasing the number of good solutions. For such f, a delicate arrangement of sublattices can be observed.[6] An algebraic number whose sequence of rational approximants allows several of its best approximants to participate in a reasonably large (i.e. computationally feasible) lattice, is the golden mean $\tau = (\sqrt{5} + 1)/2$. The hierarchy does not mean that the lattice should necessarily organize itself in that way (in fact, to find such perfect arrangement is difficult and may even be precluded due to the commensurability effects), but it does show that on scales smaller than the system size, and yet larger than the grid constant, other candidate vortex lattices can form stable domains. This, of course, is the property of every irrational number, namely that it has infinitely many rationals close to it.

The Hamiltonian of the Josephson-junction array in the presence of magnetic field perpendicular to the plane of the square array (or grid) is given by

$$H = -J \sum \cos(\theta_i - \theta_j) - J \sum \cos(\theta_i - \theta_j - 2\pi f m) \,. \tag{1}$$

θ_i represents the phase of superconducting electron wavefunction on site i of the lattice. J is the strength of Josephson coupling and is taken to be positive in Eq. (1). It also sets the energy scale; the temperatures we shall mention will be in units of J/k_B. The sums are over the nearest neighbor pairs $\langle i,j \rangle$: in the x-direction, for the first sum, and in the y-direction for the second sum. The frustrating phase $2\pi m$ appears in the second term; the vector potential has only a y-component. m is an integer which gives the x-coordinate of the vertical bond $\langle ij \rangle_y$. The symmetry mentioned above, and the evenness of cosine, give the physical property that f and $-f$ (i.e. reversal of the field) should be the same. Hence, instead of studying $55/34$ which is in the sequence of best rational approximants of τ, we could as well study $f = 13/34$, and so we do. We study a 34×34 lattice using periodic boundary conditions.

The specific problem to deal with is how the correlation of phases varies in time, as a function of temperature. In a Monte Carlo simulation, just as in the experiment, one starts at a high temperature, quenches to the temperature of interest, waits for a long-time t_w and then measures the physical quantity of interest,[7] here the correlation function,

$$q_{xy}(t) = \left\langle \frac{1}{N} \sum_{i=1}^{N} \cos(\theta_i(t_w + t) - \theta_i(t_w)) \right\rangle \,. \tag{2}$$

In Eq. (2), N is the number of lattice sites, and $\langle \, \rangle$ denotes the average over various initial conditions. t_w should be larger than the expected relaxation times for the temperature under study, in order to probe the equilibrium properties.

In the study of dynamical behavior of Ising spin glasses, Ogielski found that the dynamic correlation function $q(t)$, followed the empirical formula,[8]

$$q(t) = c \frac{\exp(-\omega t^\beta)}{t^x} \,. \tag{3}$$

This formula is composed of a stretched exponential in the numerator, and a power-law decay characterized by x. We find that the correlation data for the Josephson-junction array at $f = 13/34$ also fits this empirical function.

The stretched exponential was first found to describe the α relaxation (low-frequency response) of glasses. It is also known as the Kohlrausch–Williams–Watts formula. The exponent β is largely dominated by the long-time behavior of $q(t)$, whereas x is governed mostly by the short-time behavior, and ω is related to the average relaxation time.

It has recently been found that just as in the case of the α process predicted by the mode coupling theory of supercooled liquids, the long-time regime of the correlation function for the model (1), obeys the time–temperature super-position principle above $T^* \simeq 0.25$. In other words, the long-time regime for these temperatures fits $\exp(-(t/\tau)^{\beta'})$, with constant β'.[9]

Using the empirical formula (3), we find that the exponents β and x decrease as the temperature decreases. We consider $q(t) \sim t^{-x}Q(t/t_0)$, where $Q(t/t_0)$ depends on temperature only through t_0.[8,10] We then show that we can identify the region where the scaling holds. The relevant correlation time, with this type of scaling becomes the first moment of $q(t)$, if it is considered as a distribution. This scaling time can be calculated through the integration of the data or the empirical fit. We choose the latter, for which the scaling

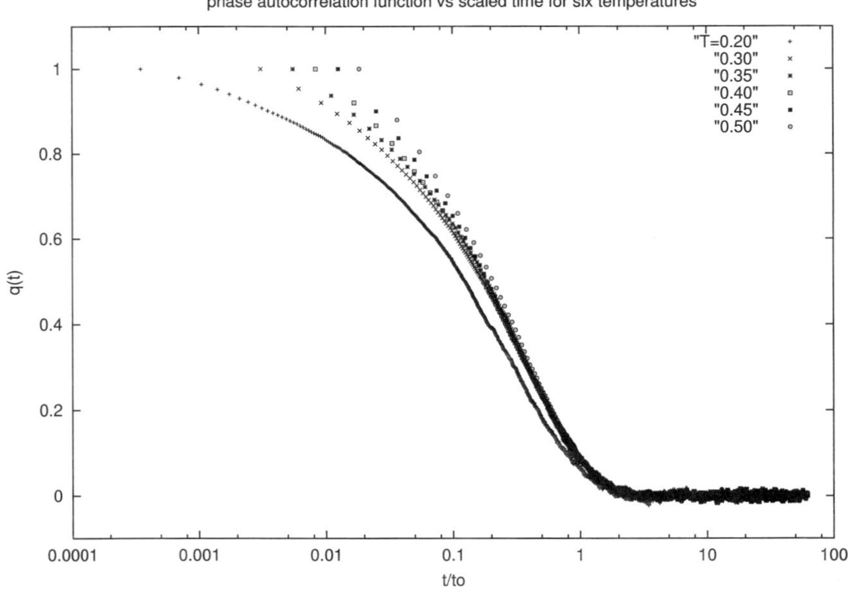

Fig. 1.   Correlation q versus scaled time t/t_0 for various $k_B T/J$ as indicated in the figure.

time t_0 becomes $\omega^{-1/\beta}\Gamma((2-x)/\beta)/\Gamma((1-x)/\beta)$, where Γ denotes the gamma function. The figure shows the outcome of plotting $q(t)$ versus the scaled time on a semilog graph.

The result shows that for temperatures, $T \geq 0.30$ the time scaled data approach each other near $t \simeq t_0(T)$. The figure indicates that for the long-time regime specified by t_0, it is possible to find a scaling function which, not surprisingly, is the stretched exponential, which dominates the long-time part of formula (3). But, in this way we have found the time region for which this scaling holds. Furthermore, the figure hints that at a temperature above 0.2 and below 0.3, the scaling does not hold anymore, as there is a sudden rise in the relaxation time. This rise can be inferred from the set of points at $q(t) = 1$. Earlier results,[9] show that the relaxation times fit the Arrhenius law above T^* and change to the Vogel–Fulcher behavior below T^*. For the scaling times t_0, we find that the Arrhenius behavior $\exp(\Delta E/T)$, is a good fit, with temperature independent $\Delta E \simeq 1.3J$.

The reason for the loss in scaling is not known for sure. A possibility we propose is that, below T^*, the system encounters a multiplicity of states in which it can get stuck.[11] The present data shows that this happens for $0.2 < T^* < 0.3$. Below T^* the metastable states are formed. A better set of data is needed to decide whether a particular form of scaling is present, just as the system is crossing into the deeper valleys of the landscape. This requires probing the correlation function for a more nearly spaced set of temperatures.

Acknowledgments

The author thanks H. Fazli for useful discussions, and also wishes to thank the Abdus Salam ICTP, where part of this work was carried out, for their hospitality.

References

1. S. Teitel and C. Jayaprakash, *Phys. Rev.* **B27**, 598 (1983).
2. M. Franz and S. Teitel, *Phys. Rev.* **B51**, 6551 (1995).
3. J. P. Straley, A. Y. Morozov, and E. B. Kolomeisky, *Phys. Rev. Lett.* **79**, 2534 (1997).
4. T. C. Halsey, *Phys. Rev. Lett.* **55**, 1018 (1985).
5. W. Götze and L. Sjögren, *Rep. Prog. Phys.* **55**, 241 (1992).
6. M. R. Kolahchi, *Phys. Rev.* **B59**, 9569 (1999).
7. W. Kob, in *Annual Reviews of Computational Physics III*, ed. D. Stauffer, (World Scientific, Singapore, 1995).

8. A. T. Ogielski, *Phys. Rev.* **B32**, 7384 (1985).
9. B. Kim and S. J. Lee, *Phys. Rev. Lett.* **78**, 3709 (1997); M. R. Kolahchi and H. Fazli (unpublished).
10. L. Sjörgen and W. Götze, *J. Non-Crys. Sol.* **131–133**, 153 (1991).
11. T. R. Kirkpatrick and P. G. Wolynes, *Phys. Rev.* **B36**, 8552 (1987).

Annual Reviews of Computational Physics VIII (pp. 307–319)
Edited by Dietrich Stauffer

MONTE CARLO SIMULATION OF
MICROSCOPIC STOCK MARKET MODELS

DIETRICH STAUFFER

*Institute for Theoretical Physics, Cologne University,
D-50923 Köln, Euroland*

Computer simulations with random numbers, that is, Monte Carlo methods, have been considerably applied in recent years to model the fluctuations of stock market or currency exchange rates. Here we concentrate on the percolation model of Cont and Bouchaud, to simulate, not to predict, the market behavior.

1. Introduction

Computational physics is more than just the numerical solution of partial differential equations. More and more are the methods of statistical physics applied to biological or sociological questions. Here, even if I talk about money,[1–4] you should not expect to get rich by listening.

It is not the physicists who have introduced physics methods like Monte Carlo simulations to economic science. When I was still a student, that means very long time ago, Harry Markowitz, the later Nobel laureate in economics, invented his own computer language for economic applications. His paper[5] with graduate student Kim, a Monte Carlo simulation of the 1987 Wall Street crash, was published several years before such studies became fashionable among physicists. (The first Monte Carlo simulation of markets which I know[5] was published by Stigler 35 years ago.) And closer to my home, the young economics professor Thomas Lux at Bonn University has cited physics research on synergetics in his economic publications and has even published now in *Nature* and *Physica A*.[6] Nevertheless, the influx of many physicists with their own ideas into what was called "Econophysics" by Gene Stanley might produce useful results, besides reinventions of the wheel. During our meeting, the European Physical Society will be holding its first econophysics conference in

Ireland, though French-speaking regions of Europe seem to dominate in this research.

Many papers deal with the phenomenological description of real markets, while this review emphasizes more on the microscopic models dealing with single investors similarly to physics simulations dealing with single atoms moving along with other atoms in a fluid. But towards the end I will give you some advice on how to get rich in the year 1999.

2. Similarity to Physics

The various microscopic models,[7] as reviewed in Refs. 4 and 8, use many investors trading with each other. The collective behavior of different investors determines the change of the price, according to supply and demand. Physicists like to treat the decisions of the single investor by random numbers, as they do in Monte Carlo simulations of physical systems, in this way hiding the fact that the details of the decision process are unknown. A better interpretation of this randomness was already given by Kim and Markowitz[5]: The investors deal with the money of their customers, and some customers withdraw money from their account to pay for a new car or because of some sickness. These reasons to withdraw or deposit money have nothing to do with the market, and are not part of the market model, and thus are approximated as random.

In physics we try to simulate systems with 10^{25} atoms, which cannot be all put onto a computer. Thus we try to extrapolate to infinite system sizes by comparing simulations with varying numbers N of atoms. Similar studies have been made for markets with N investors, and in most cases (Ref. 7: Hellthaler, Kohl, Egenter *et al.*, Busshaus *et al.*, and Farmer) the result was that for $N \to \infty$ the behavior of the simulated market became quite smooth or periodic and thus predictable, in contrast to real markets. Thus, if these models are good descriptions of markets, then real markets with their "chaotic" strong fluctuations are dominated by a rather limited number of large players. We do not have to follow conspiracy theories that a small group of criminals steer the world's market; we simply have to assume that the hundred most important investors or investment companies have been much more influential than the millions of less wealthy private investors.

The only known model where the market may stay realistic even for $N \to \infty$ seems to be the Cont–Bouchaud percolation model[8] to be explained in the next section. This model has a critical point or second-order phase transition, called the percolation threshold and is discussed elsewhere in these proceedings. At such critical points, the fluctuations are as large as the average values for the order parameter or for other quantities going to zero or infinity. In other words, the law of large numbers no longer works here. Thus even for $N \to \infty$ the prices behave in an unpredictable way, as in reality. If in the same model we move away from the critical point, we recover the usual law that the relative fluctuations go to zero if $N \to \infty$. Thus, generally, we expect market models to stay realistic for $N \to \infty$ only if they operate at a critical point. Let us now look into this model more precisely.

3. Cont–Bouchaud Model

Let us take a large lattice in d dimensions and occupy every site with probability p, while leaving it empty with probability $1 - p$. The occupied sites are people with money to invest. The groups of neighboring occupied sites are called clusters, and are interpreted as companies formed by neighboring investors. You may think that in a city, people go to a bank in their neighborhood and thus all get the same advice on how to invest their money. In this way, all sites belonging to the same cluster make the same decision at the same time, that means the single cluster acts like a single company making one decision only. This model thus approximates the herding of investors which influence each other.

The original paper of Cont and Bouchaud[9] worked with bond instead of site percolation and, more importantly, assumed that everybody interacts with everybody else using the same probability, independent of distance. This infinite-range percolation limit, also known as random graph, seems to have the same critical exponents as percolation on a Bethe lattice or in infinite dimensions and thus corresponds to the very first percolation model published by Flory nearly 60 years ago.[10] (Kauffman[11] suggested this same limit for his theory of the origin of life through autocatalytic reactions among sufficiently complex molecules.) Thus the original Cont–Bouchaud model is based on the idea that in today's computerized world of finance, everybody can communicate within seconds with a business partner on another continent. This limit can be solved exactly.[10] For p below some percolation threshold p_c only finite clusters appear, for $p > p_c$ also an infinite cluster.

The opposite extreme is what this meeting (see contributions of Erzan, Marrink *et al.*, and Stauffer and Jan) dealt with when talking about percolation: neighbors have to be nearest neighbors of the lattice. Thus on the square lattice each site has at most four occupied neighbors, and on a d-dimensional hypercubic lattice it has $2d$ neighbors. For $d > 6$ the critical exponents of this model agree with those on the Bethe lattice, but for the more realistic dimensions $d = 2$ and 3 they are slightly different. For example, the number n_s of clusters containing s sites each decays for large s as $s^{-\tau}$ right at the percolation threshold, with $\tau = 2.05$, 2.19 and 2.5 for $d = 2$, 3 and > 6. Away from p_c, the cluster numbers decay asymptotically as a simple exponential below p_c, $\log n_s \propto -s$ and as a stretched exponential above p_c, $\log n_s \propto -s^{1-1/d}$.

Whether we have an infinite range of connections, or just look at clusters of nearest neighbors, the market model from then on is the same. For each time step, each cluster randomly, with probability $2a$, decides to be active, while with probability $1 - 2a$ it remains inactive at that time. Active clusters decide, in the simplest case randomly with equal probabilities $1/2$, whether to buy or to sell. The amount of trade is proportional to the number of sites within this cluster. The total demand is the sum over all buying clusters, the total supply the sum over all selling clusters. The market price then goes up or down at this time step, proportional to the difference between demand and supply. This short description already defines the model. We see that it has only two essential parameters: the concentration p and the activity a to buy as well as to sell.

For comparison with real markets, the activity $a < 1/2$ may be assumed to be an increasing function of the time between two measurements of the market price. If we take this time interval as one minute, then most traders will not be active during this interval, and a is very small. If, on the other hand, we talk about months or years, then the probability of professional investors to act during this long interval is much higher and thus a is closer to its upper limit $1/2$.

The limit of small a is particularly simple, and was already recommended by Cont and Bouchaud who take $1/a$ proportional to the system size. For $a \to 0$ at fixed large system size, there will be in most cases no active cluster at all; only now and then, one cluster will buy or sell, and the probability of two clusters to trade simultaneously is negligibly small. (The model does not require that each buyer finds a seller and is thus similar to a grand canonical ensemble of physics.) Then the price in most cases will not change at all, and sometimes it will change by an amount proportional to the size of a single cluster. Thus the histogram $P(x)$ of price changes x, that means the probability density function, will for nonzero x be proportional to the cluster size distribution n_s mentioned above. As one can see from my excellent book[11] (do not buy those of my competitors[11]), much is known about this cluster size distribution. For large s and concentrations p below the percolation threshold p_c, this distribution decays with a simple exponential in s, while for $p > p_c$ the distribution decays as a stretched exponential, $\log(n_s) \propto -s^{1-1/d}$ in d dimensions. (This exponential function is multiplied with some power of s and other higher-order correction terms.) Thus for small activities right at the critical point we get the above power law for the distribution $P(x)$ of price fluctuations: $P \propto x^{-\tau}$, while above and below p_c the power law behavior is cut off exponentially at large x.

For finite but still small activities a not much is changed, as seen[12] in Fig. 1 for a square lattice with a million sites. At $a = 0.0005$ we see in this log–log plot that the data follow a straight line (power-law decay) up to large price changes, where a steeper decay just becomes visible. For ten times larger activity, the power-law decay again is followed for large price changes by a steeper decay, setting in somewhat earlier. Increasing the activity by a further

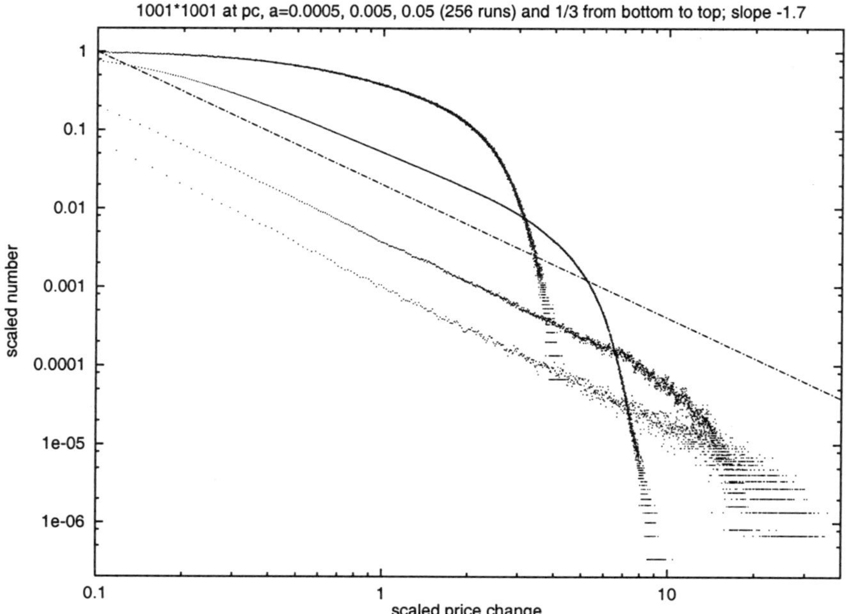

Fig. 1. Probability $P(x)$ for relative price changes x in the Cont–Bouchaud model on large critical square lattices. The activity is given in the title line and changes from small (lower right) to large (upper left). From Ref. 12.

factor 10 to 0.05 gives a power law followed by an even earlier crossover to an exponential decay. Finally, at $a = 1/3$, when with equal probability a cluster buys, sells, or sleeps, the region with a power-law decay has vanished and the data follow reasonably a Gaussian distribution. (Some people claim that the Cont-Bouchaud model is nothing but a translation of known percolation concepts to markets, but I have not yet seen them deriving this crossover to Gaussian distributions from percolation.)

In such histograms, the number of observed events depends on the computational effort, and the price change is determined only up to an arbitrary proportionality factor. Thus it is practical to normalize the height and the width to unity, as done in this figure. Also, in any quantitative comparison one has to be careful: does one want the probability $P(x) \propto x^{-\tau}$ to have a given price change x, or the probability $P(> x) \propto x^{-\tau}$ with $\tau = 1 + \mu$ that the price change is $> x$.

Reality seems to be somewhat similar[1]: A round peak near zero price change (not visible in Fig. 1 because of its logarithmic scale) is followed by a power law at intermediate price changes and an exponential cutoff at large price changes. (Perhaps[13] instead of power law and exponential cutoff, real data for intermediate x are described better by a power law with an exponent near 3 for $P(> x)$.) Only for price changes defined for very long time intervals, like a comparison from one month to the next month, may the distribution follow a Gaussian. Such month-to-month data are of course very difficult to produce with high statistics in reality, whereas the correponding computer simulations with $a \sim 1/3$ as in Fig. 1 are easy to get. (According to Chowdhury,[9] we do not get this Gaussian if instead of the Cont–Bouchaud cluster decisions we simply let the price follow a Levy flight.)

If we want to get the exponent 3 of Ref. 13, we assume[14] that a market integrates over all possible concentrations p with $0 < p < p_c$ and that, following Zhang,[15] the price change varies as the square root of the difference between demand and supply. Then the exponent for $P(> x)$ is about 2.9 on the square lattice, or $2(\tau + \sigma - 1)$ with the general percolation exponents. A variant of this model[14] gives wings in the distribution, which may correspond to outliers like the crashes of 1929 and 1987.

Historically, the first theory of Bachelier one century ago assumed the prices to follow a random walk with a Gaussian distribution of price changes. Mandelbrot, the inventor of the fractal concept, pointed out in the 1960's that Levy-stable distributions work better: they have a power law $\propto x^{-\mu} = x^{1-\tau}$

for intermediate and large x with $\mu \leq 2$ and a rounded peak at $x \to 0$. Only in the 1990's were exponential cut-off or exponents μ near 3 published in the analyses of real data known to me.[1,13,16]

If such two-dimensional simulations are repeated in three to seven dimensions, not much is changed.[12] More interesting are simulations away from the $p = p_c$ used in Fig. 1. For $p > p_c$ we neglect the contribution of the infinite cluster, which would produce enormous crashes and upturns of the market. The finite clusters at a suitably selected concentration slightly above the two-dimensional percolation threshold give, with 40 million simulated trades, an exponent $\mu \simeq 3$ in Fig. 2; however, when the computational effort is increased to 640 million data points, an order of magnitude more than the experimental statistics of Gopikrishnan *et al.*,[13] then slight curvature becomes visible in the log–log plot of Fig. 1, suggesting that this exponent $\mu \simeq 3$ is only an effective value for intermediate x while the asymptotic decay (for infinitely good statistics in infinitely large lattices) may be a (stretched) exponential like the

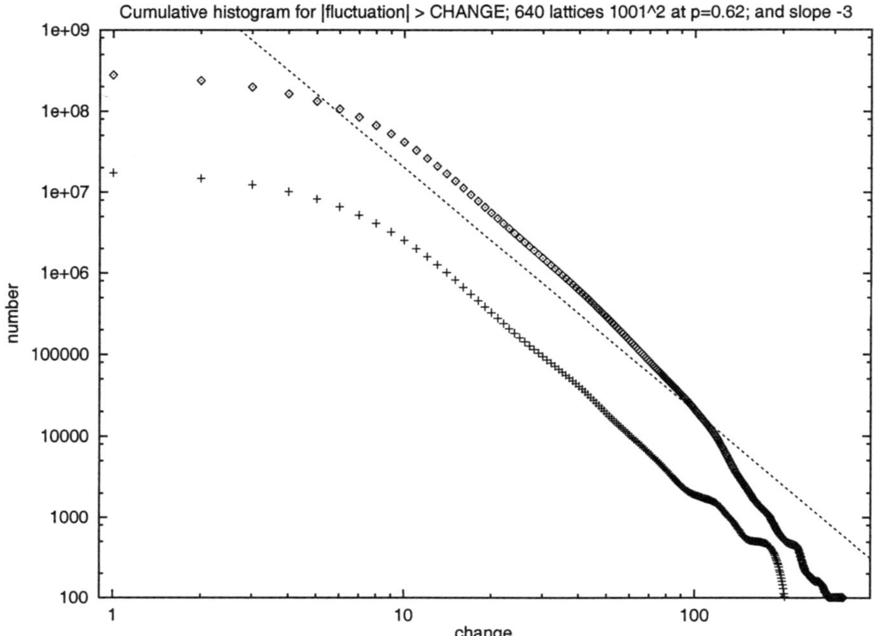

Fig. 2. Probability $P(>X)$ for ralative price changes x (arbiltrary units) in the Cont–Bouchaud model on large square lattices at $p = 0.62$ slightly above the percolation threshold. The straight line has the experimental slope of Ref 13. From Ref. 12.

cluster size distribution mentioned above. Present experimental data may not yet be good enough to see this distinction; but with better computerization of market data for very small time increments will soon become widely available such that reality can catch up with our simulations in Fig. 2.

"Volatility clustering" means that a large change on one day is likely to produce a large change also on the next day; only the sign of this change cannot be predicted. Thus, like the weather, real markets have turbulent as well as calm periods. On the computer, we can reproduce this effect by letting the Cont–Bouchaud model in the occupied sites diffuse slowly to empty nearest neighbor sites.[12] Not much is changed in the $P(x)$ histograms if we assume the traders to prefer selling at high prices and buying at low prices, and if we assume that "high" and "low" in this sense are defined relative to a fundamental price following a random walk: Chang.[9] Work on the distribution of wealth among the traders did not give nice power laws for the wealth distribution.[18]

These Cont–Bouchaud models treat buying and selling in a symmetric way, and thus the average price does not change: Inflation is subtracted (I mean inflation of stock prices, not inflation in the sense of Rivier's article in this volume). In reality, prices may rise to a sharp peak from which they fall rapidly again, while they reach a minimum much more slowly, and rise slowly from it. This asymmetry can be incorporated by letting the activity increase or decrease proportionally to the last known change in price; see Ref. 19 for related feedback models. Then[20] volatility clusters automatically, even without the above-mentioned diffusion.

(The kurtosis of the price fluctuations in the original infinite-range model was calculated in Ref. 21.)

Some technical aspects: Since the units for prices are quite arbitrary it is best to look at *relative* price changes x, fixing the initial price at unity. Then the model simulates changes in the logarithm of the price. The Cont–Bouchaud model has an arbitrary proportionality factor between the relative price change and the difference between demand and supply; the easiest is to identify one site with one unit change x, i.e. one single site deciding to buy gives $x = 1$, and one isolated seller produces $x = -1$. Now the x summed over all clusters is an integer, easy to bin for $P(x)$. The proportionality factor then enters only the axis units in plots of log(price) or of $P(x)$. The larger the lattice is, the larger are the typical x values. When we change the buying preference or the trader activities by an amount proportional to the price change, we have to normalize by the lattice size, for example, $a(t = 1) = a(t) + \alpha x/L^d$. Reference 20 used $\alpha = 10$ for the proportionality factor.

4. Other Models

Among the many other models[7] reviewed in Refs. 4 and 8 I emphasize here only those which have some similarity with biology, where selection of the fittest removes the less successful and where different species may coexist with each other. Also in the Cont–Bouchaud model, the selection of the fittest traders was studied.[18]

Caldarelli *et al.*[7] considered only one stock and its price fluctuations, but they applied Darwinian evolution to the population of traders. In the simulation each trader begins with a randomly selected strategy of buying and selling, and all traders receive the same capital. At each time step the trader with the smallest remaining capital is eliminated and replaced by a new one with a randomly selected strategy. The amount which a trader wishes to buy or sell is given in part by a complicated function which depends on the time average of the derivatives of the past market price. The coefficients entering this function for each trader are chosen randomly. In addition, the trading decision is influenced by the present amount of stocks and other wealth held by the investor who in a stable market prefers to have a certain fraction of wealth in cash. Also some noise is introduced into the decisions. The resulting curves of price versus time look realistic, and the price fluctuations from one time step to the next deviate strongly from a Gaussian. Like Levy distributions they have fat tails with a power-law decay. The wealth distribution of the traders also follows a power law, but with a different exponent.

All investors are divided by Lux[6] into three groups: the fundamentalists look at the fundamental price assumed to fluctuate randomly, and abhor speculations. The speculators, on the other hand, may be optimistic or pessimistic noise traders. People change their minds based on both herding instincts and past performance, that means depending on whether the market was going up or down. The model reproduced volatility clustering as well as $\mu \simeq 3$ and gave a lot of noise traders during turbulent periods of the market, while in calm periods most of them became fundamentalists.

Similarly, Farmer[7] assumed an "ecology" of different trading strategies. Old strategies which are no longer successful are replaced by new ones, just as more fit biological species replace less fit species. Impatient buyers drive the prices up and impatient sellers push them down according to a market impact function.

Also randomness is involved in the Lotka–Volterra-type differential equations which are used in the simulations. If too many move into one successful niche, the advantages of that niche are diminished, and a new strategy may emerge as more successful. Similarly to the biological evolution of species, a diversity of co-evolving trading styles emerges. But in contrast to Nature, this evolution does not take millions of years but at most a few decades.

Perhaps the most sophisticated models are the learning agents discussed by Palmer *et al.* and Arthur *et al.*,[7] involving bit-strings, mutations, and recombination as in biology. (They even allow for the influence of sunspots on our markets.) Each agent follows its own set of 60 rules for what is to be done under certain conditions. These conditions are coded as strings of about 75 bits. A simplified genetic algorithm is used to recombine the bit-strings so that they are changed. So the rules adapt, like living beings, to the environment and work better and better. The rules needed for success change with time. Crashes and bubbles without simple explanations are found. The wealth distribution is unequal for intermediate times, but for long periods of time those who were wealthy can become poor again, and the other way around.

5. Conclusion

For many questions on economic activities, we looked here simply at the distribution of price fluctuations. Moreover, this review, partly taken from earlier papers of the author, concentrated on the Cont–Bouchaud model for market fluctuations because it is the closest to the subject of this school and the experience of this author. By definition, you cannot get rich from this model as long as the clusters make their decisions randomly and thus let the price fluctuate up or down with equal probability. If you want to become rich, then look at Fig. 3 of Johansen and Sornette. They fitted the decay of the Japanese stock market after its peak a decade ago onto log-periodic oscillations $\propto \sin(\omega \log |t - t_c|)$ and in this way predicted[17] in January 1999 that in 1999 the Japanese stock market (Nikkei index) should go up. So far they have been right, and had you followed them at the beginning of this year, you would have become rich already. Because of the many critical questions posed to me in Zanjan, in particular by Prof. Rivier, I invested some money in Japanese stocks after my return. But who cares about money when we have citation rates to care about. (Despite Khodadad Azizi, few Cologne people at conference time care about football!)

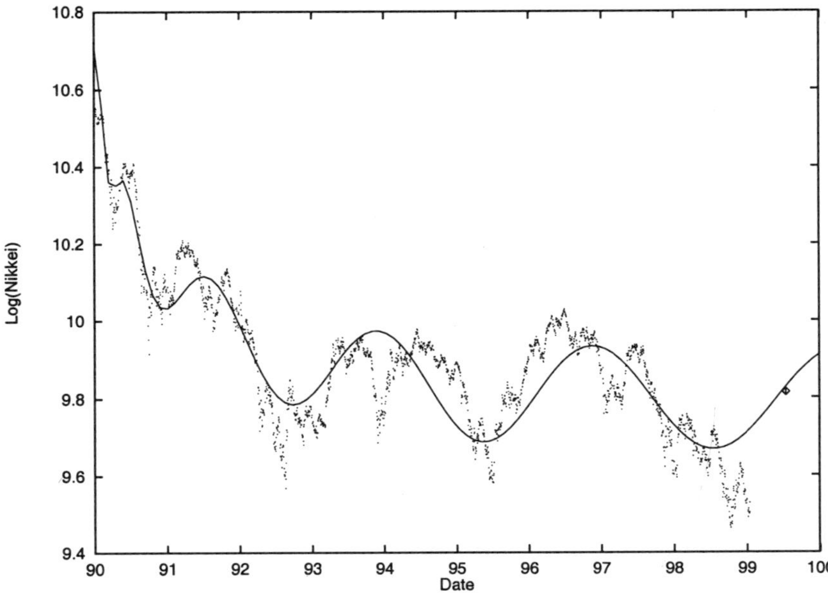

Fig. 3. Johansen–Sornette fit and prediction for the Nikkei index in Tokyo. The curves show three fits of increasing quality with an increasing number of free parameters. From Ref. 17, adapted by these authors to show the Nikkei index at the time this talk was given.

Acknowledgments

I thank Sorin Solomon for introducing me to this field and numerous others for discussions and collaboration.

References

1. J. P. Bouchaud and M. Potters, *Théorie des Risques Financieres* (Alea-Saclay/Eyrolles, Paris, 1997); English translation in www.science-finance.fr for Cambridge University Press; J. P. Bouchaud, *Physica* **A263**, 415 (1999); For a 1998 conference see *Physica* **A269**, 1–187 (1999); H. Levy, M. Levy, and S. Solomon, *Microscopic Simulation of Financial Markets* (Academic Press, New York, 2000).
2. J. Kertész and I. Kondor, *Econophysics: An Emerging Science* (Kluwer, Dordrecht, 1999), in press.
3. R. N. Mantegna and H. E. Stanley, *Econophysics: An Introduction* (Cambridge University Press, Cambridge, 2000).
4. S. Moss de Oliveira, P. M. C. de Oliveira, and D. Stauffer, *Evolution, Money, War and Computers* (Teubner, Stuttgart-Leipzig, 1999.)

5. G. W. Kim and H. M. Markowitz, *J. Portfolio Management* **16**, 45 (1989). See also G. J. Stigler, Public Regulation of the Securities Market, *J. Business* **37**, 117 (1964).

6. T. Lux, *Economic J.* **105**, 881 (1995); T. Lux and M. Marchesi, *Nature* **397**, 498 (1999).

7. H. Takayasu, H. Miura, T. Hirabashi, and K. Hamada, *Physica* **A184**, 127 (1992); M. Levy, H. Levy, and S. Solomon, *Econ. Lett.* **94**, 103 (1994) and *J. Physique* **I5**, 1087 (1995); M. Levy and S. Solomon, *Physica* **A242**, 90 (1997); T. Hellthaler, *Int. J. Mod. Phys.* **C6**, 845 (1995); R. Kohl, *Int. J. Mod. Phys.* **C8**, 1309 (1997); R. G. Palmer, W. B. Arthur, J. H. Holland, B. Lebaron, and P. Tayler, *Physica* **D75**, 264 (1994); W. B. Arthur, J. H. Holland, B. Lebaron, R. G. Palmer, and P. Tayler, in *The Economy as an Evolving Complex System II*, eds. W. B. Arthur, S. Durlauf, and D. Lane, (Addison-Wesley, Redwood City, CA, 1997); C. Tsallis, A. M. C. de Souza, and E. M. F. Curado, *Chaos, Solitons and Fractals* **6**, 561 (1995); G. Caldarelli, M. Marsili, and Y. C. Zhang, *Europhys. Lett.* **40**, 479 (1997); R. Chatagny and B. Chopard, *International Conference on High Performance Computing and Networks*, Vienna 1997; D. Sornette and A. Johansen, *Physica* **A245**, 411 (1997) and **261**, 581 (1998); K. Steiglitz, M. L. Honig, L. M. Cohen, in *Market-Based Control: A Paradigm for Distributed Resource Allocation*, ed. by S. Clearwater, Chap. 1 (World Scientific, Hong Kong, 1996); P. Bak, M. Paczuski, and M. Shubik, *Physica* **A246**, 430 (1997); G. Weisbuch, A. Kirman, and D. Herreiner, preprint for *J. Theor. Economy*. K. N. Ilinski and A. S. Stepanenko, preprint cond-mat/9806138 = *J. Complex Systems*, November 1998; J. D. Farmer, e-print adap-org/9812005 (1998); C. Busshaus and H. Rieger, *Physica* **A267**, 443 (1999); E. Egenter, T. Lux, and D. Stauffer, *Physica* **A268**, 250 (1999).

8. D. Stauffer, *Ann. Physik* **7**, 529 (1998).

9. R. Cont and J. P. Bouchaud, e-print cond-mat/9712318 = Macroeconomic Dynamics **4**, 170 (2000); D. Stauffer and T. J. P. Penna, *Physica* **A256**, 284 (1998); Ref. 12; I. Chang and D. Stauffer, *Physica* **A264**, 294 (1999); D. Chowdhury and D. Stauffer, *Eur. Phys. J.* **B8**, 477 (1999).

10. P. J. Flory, *J. Am. Chem. Soc.* **63**, 3083, 3091, 3096 (1941).

11. D. Stauffer and A. Aharony, *Introduction to Percolation Theory*, Taylor and Francis, London 1994; A. Bunde and S. Havlin, *Fractals and Disordered Systems* (Springer, Berlin-Heidelberg, 1996); M. Sahimi, *Applications of Percolation Theory* (Taylor and Francis, London, 1994).

12. D. Stauffer, P. M. C. de Oliveira, and A. T. Bernardes, *Int. J. Theor. Appl. Finance* **2**, 83 (1999).

13. T. Lux, *Appl. Financial Economics* **6**, 463 (1996); P. Gopikrishnan, M. Meyer, L. A. N. Amaral and H. E. Stanley, *Eur. Phys. J.* **B3**, 139 (1998) and *Phys. Rev.* **E60**, October (1999).

14. D. Stauffer and D. Sornette, cond-mat/9906434, *Physica* **A271**, 496 (1999).

15. Y. C. Zhang, *Physica* **A269**, 30 (1999).

16. R. N. Mantegna and H. E. Stanley, *Nature* **376**, 46 (1995).
17. A. Johansen and D. Sornette, *Int. J. Mod. Phys.* **C10**, 563 (1999) and **11**, 359 (2000); see also D. Sornette, *Phys. Rep.* **297**, 239 (1998) and **313**, 237 (1999).
18. J. Liebreich, *Int. J. Mod. Phys.* **C10**, 1317 (1999).
19. S. Focardi, S. Cincotti, and M. Marchesi, "Self-organized criticality and large craches in financial markets," Electronic conference proceedings at dibe.inige.it/wehia.
20. D. Stauffer and N. Jan, *Physica* **A277**, 215 (2000).
21. R. d'Hulst and G. J. Rodgers, *Physica* **A280**, 554 (2000).

Annual Reviews of Computational Physics VIII (pp. 321–328)
Edited by Dietrich Stauffer

A HISTORY-DEPENDENT MODEL FOR PREDATOR–PREY PROBLEM

ROUZBEH GERAMI* and MOHAMMAD R. EJTEHADI

*Institute for Studies in Theoretical Physics and Mathematics,
P.O. Box 19395-5531, Tehran, Iran*

**Department of Physics, Sharif University of Technology,
P.O. Box 11365-9161, Tehran, Iran*
E-mail: {rouzbeh,reza}@theory.ipm.ac.ir

A time-delayed stochastic model has been introduced for the predator–prey problem in which the prey are immobile edible plants and the predators are diffusing herbivors. In the stationary state, clusters of predators and preys are formed, although they do not see each other. It is shown that the interaction rate should be lowered by a correction factor because of this clustering.

1. Introduction

The time evolution of systems of interacting species modeling natural ecosystems has attracted wide attention since first studied by Lotka[1] and Volterra.[2] A much studied category of such systems is that of two interacting species, the so-called predator–prey models.[3,4] However, most of the existing models neglect the effect of time delays on the dynamics of the models. By time-delayed systems we mean systems for which their dynamics is not defined by only knowing their current state, but some information about the previous states is also required. Time delays are present in many different physical or biological systems, and are able to particularly account for many features of ecological phenomena,[5,6,7] although they have not been studied extensively. Recently we have introduced a new model of the predator–prey problem with history-dependent dynamics,[8] and here we report some results of the model.

In this model, the ecosystem consists of a (an infinite) square lattice each site of which if not empty, is occupied by either predators or a plant. The predators move randomly to one of the nearest neighbors (two-dimensional free random walk) and do not interact with each other, therefore multiple occupancy of the sites is allowed. If a predator enters a site occupied by a plant, it will eat it. However after c time steps another plant grows at that site.

To every predator an *energy* is assigned, indicating the number of steps that it can go without eating anything. As a result, the energy is lowered by one at every time step. Eating a plant raises the energy to the maximum value l, so that a predator that has not eaten anything in l steps will die. At every time step each predator reproduces with probability b. The offspring is positioned at the same site and half of the parent's energy is transferred to it.

These rules are applied in the following order. The predators are first moved in a random sequence. They eat every plant that they can, after which they reproduce with some probability and finally plant growth occurs. In the case of more than one predator entering a plant site, the early comer eats the plant. In our model, time delays enter the temporal evolution equations through the terms representing plant growth and predator death. Somewhat similar models without such time delays has also been published before.[9,10]

Simulations of the model are made on a $M \times M$ square lattice with $M = 100$ and with periodic boundary conditions. As initial conditions, predators and plants are distributed randomly and the value of l is assigned to the energy of every predator. The sites that are initially plant-free must be filled with plants in the first c steps, so a random integer τ, $0 < \tau < c$, is assigned to every such site, and a plant occupies that site at time $t = \tau$.

Let $P(\mathbf{x}, t)$ and $N(\mathbf{x}, t)$ denote predator and plant the respective local densities and $p(t)$ and $n(t)$ be their respective spatial mean values, that is, $p(t) = \langle P(\mathbf{x}, t) \rangle$ and $n(t) = \langle N(\mathbf{x}, t) \rangle$ (where $\langle \cdot \rangle$ stands for spatial averaging). $P(\mathbf{x}, t)$ is an integer number including 0, while $N(\mathbf{x}, t)$ is either 0 or 1. $p(t)$ and $n(t)$ are assumed to be equal to the probability of predators and plants occupying a lattice site (assuming that p does not become larger than one).

As expected, time evolution of the two species can lead to a stationary state (Fig. 1) in which both n and p fluctuate about their (time-independent) mean values, and the fluctuations are predominantly anticorrelated (in the sense that when n goes up p comes down and vice versa). Therefore by averaging $n(t)$ and $p(t)$ over many realizations of the system we find a fixed point in the $(\langle\langle p \rangle\rangle, \langle\langle n \rangle\rangle)$ phase space (Fig. 2) (where $\langle\langle \cdot \rangle\rangle$ represents the expectation value found by averaging over different realizations).

Trivially, $(n, p) = (1, 0)$ is also a fixed point (*extinction* state). In a wide range of parameters this is unstable, and there exists the just described active

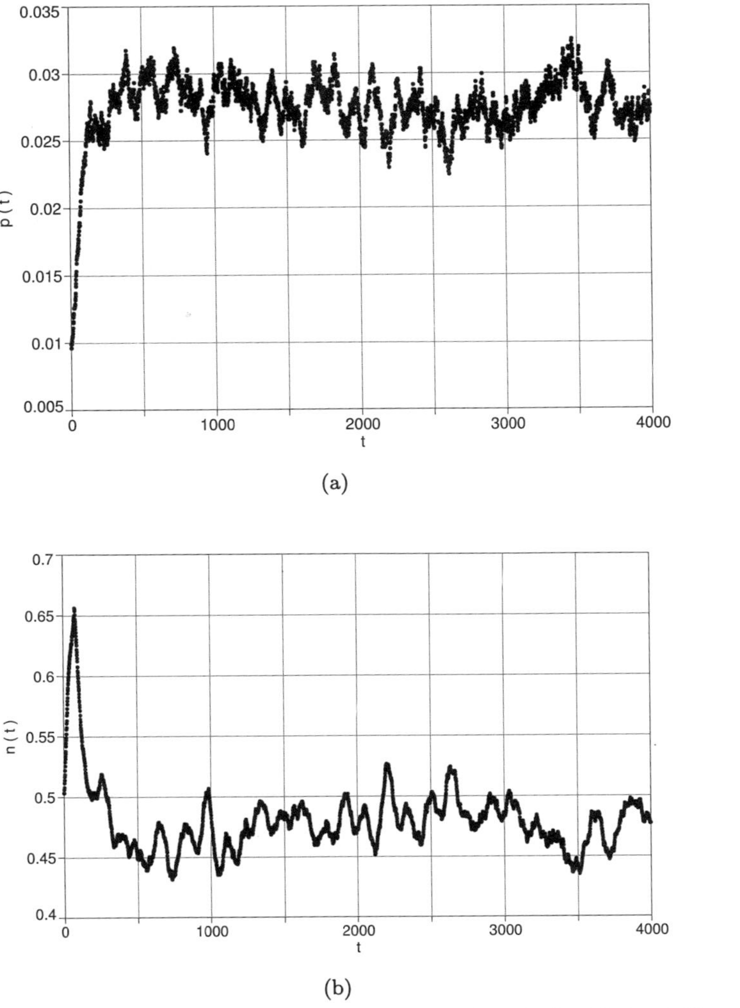

(a)

(b)

Fig. 1. Predator and plant mean densities with respect to time for $l = 20$, $b = 0.02$, $c = 80$, $p(0) = 0.01$ and $n(0) = 0.75$. (a) $p(t)$, (b) $n(t)$.

oscillatory state with a $(\langle\langle p\rangle\rangle, \langle\langle n\rangle\rangle)$ stable fixed point. But in a large region in the parameter space of l, b and c, the point $(1, 0)$ is stable and there is no nonextinction stationary state. This is the case for sufficiently large c (low growth rate for the plants), low l (low energy content of a plant) or low b (low predator birth rate). Even as an unstable fixed point, $(1, 0)$ can be reached (in transient

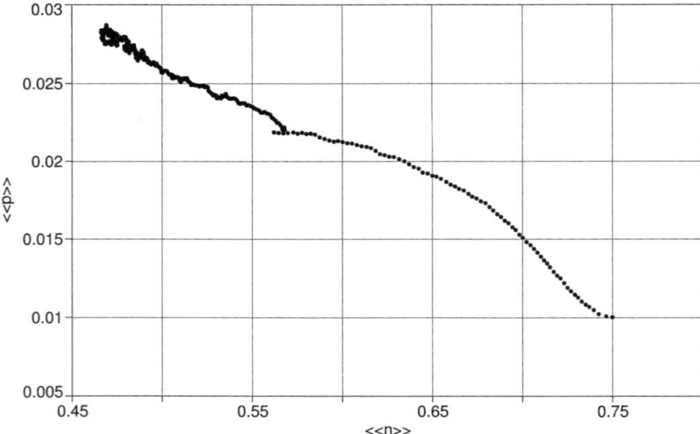

Fig. 2. Fixed point in the phase space of expectation values of predator and plant mean densities $(\langle\langle p\rangle\rangle, \langle\langle n\rangle\rangle)$, for the same parameter set as in Fig. 1.

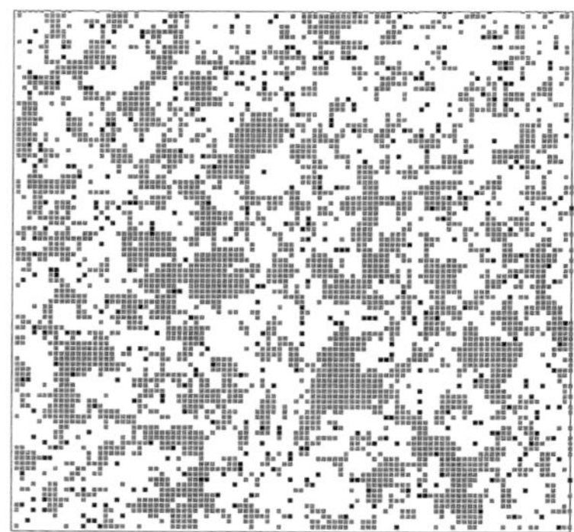

Fig. 3. Distribution of predators and plants in a 100×100 lattice for $l = 20$, $b = 0.02$, $c = 60$ and $t = 500$. Predators are represented by black dots and plants by gray.

region) by specific initial conditions that are large $p(0)$ or large $n(0)$. In the latter case, the initial high density of plants increases p and decreases n very much and consequently all the predators die of starvation. In the following we consider the nontrivial (nonextinction) stationary state.

Although the predators (plants) have no interaction with each other, the spatial distributions of $N(\mathbf{x})$ and $P(\mathbf{x})$ are not uniform in the stationary state. This is due to the rules of the game that are random motion of predators and the laws of birth and death.[11] As a typical pattern, Fig. 3 shows the emergence of clusters of predators and plants for $l = 20$, $b = 0.02$, $c = 60$ and $t = 500$, when the system is in its stationary state. Formation of the clusters is characterized quantitatively by the predator or plant autocorrelation functions defined by

$$C_n(\mathbf{d}) = \frac{\langle N(\mathbf{x}+\mathbf{d})N(\mathbf{x})\rangle - n^2}{n^2} \,, \tag{1}$$

$$C_p(\mathbf{d}) = \frac{\langle P(\mathbf{x}+\mathbf{d})P(\mathbf{x})\rangle - p^2}{p^2} \,. \tag{2}$$

These clusters form separately, since for a plant at a site no predators can be at the same site. This is shown by the predator–plant correlation function:

$$C_{np}(\mathbf{d}) = \frac{\langle N(\mathbf{x}+\mathbf{d})P(\mathbf{x})\rangle - np}{np} \,. \tag{3}$$

Figure 4 shows $\langle\langle C_n(\mathbf{d})\rangle\rangle$, $\langle\langle C_p(\mathbf{d})\rangle\rangle$ and $\langle\langle C_{np}(\mathbf{d})\rangle\rangle$ as functions of d along the lattice axis, for the same parameter set as in Fig. 3. They all vanish as d increases, but while C_n and C_p are positive functions for small d, representing formation of the clusters, C_{np} is negative since the probability that a plant

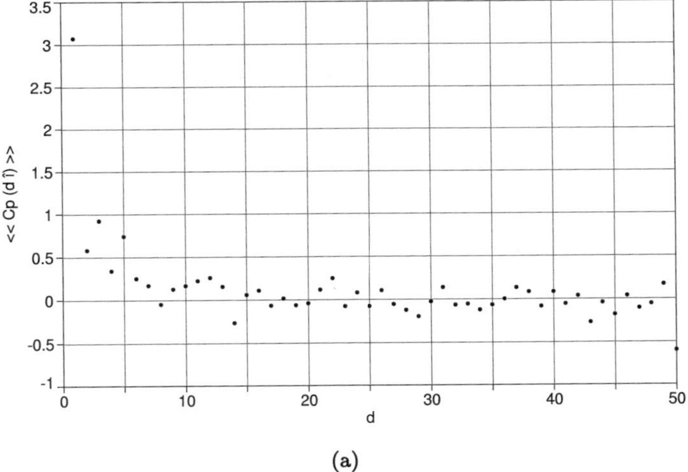

(a)

Fig. 4. Expectation values of (a) autocorrelation function of predators (b) autocorrelation function of plants (c) predator–plant correlation function, as a function of d along the lattice axis for $l = 20$, $b = 0.02$, $c = 150$ and $t = 500$. An exponential function best fits to C_n with correlation length increasing with c.

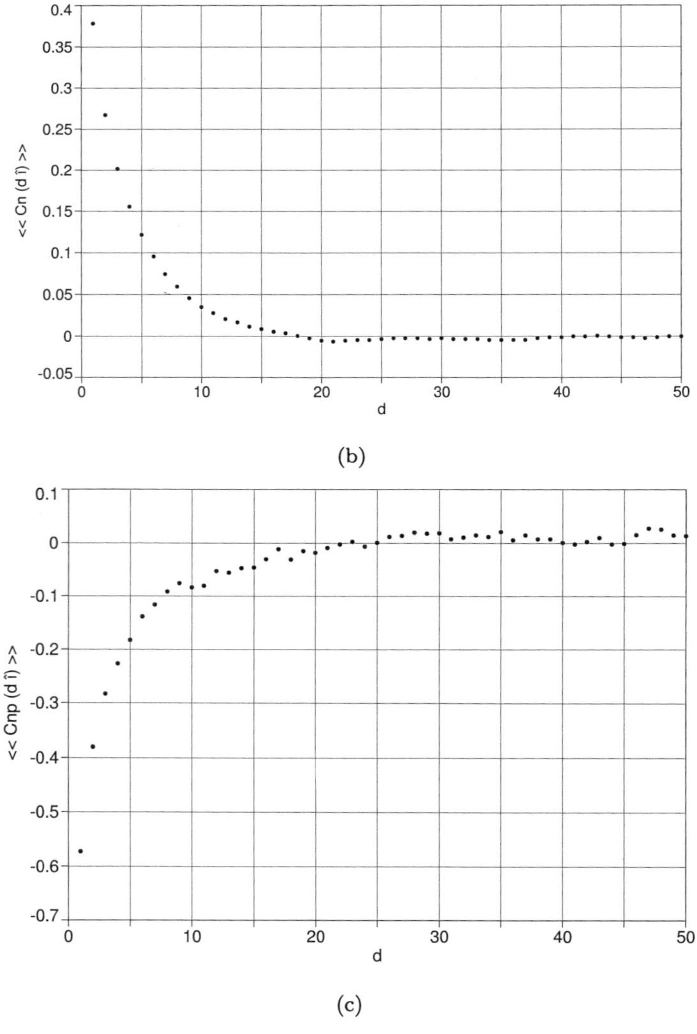

(b)

(c)

Fig. 4. (*Continued*)

occupies a site decreases if there is a predator in the neighborhood. Diffusion of the predators increases the fluctuations in C_p and C_{np}. An exponential function best fits to C_n with correlation length increasing with c.

If the probabilities that a site is occupied by a predator or a plant were independent, the density of the eaten plants at every time step would be given by

$$\Delta_- n(t) = n(t)p(t)\,, \tag{4}$$

that is, $\Delta_- n(t)$ is the probability that a site is simultaneously occupied by both a predator and a plant. To take into account the just-described correlations, we modify this expression, by writing it as

$$\Delta_- n(t) = rn(t)p(t)\,, \tag{5}$$

where $0 < r < 1$, and this also can be thought of as a rate. Stronger correlations imply larger clusters which lowers the value of r. Introduction of $r < 1$ rate, can also be justified in this way: since predators move randomly, a predator lowers its food-eating chance by repeatedly coming back to the sites which had previously been occupied by itself and it had already eaten the plants in such sites.

To calculate $r(t)$ by simulation, we enumerate the total number of the eaten plants at time t and divide it by $M^2 n(t)p(t)$. Figure 5 represents as a function of time, the value of $\langle\langle r(t)\rangle\rangle$ for $l = 20$, $b = 0.02$ and $c = 60$ which indicates that it becomes essentially a constant at about $\langle\langle r(t)\rangle\rangle \simeq 0.54$ in the stationary state. In fact $\langle\langle r(t)\rangle\rangle$ varies slightly as a, b and c change. The value of r can also be read from the correlation function (Fig. 4(c)). Since the probability

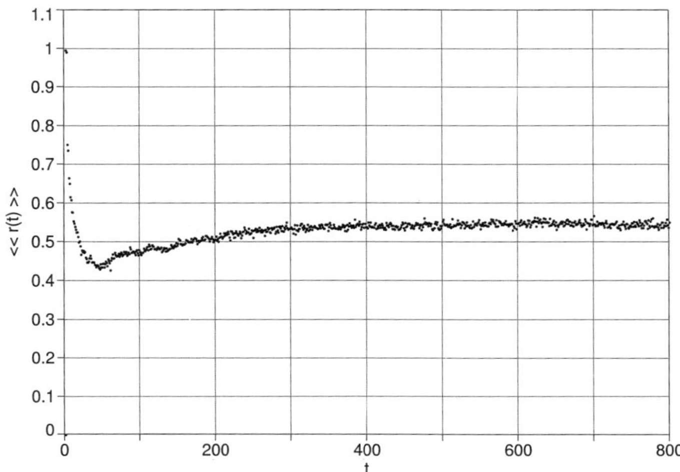

Fig. 5. Numerically calculated expectation value of $r(t)$ as a function of t for $l = 20$, $b = 0.02$ and $c = 60$, with 0.54 mean value for $t > 400$.

that a plant is eaten is one quarter of probability that a predator and a plant are nearest neighbors, and this probability is equal to the probability of finding a predator and a plant within a unit distance $\mathbf{i}$, we have

$$\Delta_- n = rnp = \langle n(\mathbf{x}+\mathbf{i})n(\mathbf{x})\rangle\,, \tag{6}$$

then

$$C_{np}(\mathbf{i}) = r - 1\,. \tag{7}$$

From Fig. 4(c) we find that $C_{np}(\mathbf{i}) \simeq 0.46$ and $r = 0.54$ in complete agreement with the independently calculated value of r (Fig. 5).

We have shown[8] that r is an important parameter that controls the ability of the mean-field equations to have chaotic solutions. We found that the time-evolution equations with a realistic value of $r(l, b, c)$ which is obtained from the simulation, do not exhibit chaotic behavior. This is similar to what occurs in the simulation for which there always exists a stationary state. However, setting $r = 1$ artificially, can produce a chaotic behavior. This has an interesting interpretation: formation of the clusters and emergence of correlations remove the chaotic regime.

Acknowledgments

The authors offer their special thanks to N. Hamedani, V. Shahrezaei, H. Seyed-Allaei and S. E. Faez for invaluable discussions and to D. Stauffer, M. Sahimi, A. Erzan and A. Aghamohammadi for careful reading of the manuscript, and useful comments.

References

1. A. J. Lotka, *Proc. Natl. Acad. Sci. U.S.A.* **6**, 410 (1920).
2. V. Volterra, *Mem. Accad. Nazionale Lincei 2* **6**, 31 (1926).
3. N. Boccara, O. Roblin, and M. Roger, *Phys. Rev.* **E50**, 4531 (1994).
4. J. Satulovsky and T. Tome, *Phys. Rev.* **E49**, 5073 (1994).
5. J. D. Murray, *Mathematical Biology* (Springer Verlag, NY, 1993).
6. J. Faro and S. Velasco, *Physica* **D110**, 313 (1997).
7. V. Mendez and J. Camacho, *Phys. Rev.* **E55**, 6476 (1997).
8. R. Gerami and M. R. Ejtehadi, *Eur. Phys. J. B*, to appear.
9. W. F. Wolff, in *Ecodynamics — Contributions to Theoretical Ecology*, eds. W. Wolff, C. J. Soeder, and F. R. Drepper (Springer, Berlin-Heidelberg, 1988), p. 285.
10. I. Mroz and A. Pekalski, *Eur. Phys. J.* **B10**, 181 (1999).
11. M. Meyer, S. Havlin, and A. Bunde, *Phys. Rev.* **E54**, 5567 (1996).

is comparable to this characteristic time, namely $R/c \sim \tau$. Hence, taking into account the retardation effect, the interaction is

$$\mathcal{V}_r(R) = -\hbar\omega_0 \times \frac{V_1 V_2}{R^6} \times f\left(\frac{R\omega_0}{c}\right). \tag{17}$$

The crossover function $f(x)$ is a constant for $x \to 0$, and vanishes as $1/x$ for $x \to \infty$. It is clear that in the latter limit the dependence on ω_0 vanishes, and the Casimir–Polder result of Eq. (16) is recovered.

There are also interactions between spheres immersed in a critical fluid, which have been studied both analytically[43] and experimentally.[44] Surprisingly, the interaction does not have the same power law forms as in Eq. (15); its exponents depending on the critical exponents, as well as boundary conditions.[45]

2.2. *Inclusions on membranes*

Dispersion forces are not limited to particles in three-dimensional space, but also occur for inclusions on films and membranes, the latter is of potential importance for understanding the interactions between proteins floating on a cell membrane. In addition to their structural role of forming the exterior frames of the cell and its interior organelles and vesicles, lipid bilayers act as the host and regulator of many biophysical and biochemical reactions.[46,47] Inter- and intra-cellular recognition and transport, adhesion, regulation of ion concentrations, and energy conversion, are but a few of the processes taking place at the membrane. These tasks are carried out by a variety of proteins, glycolipids, and other macromolecules that move through the many different lipids that make up the bilayer. It is thus essential to understand how inclusions affect the physical properties of the membrane, and how the membrane in turn contributes to the interactions between inclusions.

There are a number of different forces which act within membranes[48]: van der Waals interactions fall off with separation R, as $1/R^6$ at long distances; the Coulomb interaction is strongly screened under physiological conditions (typical ion concentrations are a few mM, giving a screening length of less than 10 Å); hydration and structural forces are also short-ranged. There are additional interactions between inclusions which are mediated by the membrane; the inclusion disturbs the lipid bilayer and this disturbance propagates to neighboring inclusions (c.f. Refs. 47–51 and references therein). When macroscopic thermal fluctuations are unimportant (we refer to this case as $T = 0$), the resulting interactions tend to be short-ranged, falling off exponentially with a characteristic length corresponding to the distance over which the lipid disturbance "heals." For example, if in the region around an inclusion the membrane is forced to deviate from its preferred thickness (~ 40 Å), then the resulting disturbance decays exponentially with a length comparable to this thickness.[51]

If the membrane is to mediate long-ranged interactions, then in the long-distance limit it should be possible to neglect the membrane thickness and details concerning lipid structure. In this limit, the membrane is well described by the elastic Hamiltonian,[52]

$$\mathcal{H} = \int \mathrm{d}S \left[\sigma + \frac{\kappa}{2} H^2 + \bar{\kappa} K \right] , \qquad (18)$$

where $\mathrm{d}S$ is the surface area element, and H, K are the mean and Gaussian curvatures respectively. The elastic properties of the surface are described by the tension σ, and the bending rigidities κ and $\bar{\kappa}$. A finite surface tension is the most important coupling in $\mathcal{H}$ and dominates the bending terms at long wavelengths. This is the case for films on a frame, interfaces at short distances, and possibly closed membranes in the presence of osmotic pressure differences between their interior and exterior. On the other hand, for closed bilayers in the absence of osmotic stress, as well as for microemulsions, the surface tension is effectively zero.[53–55] In these cases, the energy cost of fluctuations is controlled by the rigidity terms. For simplicity we shall refer to surface tension dominated surfaces as films, and to rigidity controlled ones as membranes.

The long distance interactions between inclusions in a membrane were examined in Ref. 50. It was shown that if the inclusions are asymmetric across the bilayer and impose a local curvature, $\mathcal{H}$ gives rise to a repulsive ($T = 0$) interaction that is long-ranged, falling off at a distance of $1/R^4$. The energy scale of the interaction is set by κ and $\bar{\kappa}$. If thermal fluctuations of the membrane are included ($T \neq 0$), on the other hand, there is a $1/R^4$ interaction for *generic* inclusions, as long as the rigidity of the inclusion differs from that of the ambient membrane.[50] In particular, if the inclusions are much stiffer than the membrane, the fluctuation-induced potential is

$$\mathcal{V}(R) = -k_B T \times \frac{A^2}{R^4} \times \frac{6}{\pi^2} , \qquad (19)$$

where A is the area of each inclusion. (In this formula, the result in Ref. 50 has been corrected by a factor of $1/2$.[56]) The interaction is attractive and

Annual Reviews of Computational Physics VIII (pp. 329–339)
Edited by Dietrich Stauffer

BIOLOGICAL AGEING IN THE 20TH CENTURY

DIETRICH STAUFFER

Institute for Theoretical Physics, Cologne University, 50923 Köln, Euroland

The phenomenological ageing theory of Azbel as well as many Monte Carlo simulations based on Penna's mutation accumulation model are reviewed. We emphasize the work of Suzana and Paulo Murilo de Oliveira on the growth of life expectancy in the 20th century.

1. Introduction

Simulations of biological ageing were reviewed by Bernardes,[1] more superficially by this writer,[2] and in detail by Suzana Moss de Oliveira *et al.*[3] Thus we start with Azbel's[4] phenomenological theory in the next section, followed by Monte Carlo simulations. Age is defined here not in terms of beauty or scientific originality but very simply and quantitatively in terms of the mortality, the probability to die within the following time interval. Most of the review is taken from Refs. 3 and 18.

2. Azbel Theory

The Gompertz law of the 19th century tells us that the mortality $q(x)$ of humans increases exponentially with age x, after childhood troubles have been overcome. Figure 1 shows recent French data indicating roughly

$$q \propto \exp(bx), \tag{1}$$

for both males and females. More precisely, child mortality is very high and does not follow this Gompertz law, and men have a mortality roughly twice as high as women, except at very old age when men finally achieve equality.

Traditionally, the mortality q is defined as the fraction at age x which survives until age $x + 1$:

$$q(x) = [N(x) - N(x + 1)]/N(x). \tag{2}$$

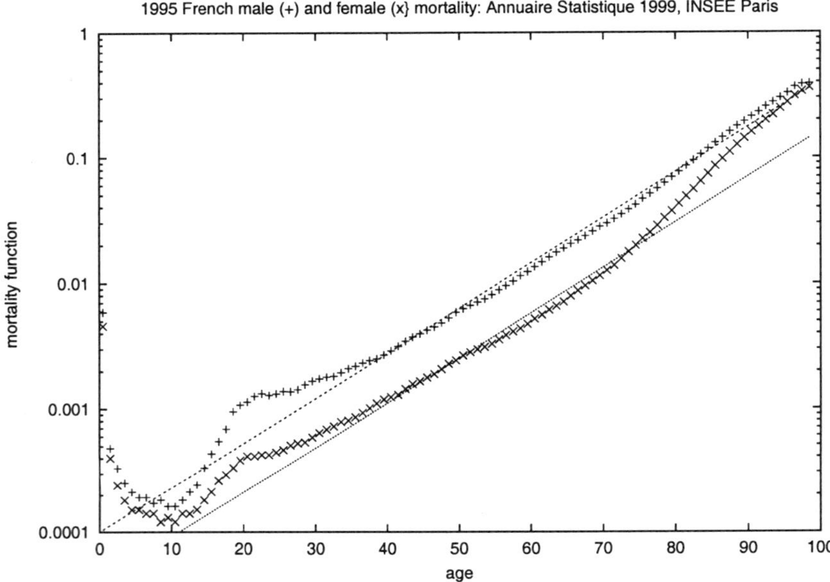

Fig. 1. French mortality: "+" for males, "x" for females. The straight line shows the Gompertz law with slope $b = 1/(12 \text{ years})$.

Here $N(x)$ is the number of survivors at age x in a stationary population, and the age x is measured in suitable units, like a year for humans and a day for fruitflies. Obviously, this definition leads to $q \leq 1$ independent of the choice of the time unit, and it thus contradicts Eq. (1) where q goes to infinity for age x going to infinity. This definition problem is solved by replacing Eq. (2) with

$$q(x + 1/2) = \ln[N(x)/N(x+1)], \tag{3}$$

a quantity also called hazard function, mortality function or force of mortality μ. This quantity can increase beyond unity as required by the Gompertz law (1). (Basically, one should define $q = -d\ln N/dx$; approximating this derivative by a constant in the age interval from x to $x + 1$ leads to Eq. (3).) This definition is already used in Fig. 1, though q there does not yet surpass unity.

More precisely, Ref. 4 defines a dimensionless mortality $q(x)/b \propto \exp(bx)$ independent of the time unit, a characteristic age X and a proportionality factor A which both are the same for the whole species, while b may differ for different groups:

$$q(x)/b = A \exp(b(x - X)) . \tag{4a}$$

With dimensionless time units $y = bx$, $Y = bX$ and thus $q/b \simeq -d \ln N/dy$ this equation reads

$$-d \ln N/dy = A \exp(y - Y) . \tag{4b}$$

In other words, if the logarithm $\ln(q/b)$ of the dimensionless mortality is plotted versus age x then all straight-line fits for humans go through the same point $\ln A$ at $x = X$, even though their slope b changes drastically with country and century. Human lifetables from Sweden, Japan and Germany in the last two centuries, involving more than 10^9 people and a thousand-fold change in the mortality, follow at intermediate ages roughly this law with $X = 103$ years and A around 11, while b ($1/b \simeq 11$ years in Fig. 1) changes with improved living conditions and health care. Also Thatcher[8] noted, going back even thousands of years to Chinese data, that at age 99 the mortality function is always about $1/2$. Thus X seems to be determined by our genes, while b is influenced by the environment (if we count all deaths) and also may differ for different individuals. Similar fits were successful for medflies where X is measured in days instead of years. When we discard all deaths due to infections, accidents, etc as "premature" deaths and count only those due genetic factors (which we can do easily in the computer model of the next section), then also the factor b is determined completely by our genes.

Other than humans, only some mediterranean fruitflies were studied with more than a million individuals.[5] Here, an interesting effect was observed for very old age: The mortality saturated or even decayed with increasing age. If humans would learn such survival techniques from these medflies, a small fraction of us might live a thousand years. To explain such and other data, Azbel[4] introduced a distribution of Gompertz factors b, Eq. (1), in the population. This heterogeneity within one species means that some genetically well endowed but rare families live much longer than the vast majority.[10] And this heterogeneity can then explain such deviations from the Gompertz law.

These fly observations lead to a lot of exaggeration in the literature about human mortality, like for flies, having a maximum: "Humans who make it to 110 years of age appear to have truly better survival rates than those who make it to 95 or 100" (page 14 of Ref. 6); "beyond 85 years, the mortality rate stops increasing exponentially and becomes constant, or actually decreases" (page 122 of Ref. 6); "mortality decelerates at older ages. . . . the rate of increase

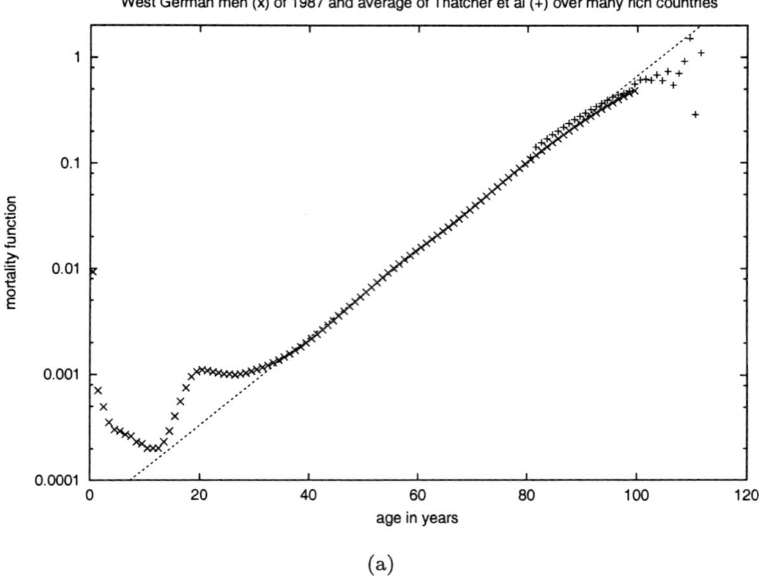

(a)

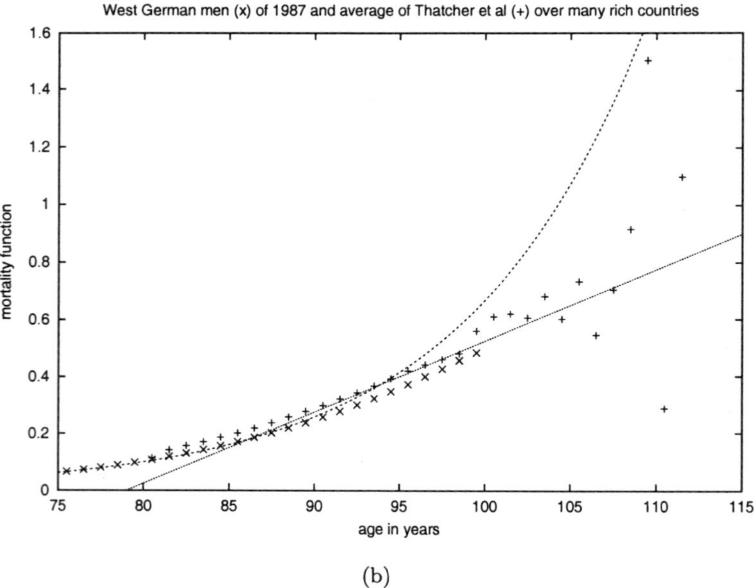

(b)

Fig. 2. Comparison of real data with Gompertz law, mortality function $\propto \exp(0.095x)$, on a logarithmic (top) and linear (middle) scale.

slows down" (page 18 of Ref. 6); "mortality continues to rise throughout adult life, but at a decreasing rate after the age of 75 or 80" (page 47 of Ref. 6). This latter deceleration has also been claimed on pages 19, 24 and 25 in Ref. 6. The reality[8] is different: Fig. 2 top shows that German men obey quite well the Gompertz law, Eq. (1), for middle age (straight line); and the expanded plot in Fig. 2 bottom (linear instead of logarithmic scale) shows that the deviation from the Gompertz law for old age is compatible with a mortality function linear in age (straight line) instead of exponential (curve). The plus signs there are presumably the world's best data from Ref. 8, based on Japan and Western Europe together. Within their scattering there is no clear evidence for an acceleration (upward curvature) or deceleration (downwards curvature). They allow also an extrapolation to a plateau value or even a maximum, but if we take into account systematic errors from people misstating their age, perhaps these data are also compatible with the Gompertz curve.

Some of the claimed decelerations come from looking at women's instead of men's mortality for age above 80 only and ignoring the opposite deviations from the Gompertz law seen if female mortality below 80 years is included; Fig. 1. Others may come from faulty statistics: USA show a clear maximum of human

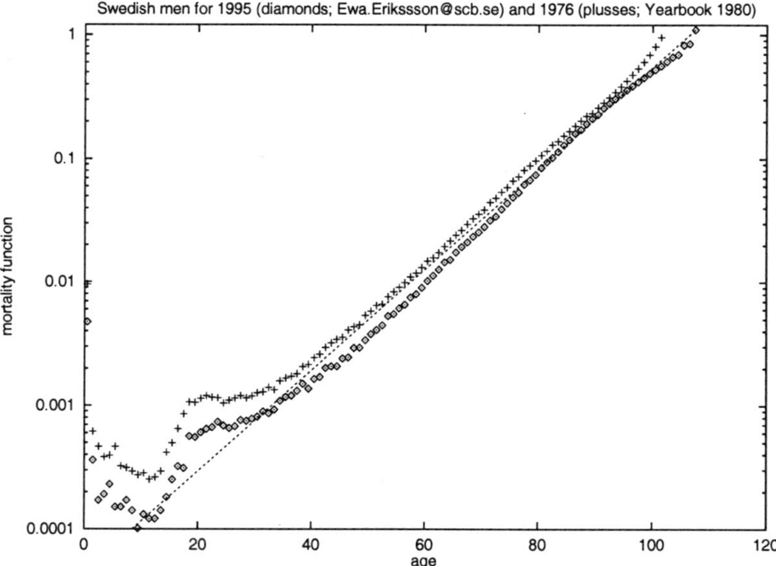

Fig. 3. Older and recent male annual mortalities from the biggest Scandinavian country.

mortality at old age and humans living beyond 130 years[9]; this is not believed by experts. Western Europe and Japan have better statistics and their is no clear mortality maximum. The Scandidavian data seem best[8] and show no clear downward deviation from the Gompertz extrapolation of male mortality functions, Fig. 3. Thus, the more reliable the statistics is, the smaller are the deviations from the Gompartz law for old men. (If plotted not logarithmically as in Fig. 3, but linearly versus age, the Swedish mortality function of Fig. 3 still shows an acceleration, $d^2q/da^2 > 0$ and not a deceleration at old age.)

All these questions have great social implications: Who will pay for my retirement if life expectancy in Germany has doubled within 110 years? At the beginning of the 20th century the German retirement age was mostly 70 years, and now it is 65. The Californian lecturer at the school had already mentioned that the official retirement age there has been increased to 67 years. Recent male mortality tables for the USA in the Berkeley Mortality Database of Wilmoth give excellent agreement with the Gompertz law even for centenarians.

Thus we now study a more microscopic model, in line with the tradition of physics to look for explanations via basic particles, beyond phenomenological theory. Atoms and quarks then correspond to individuals and their genome.

3. Penna Model

In Statistical Physics, energy and entropy counteract each other with the result that neither the energy nor the negative entropy but the free energy is minimal. Similarly, the mutation accumulation hypothesis[7] explains ageing from the counterbalance of Darwinian selection, which prefers perfect individuals, and random detrimental mutations, which cause disorder and disease. These mutations can be inherited and can happen with equal probability at any stage in life. Thus at first one may think that they increase the mortality $q(x)$ equally for all ages x. But this is not so once we take into account the effects of reproduction: A dangerous mutation killing us before we have children will die with us; a mutation killing us in old age will be given on to our children and thus may remain forever in the population. Thus in a stationary equilibrium of new mutations versus selection pressure, mutations affecting us at old age will stay in the population much more often than mutations already afflicting young people. Therefore hereditary diseases increase the mortality much more in old than in young age, as required by Eq. (1).

The Penna model[11] of 1994 is now by far the most widely used computer simulation technique to predict $q(x)$ in ageing populations, and to my knowledge the first one reproducing approximately Eq. (1); see Ref. 12 for later models also succeeding in this aspect. The maximum age is divided into 32 intervals, called years, on a computer with 32-bit words. In each interval, one bad hereditary disease, caused by an earlier mutation, may start to affect the individual until the end of its life. Three or more such active diseases kill us. Each bit corresponds to one year and is set if and only if starting from the year to which the bit position belongs a dangerous hereditary disease threatens our life. The human genome has 10^5 and not just 32 genes, but most of these genes do not cause death if mutated. Besides, qualitatively there was not much difference between 32, 64 and 128 bits taken into account. After a minimum age of reproduction, typically 8 years, each adult individual typically gets one child per year until he or she dies. This child can differ at one randomly selected bit position from the genome of the parent: If the parent bit was not yet set the child bit is now set; if the parent bit was already set before, the child bit remains set (or is set back to zero). Besides these genetic deaths due to bad mutations, there are also deaths due to restrictions of food and space which are taken into account by an additional death probability N/N_{max} (Verhulst factor), where N is the total population and N_{max} a maximum population.

Figure 4 from Ref. 3 shows that this model gives a rough exponential increase of mortality q with age x, as required by Eq. (1). However, in order to find this we had to subtract the "premature"[3] deaths due to the Verhulst factor from the genetically determined deaths due to mutations. More easily, one can do this approximately after the simulation by the Oliveira trick: Subtract that at very young age from the total mortality since the latter is hardly due to mutations. Thus the diamonds in Fig. 4 represent the total mortality, equal to about 0.145 at young age from the Verhulst factor, and the plusses are the same data with this number subtracted. In rich countries at peace, death by hunger and infectious diseases should be relatively rare (the Verhulst factor being replaced by birth control), and then the plusses should correspond to reality except for traffic accidents, etc.

An important aspect of asexual reproduction which is kept also by its sexual variant[13] is that after sometime all survivors are offspring of the same mother and father. Thus they all may share bits set at the same years in youth, causing fluctuations which do not go away if we simulate over longer time. This biblical effect also causes problems in parallel computing[14] since after some

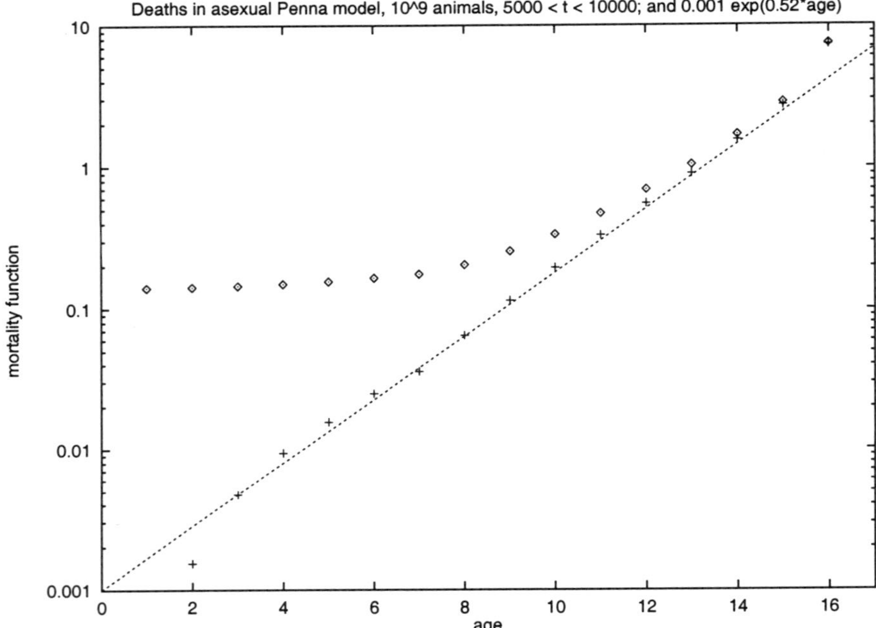

Fig. 4. Mortality in the asexual Penna model with (higher values) and without (lower values) Verhulst deaths.

time all survivors of asexual reproduction are located on the same processor if no load-balancing is made.

How can we now simulate the medical progress, due to which, for example, in Germany the average human life expectancy has doubled within 110 years, ignoring wars. Suzana Moss de Oliveira and Paulo Murilo de Oliveira[18] took into account for earlier centuries a probability of 30% for every time interval, that the age of genetic death is diminished by one time unit, due to infectious diseases, etc. Then, for the 20th century, this effect is omitted and they came back to the standard Penna model. Figure 5 shows that indeed one finds different mortality functions crossing each other at an Azbel–Thatcher fixed point, similar to reality.

The usual Penna model gives a maximum lifespan; this can be avoided by replacing the deterministic genetic death (three mutations kill) with a probabilistic one similar to a Fermi function. Then mortalities similar to the recent Swedish data in Fig. 3 can be obtained.[19]

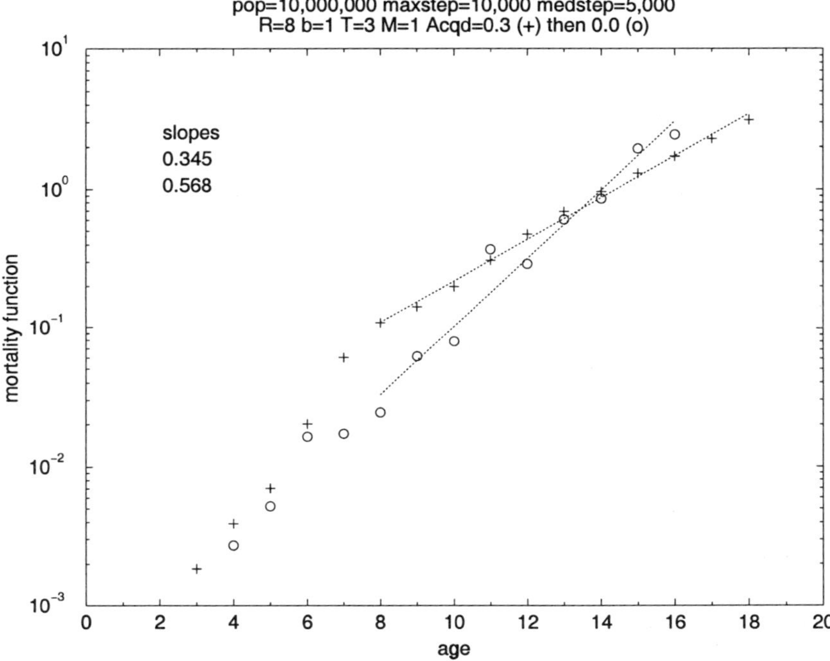

Fig. 5. Mortality in the Penna model with (smaller slope) and without (steeper slope) deaths due to infections, etc.

4. Discussion

This paper presented a phenomenological and a more microscopic description of biological ageing. These genetic approaches seem to be what most recent papers emphasize. That does not prove, however, that this approach is correct. Perhaps instead, we get old because of wear and tear in faculty meetings, similar to the bodies of top athletes. Or somatic mutations are the dominating factor, coming from oxygen radicals arising from metabolism. (Tea drinking Iranians may wish to work on that hypothesis since tea is said to keep oxygen radicals in check.) Or programmed cell death, which surely is crucial for the development of the embryo, also tells our heart to stop beating after 10^2 years, via shortening of the telomeres. The observation that average lifespan corresponds roughly to about ten oxygen molecules consumed per body atom[15] of an animal suggests a rather common cause for death. But the different lifestyles of Pacific Salmon[16] and of California Redwood Trees[17] indicate that not all living beings are the same, even if both extremes have been described by the Penna model. Besides, bacteria have shown a way to immortality, provided we do not define cell division as death.

Will future simulations cure ageing? Presumably not, but understanding nature may give hints for improvement. Similarly, mankind's problem with energy supply was not solved when the conservation of mechanical energy was proven by Newton's laws, or its equivalence to heat understood. But understanding the Carnot cycle in thermodynamics leads to an understanding of steam engine efficiencies and their improvement. Presently, we are not yet at Newton's stage in ageing theories, but we rather behave like the alchemists of science hundreds of years ago. On their basis, modern chemistry developed 200 years ago, while quantum chemistry came much later.

Acknowledgments

The author thanks A. T. Bernardes, S. Cebrat, N. Jan, S. Moss de Oliveira, P. M. C. de Oliveira, T. J. P. Penna, A. R. Thatcher and many others for their cooperation.

References

1. A. T. Bernardes, Monte Carlo simulations of biological ageing , in *Annual Reviews of Computational Physics*, Vol. IV, ed. D. Stauffer (World Scientific, Singapore 1996), p. 359.
2. D. Stauffer, *Computers in Physics* **10**, 341 (1996).
3. S. Moss de Oliveira, P. M. C. de Oliveira and D. Stauffer, *Evolution, Money, War and Computers* (Teubner, Stuttgart-Leipzig, 1999).
4. M. Ya. Azbel, *Proc. Roy. Soc. London* **B263**, 1449 (1996); *Phys. Repts.* **288**, 545 (1997); *Physica* **A249**, 472 (1998); *Proc. Natl. Acad. Sci. USA* **95**, 9037 (1998), and **96**, 3303 (1999).
5. J. R. Carey, P. Liedo, D. Orozco and J. W. Vaupel, *Science* **258**, 457 (1992).
6. K. W. Wachter and C. E. Finch, *Between Zeus and the Salmon. The Biodemography of Longevity* (National Academy Press, Washington DC, 1997); see also the many articles by many authors in *La Recherche* **322** (July/August 1999).
7. L. Partridge and N. H. Barton, *Nature* **362**, 305 (1993); B. Charlesworth, *Evolution in Age-Structured Populations*, 2nd edition (Cambridge University Press, Cambridge, 1994); M. Rose, *Evolutionary Biology of Aging* (Oxford University Press, New York, 1991).
8. A. R. Thatcher, V. Kannisto and J. W. Vaupel, *The Force of Mortality at Ages 80 to 120* (Odense University Press, Odense 1998); A. R. Thatcher, *J. Roy. Statist. Soc.* **A162**, 5 (1999) and private communication.
9. P. Klement and S. Doubal, *Mech. Aging Dev.* **98**, 167 (1997).

10. T. T. Perls *et al.*, *Lancet* **351**, 1569 (1998); P. M. C. de Oliveira *et al.*, *Lancet* **252**, 911 (1998).
11. T. J. P. Penna, *J. Stat. Phys.* **78**, 1629 (1995).
12. L. D. Mueller and M. R. Rose, *Proc. Natl. Acad. Sci. USA* **93**, 15249 (1996); S. D. Pletcher and J. W. Curtsinger, Evolution 52, 454 (1998); A. Pekalski, *Eur. Phys. J.* **B13**, 791 (2000); S. D. Pletcher and C. Neuhauser, *Int. J. Mod. Phys.* **C11**, 525 (2000).
13. P. M. C. de Oliveira, S. Moss de Oliveira and D. Stauffer, *Theory in Biosciences* **116**, 65 (1997). See also Y. C. Zhang, M. Serva and M. Polykarpov, *J. Stat. Phys.* **58**, 849 (1990).
14. F. Meisgen, *Int. J. Mod. Phys.* **C8**, 575 (1997).
15. M. Ya. Azbel, *Proc. Natl. Acad. Sci. USA* **91**, 12453 (1994).
16. T. J. P. Penna, S. Moss de Oliveira and D. Stauffer, *Phys. Rev.* **E52**, R3309 (1995).
17. M. Argolo de Menezes, A. Racco and T. J. P. Penna, *Physica* **A233**, 221 (1996).
18. P. M. C. de Oliveira, S. Moss de Oliveira, D. Stauffer and C. Cebrat, *Proceedings of the Max Born Symposium*, May 1999, Wroclaw, Poland, *Physica* **A273**, 145 (1999); E. Niewczas, S. Cebrat and D. Stauffer, *Theory in Biosciences* **119**, 122 (2000).
19. D. Stauffer, "Probabilistic generalization of the Penna ageing model and the oldest old", *Int. J. Mod. Phys.* **C10**, 1363 (1999).